Nature Based Strategies for Urban and Building Sustainability

Nature Based Strategies for Urban and Building Sustainability

Edited by

Gabriel Pérez
University of Lleida, Spain

Katia Perini
University of Genoa, Italy

Butterworth-Heinemann
An imprint of Elsevier

Butterworth-Heinemann is an imprint of Elsevier
The Boulevard, Langford Lane, Kidlington, Oxford OX5 1GB, United Kingdom
50 Hampshire Street, 5th Floor, Cambridge, MA 02139, United States

British Library Cataloguing-in-Publication Data
A catalogue record for this book is available from the British Library

Library of Congress Cataloging-in-Publication Data
A catalog record for this book is available from the Library of Congress

ISBN: 978-0-12-812150-4

For Information on all Butterworth-Heinemann publications
visit our website at https://www.elsevier.com/books-and-journals

Working together
to grow libraries in
developing countries

www.elsevier.com • www.bookaid.org

Publisher: Joe Hayton
Acquisition Editor: Ken McCombs
Editorial Project Manager: Andrae Akeh
Production Project Manager: Anitha Sivaraj
Cover Designer: Greg Harris

Typeset by MPS Limited, Chennai, India

Contents

Section II
Nature Based Strategies and Technologies

Section III
Nature Based Strategies: Benefits and Challenges

Section IV
Nature Based Strategies: Social, Economic and Environmental Sustainability

List of Contributors

Eleanor Atkins Staffordshire University, Stoke-on-Trent, United Kingdom

Thomas Auer Technische Universität München, München, Germany

Tijana Blanusa Royal Horticultural Society, London, United Kingdom

Luisa F. Cabeza University of Lleida, Lleida, Spain

Inma R. Cantalapiedra Universitat Politècnica de Catalunya, Barcelona, Spain

Ata Chokhachian Technische Universität München, München, Germany

Philippe Clergeau Muséum national d'Histoire naturelle, Paris, France

Julià Coma Universitat Politècnica de Catalunya, Barcelona, Spain; University of Lleida, Lleida, Spain

John W. Dover Staffordshire University, Stoke-on-Trent, United Kingdom

Haibo Feng The University of British Columbia, Kelowna, Canada

Rafael Fernández-Cañero University of Seville, Seville, Spain

Ilaria Gnecco University of Genova, Genova, Italy

Gary Grant Green Infrastructure Consultancy, London, United Kingdom

Kasun N. Hewage The University of British Columbia, Kelowna, Canada

Manfred Köhler Neubrandenburg University of Applied Sciences, Neubrandenburg, Germany

Benz Kotzen University of Greenwich, London, United Kingdom

Kelly Ksiazek-Mikenas Northwestern University, Evanston, IL, United States

Ana M. Lacasta Universitat Politècnica de Catalunya, Barcelona, Spain

Stefano Lazzari University of Genoa, Genoa, Italy

Fréderic Madre Muséum national d'Histoire naturelle, Paris, France

Adriano Magliocco University of Genoa, Genoa, Italy

Flavie Mayrand Muséum national d'Histoire naturelle, Paris, France

Sarah Milliken University of Greenwich, London, United Kingdom

Panayiotis A. Nektarios Agricultural University of Athens, Athens, Greece

Anna Palla University of Genova, Genova, Italy

Angelina Peñaranda Universitat Politècnica de Catalunya, Barcelona, Spain

Gabriel Pérez University of Lleida, Lleida, Spain

Luis Pérez-Urrestarazu University of Seville, Seville, Spain

Katia Perini University of Genoa, Genoa, Italy

Enrica Roccotiello University of Genoa, Genoa, Italy

Paolo Rosasco University of Genoa, Genova, Italy

Bradley Rowe Michigan State University, East Lansing, MI, United States

Paola Sabbion University of Genoa, Genoa, Italy

Miguel Urrestarazu University of Almeria, Almeria, Spain

Timothy Van Renterghem Ghent University, Gent, Belgium

Madalena Vaz Monteiro Forest Research, Farnham, United Kingdom

Alan Vergnes Université Paul-Valéry Montpellier, Montpellier, France

Introduction

More than 50% of the world's population lives in urban settlements, with an increasing trend that will reach 60% by 2030 (United Nations, 2016). Most of the people will globally live in cities with more than half a million inhabitants. Therefore, the state of the urban environment will directly influence the quality of life of these citizens. Nature-based solutions can significantly improve the environmental quality of dense urban areas providing "environmental, social, and economic benefits and help build resilience" (European Commission, 2016).

In this field many recent researches have been developed all over the world in order to quantify the wide range of ecosystem services provided by urban greening.

In the present book the most relevant researchers worldwide provide an overview on the current state of the art in research regarding nature-based (NB) strategies and their potential role as means to improve cities conditions. Indeed green roofs, vertical greening systems, and green streets act on the causes of ecological and environmental issues deriving from anthropogenic activities and also on their effects at both district-city scale (i.e., when a wide surface in the same area is greened) and building scale.

Nature Based Strategies for Urban and Building Sustainability provides to academics and industry a description of the current state of technology, concerning systems and methodologies, available to integrate nature in the built environment. The most significant research results regarding green roofs, vertical greening systems, and green streets on operation and ecosystem services are featured. At the same time, the future challenges and the critical aspects are also presented, in order to provide hints to exploit nature-based strategies in terms of environmental and economic sustainability. In fact, while the potential benefits can be very interesting, the still existing barriers (technological, social, economic, environmental, etc.) could reduce the diffusion of greening systems. The book considers all these aspects with a holistic approach, thanks to the contribution of authors working in different fields, e.g., engineering, architecture, botany, landscape architecture, etc.

The book is divided in four sections. The first one describes the general and most relevant aspects related to NB strategies in urban areas. Chapter 1.1 provides an introduction to urban sustainability issues and urban ecosystems. Considering the important role of ecosystem services in enhancing the resilience of cities to climate change, a synthesis of recent literature on the ones provided by nature-based strategies in urban environments is presented in Chapter 1.2. Finally, an overview on the ways that governments, local authorities, and non-government organizations use in order to increase the amount of greening in cities, through incentives, is given in Chapter 1.3.

There are several possible integration modalities of green elements in architecture. These can have a major or a minor influence on a design project conception and on formal and functional characteristics. Section 2, "Nature-based strategies and technologies" provides not only the main classifications but also an overview on the main plant species used and their characteristics, substrates, irrigation and maintenance of vertical greening systems, green roofs, and green streets.

Vertical greening systems (Chapters 2.1 and 2.2) are made by simple climbing plants, supporting structures for their growth or planter boxes placed at several heights with a shading function; others provide the possibility to cultivate species naturally not suitable for growing on vertical surfaces, thanks to the disposition of pre-vegetated panels, defined as living wall systems. These systems entail very different maintenance needs. Green roofs (Chapters 2.3 and 2.4) may use different plant species, for both their influence on architectural aesthetics and for microclimatic improvements obtainable. The many products available on the market propose several integrated solutions for proper drainage, waterproofing, and roof protection depending on the vegetation type, such as grass and bigger or smaller shrubs.

Finally, an overview on street trees, planting compositions of urban gardens and rain gardens, along with the substrates, maintenance and irrigation requirements and practices is presented (Chapter 2.5).

The third section includes 15 Chapters providing a review on the most relevant research results on micro-scale (quantifiable) benefits, issues, and challenges of nature-based strategies.

Interesting energy savings for buildings due to the application of these systems (vertical greenery—Chapter 3.1, and green roofs—Chapter 3.2) have been achieved, becoming one of the most promising strategies for a more sustainable future for the building sector. Also, green streets can play an important role in improving outdoor thermal comfort. Chapter 3.3 provides an overview on these effects and the most influencing parameters and the possible use of modeling.

The bad air quality of many cities is an environmental and also a very relevant heath issue (WHO, 2013).

Urban greening, along with other environmental benefits, can mitigate air pollutants and improve air quality, adsorbing fine dust particles and uptaking gaseous pollutants. Chapters 3.4, 3.5, and 3.6 discuss these effects—the main parameters to consider, the most promising strategies, and the most suitable plant species characteristics—in order to provide an overview on the effects of vertical greening, green roofs, and green streets.

At building scale it's also to highlight the acoustic insulation potential for the building of these systems, provided especially by the absorption capacity of substrates as well as due to the sound diffraction capacity of plants (as discussed in Chapters 3.7–3.9).

The effect of water falling as rain on land covered by vegetation and on the hard surface of built-up areas is very different. The vast majority of precipitation on vegetation is absorbed by the soil; some is absorbed by plants and through them it is transpired back into the atmosphere. Green roofs (Chapter 3.11) and green streets (Chapter 3.12) especially, but also vertical greening systems (when specific design strategies are considered, Chapter 3.10), are able to significantly reduce storm water runoff generation and an interesting alternative to more conventional building practices.

Finally, the potential to enhance urban biodiversity, deriving from the integration of nature-based strategies, is discussed. Chapters 3.13–3.15 consider the role of plants used, respectively, for vertical greening, green roofs, and green streets, in increasing the ecological value and the main drivers for urban biodiversity in cities.

The fourth and last section discusses social, economic, and environmental sustainability of nature-based strategies. Urban greening is widely recognized for increasing the liveability of cities with consequences on human health, as highlighted in Chapter 4.3, which discusses the social and aesthetic aspects of green streets. Also, greening systems for the building envelope can improve the environmental quality of cities (as described in Chapters 4.1. and 4.2). Green roofs can also provide highly valuable recreational spaces in dense cities. In the case of vertical greening systems, however, social perception can be critic and the diffusion of such systems requires a good acceptance level, by residents, designers, entrepreneurs, and building owners.

In order to ensure a wide diffusion of nature based strategies for urban and building sustainability, the economic impact has to be considered. Cost benefit analysis allows measuring the economic sustainability of greening systems, to show if they can be cost-effective strategies from a social and personal point of view. Chapters 4.4 and 4.5 consider different green roofs and vertical greening systems highlighting the most relevant aspects related to production, constriction, use, and disposal and the cost savings which can be taken into account in order to evaluate the payback period. Chapter 4.6 discusses the main economic benefits of green streets (deriving from the several ecosystem services provided), evaluated by local and national authorities all over the word.

As discussed, especially in the first and third sections of the book, greening systems can provide several environmental benefits in urban areas. However, green roofs and especially vertical greening systems entail different environmental costs, which could overcome the above mentioned benefits. Indeed, the environmental performance of greening systems is influenced by the types of materials and plants chosen for the systems, as well as the external factors, such as climate and building type. The last two chapters of the book (4.7 and 4.8) discuss the role of life cycle assessment methodology to evaluate the environmental sustainability of vertical greening systems and green roofs, balancing the environmental burden produced during the life span of a system and the environmental benefits, mainly related to the energy savings for air conditioning at the building scale.

Katia Perini and Gabriel Pérez

REFERENCES

European Commission, 2016. Nature-Based Solutions [WWW Document]. URL https://ec.europa.eu/research/environment/index.cfm?pg=nbs (accessed 10.03.16).
United Nations, 2016. The World's Cities in 2016.
WHO, 2013. Health risks of air pollution in Europe − HRAPIE project.

Section I

Nature Based Strategies for Urban Environment

Chapter 1.1

Introduction to Urban Sustainability Issues: Urban Ecosystem

John W. Dover

Chapter Outline

INTRODUCTION: THE URBAN ECOSYSTEM

Humans, like most social species, engineer their habitat and farms; hamlets, villages, towns, and cities can be seen as a progressive modification of natural ecosystems along an urban or human ecosystem gradient; and one, moreover, with a vast ecological footprint that relies on huge inputs of materials and energy from geographically dispersed areas (Rees, 1997). Ecosystems are composed of biotic and abiotic components and their interactions; and are increasingly modified by anthropogenic activities. The urban ecosystem results from the interaction of the human social system (in its widest sense: including culture, behavior, and economics) and built elements with the other main ecosystem processes of energy flow, flow of information (evolutionary origin and change), and materials cycling (Grimm et al., 2000).

The term, "ecosystem services," has become popular to describe the goods and services provided by biodiversity to humans, and human wellbeing (Braat and de Groot, 2012) (Fig. 1; and Chapter 1.2: Ecosystem Services in Urban Environments). Urban vegetation, and other associated biodiversity components, often referred to as green infrastructure, can be

Nature Based Strategies for Urban and Building Sustainability.
DOI: https://doi.org/10.1016/B978-0-12-812150-4.00001-X

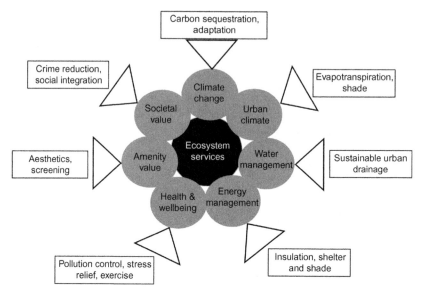

FIGURE 1 Multiple ecosystem services can be delivered by green infrastructure to address a range of challenges facing urban areas.

seen as a delivery mechanism for ecosystem services. Initially green infrastructure was seen purely as a strategic network of open green spaces and its links; this has evolved over time to embrace a wider set of features:

Green infrastructure is the sum of an area's environmental assets, including stand-alone elements and strategically planned and delivered networks of high quality green spaces and other environmental features including surfaces such as pavements, car parks, driveways, roads, and buildings (exterior and interior) that incorporate biodiversity and promote ecosystem services.

Dover (2015).

The concentration of human populations into urban areas from a dispersed, rural pattern has obvious advantages such as improved cultural development, and resources can be distributed more efficiently whether this is food, energy, goods, services, or transport. There are, equally, downsides such as the large impact resulting from even temporary disruption to those same resources, the easy transmission of disease, the need for continuous disposal of large quantities of waste and the release into the environment of toxic by-products, the stresses of large numbers of people living in close proximity, and the extreme modification of soils, the land surface, and water bodies and its concomitant impacts (Rees, 1997). As higher proportions of the global population become urbanized and densification increases, the need to mitigate the impact of the downsides becomes more important (Haaland and van den Bosch, 2015). The challenge of creating a more sustainable urban environment is substantial in the face of these requirements, but the predicted impact of global climate change makes meeting these challenges both more difficult and more pressing (Haines et al., 2006).

The remainder of this chapter will expand on *some* of the major issues confronting urban areas, and how green infrastructure helps ameliorate them. It is not intended to be exhaustive or comprehensive, and a wider exploration can be found elsewhere (see Dover (2015) and references therein).

MENTAL HEALTH AND STRESS

Urban upbringing and living brings with it the potential for negative impacts on mental health and well-being (Guite et al., 2006), in particular mood and anxiety disorders, increased incidence of schizophrenia, and mental fatigue (Paykel et al., 2000; Watts et al., 2010). Poor mental health may also result in poorer physical health (Hert et al., 2011).

Exposure to vegetation, or even views of vegetation, is known to have a number of positive outcomes including reductions in: mental distress, stress levels, tension and anxiety, health inequalities, postoperative recovery time, and use of painkillers (Grahn and Stigsdotter, 2003; Mitchell and Popham, 2007; Ulrich, 1984; White et al., 2013). Improved feelings of well-being (Luck et al., 2011), cognitive development of children, (Dadvand et al., 2015) and general health and exercise are also associated with access to green space (Bird, 2004).

CLIMATE CHANGE

Global climate change impacts vary depending on geographical position, with increased rain in some places and drought in others. While all areas are anticipated to increase in average temperature, some areas will be harder hit (IPCC, 2015). The increased frequency and severity of heat waves and flooding is expected to increase mortality and property/infrastructure damage (Patz et al., 2005). It is also likely to exacerbate the impact of air pollution (Jacob and Winner, 2008) and health issues such as asthma (D'Amato et al., 2010). Green infrastructure is identified as one of the basket of approaches that can be used to adapt societies to climate change (IPCC, 2015).

Urban vegetation removes the climate change gas carbon dioxide from the air, sequestering it (Pataki et al., 2011). If subsequently incorporated into the soil, or into long-lived species such as trees, then its rerelease can be considerably delayed. Nowak & Heisler (2010) estimated that park trees in the United States stored 75 million tons of carbon and removed 2.4 million tons annually, with soils holding a further 102 million tons. Setälä et al. (2016) in Finland estimated that soil under evergreen trees in established parks were particularly good, holding 35.5 kgC m^{-2}, compared with deciduous trees (23.4 kgC m^{-2}); both were better than new parks without trees

$(14.9\,\text{kgC m}^{-2})$. Sequestration by four parks in Rome was equivalent to 3.6% of the total CO_2 emissions for the city in 2010, and trees in woods, along avenues, and even hedges were found to be important (Gratani et al., 2016). Green roofs can also store carbon; Getter et al. (2009) found a *Sedum* roof, sequestered about $375\,\text{gC m}^{-2}$; this, with energy efficiency gains, was expected to exceed the embedded carbon debt used in construction of the roof after 9 years.

URBAN/CITY CLIMATE

Urban areas have more clouds, rain, and snow than the surrounding countryside (Shepherd et al., 2002; Taha, 1997). The main driver is the difference in the nature of the surface of towns and cities compared to the countryside, although factors also include local topography and the increased density of heat sources (Nitis et al., 2005; Taha, 1997) including air conditioners in hot countries with high population density areas (Shahmohamadi et al., 2011).

Urban areas typically have very large areas of concrete and tarmac (sealed surfaces), which absorb heat from the sun and then release it slowly over night. An additional issue in areas with many tall buildings is a reduced sky view factor (Oke, 1981); there is less sky to radiate heat out to. Urban areas are therefore hotter than rural areas, especially at night, and the effect is greater in larger settlements (Oke, 1973). Typical differences between rural and urban temperatures are in the order of $8-9°C$ (Stülpnagel et al., 1990), the urban heat island (UHI) effect (Rosenthal et al., 2008). Impacts of UHIs include increased mortality during heat waves (Li and Bou-Zeid, 2013), and causing/exacerbating other complaints such as asthma, heat stress, circulatory illnesses, and insomnia (Stülpnagel et al., 1990; Tomlinson et al., 2011).

Vegetated areas are cooler than sealed surfaces because plants evaporate water from their stomata (evapotranspiration). Rosenzweig et al. (2006) noted that UHIs are really a mosaic of hot and cool spots and suggested that UHI archipelago was a more appropriate term. Handley and Gill (2009) found areas of Manchester (UK) with a greater proportion of sealed surface had much higher maximum surface temperatures than areas with a higher proportion of vegetation (e.g., town centres $>30°C$ cf. woodlands $<20°C$). UHIs are not static phenomena, being affected by seasonal and weather changes, which, e.g., can move them downwind (Gedzelman et al., 2003).

Grass, on its own, has a substantially lower surface temperature compared with a sealed surface, but is even lower in combination with tree shade (Armson et al., 2012). Pergolas (Watanabe et al., 2014) and green façades (walls) are also useful in heat shading and reduction and, while green roofs may not have a direct impact on people at street level, like trees (O'Neill et al., 2009), they do have the potential to help reduce the UHI and the temperature regime of urban areas overall (Alexandri and Jones, 2008; Zölch et al., 2016).

WATER MANAGEMENT

The increased area of sealed surface in urban areas reduces the infiltration of rainwater into the soil (Arnold and Gibbons, 1996); rainfall is, hence, diverted to drains and attempts to enter the sewer systems. During storms, the resulting rush of water can cause flash flooding with feces and other contaminants flushed-out into streets, houses, and businesses (Coulthard and Frostick, 2010). The resulting financial, social, and health issues are obvious and potentially devastating (Rietveld et al., 2016). There's also expected to be significant impacts on waterbodies, reducing water and habitat quality (Whitehead et al., 2009). With climate change predictions including increased precipitation for some areas, combined with an increase in the frequency of extreme events, it is clear that stormwater management is a crucial and pressing matter.

Sustainable urban drainage systems are used to reduce the amount of water entering sewers. The concept is simple: infiltrate rain where it falls, if that is not possible, capture it, contain it, and release it slowly. Vegetation is an integral part of the concept, it captures rainfall, slowing its descent to the ground, what remains on leaves may be evaporated off. On reaching ground level it enters the soil and some is transpired off. Urban vegetation, whether it is at ground level, on a roof (Liptan and Strecker, 2003) or on a wall (Ostendorf et al., 2011), will contribute to a reduction in flooding. The more complex the vegetation structure, the greater will be the delaying action (Livesley et al., 2014). Green roofs act in exactly the same way as ground-level vegetation, storing some water in the growing medium with the excess being slowly directed to ground level (Hutchinson et al., 2003; Simmons et al., 2008). This water can itself be stored in a pond or rain garden for later use or infiltration (Schmidt, 2009). Various structures such as permeable pavements, swales (vegetated ditches), rain gardens, retention ponds, and constructed wetlands can all incorporate vegetation which will help slow water flow, promote infiltration, and may also remove toxic compounds (Jackson and Boutle, 2008).

ENERGY MANAGEMENT

Energy from fossil fuel burning releases carbon dioxide (CO_2). Reducing energy consumption and energy conservation reduces CO_2 emissions and the anthropogenic contribution to UHIs and air pollution. Extreme events linked to climate change will probably increase energy consumption through additional heating when the weather is cold and the use of air conditioning when hot. The latter will also increase UHIs as the hot air from buildings is expelled into the local environment.

Shelter from wind (using trees, tall shrubs, or free-standing green walls) reduces its cooling effect on buildings and the intensity of draughts that penetrate them (Akbari, 2002; Heisler, 1986). Vegetation can also provide direct insulation using green walls and green roofs (Cheng et al., 2010; Simmons et al., 2008). Green roofs are easier to fit to buildings during construction, rather than retrofitting, when any additional structural support can be incorporated. Green walls are straightforward to retrofit and can be free-standing with no loading implication for buildings. Shading windows with deciduous trees or climbers (Dimoudi and Nikolopoulou, 2003; Ip et al., 2010) can reduce air conditioning costs in the summer and yet allow sunlight through to provide solar heating in the winter. Shading air conditioning units has been claimed to reduce energy consumption by up to 10% (Parker, 1983), though Parker et al. (1996) suggested that gains may be more modest and would require a larger area to be cooled than simply the air conditioning unit itself. Building orientation affects energy saving; in the northern hemisphere shade trees planted on the west and south sides of buildings save energy in the summer but on the north side may actually increase it (Donovan and Butry, 2009).

AIR POLLUTION

The health effects of outdoor air pollution are increasingly under scrutiny, and RCP (2016) concluded that its impact is a whole life issue, starting in the womb by affecting growth, development and intelligence, and continuing throughout life. Mortality in the United Kingdom was estimated at 40,000 deaths per year and economic loss £20 billion per year.

The main pollutants of concern are particulate matter (PM) primarily from combustion sources, but also from wear and tear of tires, clutches, brakes, catalytic converters, etc., and the nitrogen oxides (NO_2 and NO) collectively known as NOx. It is typical for studies of air-pollution and health to have difficulties in isolating whether particulates or NOx are the primary actors for some conditions. Nitrogen oxides also lead to the formation of a secondary pollutant, low-level ozone (O_3), in a photochemical reaction with VOCs (volatile organic chemicals). Outdoor VOCs may be anthropogenic pollutants, e.g., from unburnt fuels, or chemicals released from industrial processes (Dore et al., 2008) or derived from natural sources (Corchnoy et al., 1992).

Inspired PM of $10\,\mu m$ in aerodynamic diameter and below is considered to be of major importance in relation to human health. Smaller particles penetrate deeper into the lungs, with the smallest being able to pass across the lungs and enter the circulatory system and to the major organs of the body. These ultrafine particulates can also enter the olfactory nerves via the nose, and thus be transported directly to the brain (Maher et al., 2016; Solomon et al., 2012). The names and different size ranges typically quoted reflect their penetrability into the body:

- PM_{10} Thoracic/Coarse (can reach the bronchioles)
- $PM_{2.5}$ Respirable/Fine (can reach the alveoli)
- $PM_{0.1}$ Ultrafine (can enter organs)

  ($1\,\mu m = 0.001\,mm$; $0.1\,\mu m = 100\,nm$)

Particulates can be composed of carbon, from combustion, but may also include a wide range of metals (Espinosa et al., 2001), and carry carcinogens such as polyaromatic hydrocarbons (Burkart et al., 2013). Health effects include hypertension, impaired lung function, sudden infant death syndrome, asthma, respiratory and cardiovascular disease, lung cancer, and dementia (RCP, 2016). In terms of mortality, globally, the impact of $PM_{2.5}$ is estimated at 3.15 million deaths per year (Lelieveld et al., 2015).

Nitrogen oxides, and in particular nitrogen dioxide, emitted from vehicle exhausts (primarily diesel) have been shown to have a range of health effects including increased mortality risk (Chiusolo et al., 2011), asthma (Brauer et al., 2002), decreased lung function in children and adults (Bowatte et al., 2016; Ierodiakonou et al., 2016), low birth weight (Brauer et al., 2008), and cognitive development (Freire et al., 2010). Increased mortality is a feature also associated with exposure to ozone (Brunkeef and Holgate, 2002) together with pneumonia, chronic obstructive pulmonary disease, adverse impacts on asthma sufferers, and decreased lung function (Berman et al., 2012).

It has been known for some time that vegetation can help remove pollutants from the air. Leaf surfaces can collect particulates and some gasses/volatiles (e.g., polychlorinated biphenyls, PCBs), can be directly absorbed through the cuticle. Ultrafine particulates and gasses (e.g., NO_2) can enter via the stomata (Barber et al., 2002; Fowler, 2002). As might be expected, the volume of vegetation, e.g., (tall vs small trees), their spatial relationships, and the type of vegetation will affect pollutant uptake, and different plants and even varieties are likely to have different uptake profiles (Lovett, 1994; Nowak, 1994; Stülpnagel et al., 1990).

CONCLUSION

The intensively modified urban habitat that humans have carved-out of the natural environment is underpinned not only by the normal underlying biotic and abiotic factors and associated ecosystem processes, but also by the human social system and a massive ecological footprint. The failure of the system to cope with the waste streams resulting from that footprint have resulted in significant disbenefits that impact on humans at a range of scales from the very local to the global. The extreme levels of vegetation removal and consequent surface sealing from buildings, roads, and other infrastructure have impacted on the ability of urban areas to self-regulate. Nevertheless, the ecosystem services delivered by vegetation (green infrastructure) have the potential to help us address some of the major environmental and social challenges of our times. Unfortunately, given the density of buildings and the expense of land, it is often difficult to create new areas for planting at ground level in towns and cities, and this is exacerbated by the complexity of below-ground infrastructure. Fortunately there is still the vertical domain, and recent technological innovations give us the opportunity to employ the various services provided by greenery by, almost literally, greening the epidermis of built structures (Sitta, 1983).

REFERENCES

Akbari, H., 2002. Shade trees reduce building energy use and CO_2 emissions from power plants. Environ. Poll. 116, S119—S126.

Alexandri, E., Jones, P., 2008. Temperature decreases in an urban canyon due to green walls and green roofs in diverse climates. Build. Environ. 43, 480—493. Available from: https://doi.org/10.1016/j.buildenv.2006.10.055.

Armson, D., Stringer, P., Ennos, A.R., 2012. The effect of tree shade and grass on surface and globe temperatures in an urban area. Urban For. Urban Gree. 11, 245—255. Available from: https://doi.org/10.1016/j.ufug.2012.05.002.

Arnold, C.L., Gibbons, C.J., 1996. Impervious surface coverage - the emergence of a key environmental indicator. J. Am. Plan. Assoc. 62, 243—258. Available from: https://doi.org/10.1080/01944369608975688.

Barber, J.L., Kurt, P.B., Thomas, G.O., Kerstiens, G., Jones, K.C., 2002. Investigation into the importance of the stomatal pathway in the exchange of PCBs between air and plants. Environ. Sci. Technol. 36, 4282—4287. Available from: https://doi.org/10.1021/es025623m.

Berman, J.D., Fann, N., Hollingsworth, J.W., Pinkerton, K.E., Rom, W.N., Szema, A.M., et al., 2012. Health benefits from large-scale ozone reduction in the United States. Environ. Health Perspect. 120, 1404—1410. Available from: https://doi.org/10.1289/ehp.1104851.

Bird, W., 2004. Natural fit. RSPB, Sandy.

Bowatte, G., Lodge, C.J., Knibbs, L.D., Lowe, A.J., et al., 2016. Traffic-related air pollution exposure is associated with allergic sensitization, asthma, and poor lung function in middle age. J. Allergy Clin. Immunol. 139, 122—129. Available from: https://doi.org/10.1016/j.jaci.2016.05.008.

Braat, L.C., de Groot, R., 2012. The ecosystem services agenda:bridging the worlds of natural science and economics, conservation and development, and public and private policy. Ecosystem Serv. 1, 4—15. Available from: https://doi.org/10.1016/j.ecoser.2012.07.011.

Brauer, M., Hoek, G., Vliet, P.V., Meliefste, K., Fischer, P.H., Wijga, A., et al., 2002. Air Pollution from traffic and the development of respiratory infections and asthmatic and allergic symptoms in children. Am. J. Respirat. Critical Care Med. 166, 1092—1098.

Brauer, M., Lencar, C., Tamburic, L., Koehoorn, M., Demers, P., Karr, K., 2008. A cohort study of traffic-related air pollution impacts on birth outcomes. Environ. Health Perspect. 116, 680—686.

Brunkeef, B., Holgate, S.T., 2002. Air pollution and health. Lancet 360, 1233—1242.

Burkart, K., Nehls, I., Win, T., Endlicher, W., 2013. The carcinogenic risk and variability of particulate-bound polycyclic aromatic hydrocarbons with consideration of meteorological conditions. Air Quality Atmosp. Health 6, 27—38. Available from: https://doi.org/10.1007/s11869-011-0135-6.

Cheng, C.Y., Cheung, K.K.S., Chu, L.M., 2010. Thermal performance of a vegetated cladding system on facade walls. Build. Environ. 45, 1779—1787.

Chiusolo, M., Cadum, E., Stafoggia, M., Galassi, C., Berti, G., Faustini, A., et al., 2011. Short-term effects of nitrogen dioxide on mortality and susceptibility factors in 10 Italian cities: the EpiAir study. Environ. Health Perspect. 119, 1233—1238.

Corchnoy, S.B., Arey, J., Atkinson, R., 1992. Hydrocarbon emissions from 12 urban shade trees of the Los-Angeles, California, air basin. Atmosp. Environ. Part B-Urban Atmosp. 26, 339–348. Available from: https://doi.org/10.1016/0957-1272(92)90009-h.

Coulthard, T.J., Frostick, L.E., 2010. The Hull floods of 2007: implications for the governance and management of urban drainage systems. J. Flood Risk Manage. 3, 223–231. Available from: https://doi.org/10.1111/j.1753-318X.2010.01072.x.

Dadvand, P., Nieuwenhuijsen, M.J., Esnaola, M., Forns, J., Basagana, X., Alvarez-Pedrerol, M., et al., 2015. Green spaces and cognitive development in primary schoolchildren. Proc. Natl. Acad. Sci. USA 112, 7937–7942. Available from: https://doi.org/10.1073/pnas.1503402112.

Dimoudi, A., Nikolopoulou, M., 2003. Vegetation in the urban environment: microclimatic analysis and benefits. Energy Build. 35, 69–76.

Donovan, G.H., Butry, D.T., 2009. The value of shade: estimating the effect of urban trees on summertime electricity use. Energy Build. 41, 662–668. Available from: https://doi.org/10.1016/j.enbuild.2009.01.002.

Dore, C.J., Murrells, T.P., Passant, N.R., Hobson, M.M., Thistlethwaite, G., Wagner, A., et al., 2008. UK Emissions of Air Pollutants 1970 to 2006. AEA Technology, Didcot.

Dover, J.W., 2015. Green Infrastructure: Incorporating Plants and Enhancing Biodiversity in Buildings and Urban Environments. Earthscan\Routledge, Abingdon.

D'Amato, G., Cecchi, L., D'Amato, M., Liccardi, G., 2010. Urban air pollution and climate change as environmental risk factors of respiratory allergy: an update. J. Investigat. Allergol. Clin. Immunology 20, 95–102.

Espinosa, A.J.F., Rodriguez, M.T., de la Rosa, F.J.B., Sanchez, J.C.J., 2001. Size distribution of metals in urban aerosols in Seville (Spain). Atmosp. Environ. 35, 2595–2601. Available from: https://doi.org/10.1016/s1352-2310(00)00403-9.

Fowler, D., 2002. In: Bell, J.N.B., Treshow, M. (Eds.), Pollutant Deposition and Uptake by Vegetation.: Air Pollution and Plant Life. Wiley, Chichester, pp. 43–67.

Freire, C., Ramos, R., Puertas, R., Lopez-Espinosa, M.J., Julvez, J., Aguilera, I., et al., 2010. Association of traffic-related air pollution with cognitive development in children. J. Epidemiol. Community Health 64, 223–228. Available from: https://doi.org/10.1136/jech.2008.084574.

Gedzelman, S.D., Austin, S., Cermak, R., Stefano, N., Partridge, S., Quesenberry, S., et al., 2003. Mesoscale aspects of the Urban Heat Island around New York City. Theoret. Appl. Climatol. 75, 29–42. Available from: https://doi.org/10.1007/s00704-002-0724-2.

Getter, K.L., Rowe, D.B., Robertson, G.P., Cregg, B.M., Andresen, J.A., 2009. Carbon sequestration potential of extensive green roofs. Environ. Sci. Technol. 43, 7564–7570. Available from: https://doi.org/10.1021/es901539x.

Grahn, P., Stigsdotter, U.A., 2003. Landscape planning and stress. Urban For. Urban Gree. 2, 1–18. Available from: https://doi.org/10.1078/1618-8667-00019.

Gratani, L., Varone, L., Bonito, A., 2016. Carbon sequestration of four urban parks in Rome. Urban For. Urban Gree. 19, 184–193.

Grimm, N.B., Grove, J.G., Pickett, S.T.A., Redman, C.L., 2000. Integrated approaches to long-term studies of urban ecological systems: urban ecological systems present multiple challenges to ecologists—pervasive human impact and extreme heterogeneity of cities, and the need to integrate social and ecological approaches, concepts, and theory. BioScience 50, 571–584. Available from: https://doi.org/10.1641/0006-3568(2000)050[0571:iatlto]2.0.co;2.

Guite, H.F., Clark, C., Ackrill, G., 2006. The impact of the physical and urban environment on mental well-being. Public Health 120, 1117–1126. Available from: https://doi.org/10.1016/j.puhe.2006.10.005.

Haaland, C., van den Bosch, C.K., 2015. Challenges and strategies for urban green-space planning in cities undergoing densification: a review. Urban For. Urban Gree. 14, 760–771. Available from: https://doi.org/10.1016/j.ufug.2015.07.009.

Haines, A., Kovats, R.S., Campbell-Lendrum, D., Corvalan, C., 2006. Climate change and human health: impacts, vulnerability and public health. Public Health 120, 585–596.

Handley, J.F., Gill, S.E., 2009. Woodlands helping society to adapt: combating climate change – a role for UK forests. In: Read, D.J., Freer-Smith, P.H., Morison, J.I.L., Hanley, N., West, C.C., Snowdon, P. (Eds.), An Assessment of the Potential of the UK's Trees and Woodlands to Mitigate and Adapt to Climate Change. The Stationary Office, Edinburgh, pp. 180–194.

Heisler, G.M., 1986. Effects of individual trees on the solar radiation climate of small buildings. Urban Ecol. 9, 337–359.

Hert, M., Correll, C.U., Bobes, J., Cetkovich-Bakmas, M., Cohen, D.A.N., Asai, I., et al., 2011. Physical illness in patients with severe mental disorders. I. Prevalence, impact of medications and disparities in health care. World Psych. 10, 52–77. Available from: https://doi.org/10.1002/j.2051-5545.2011.tb00014.x.

Hutchinson, D., Abrams, P., Retzlaff, R., Liptan, T., 2003. Stormwater monitoring two ecoroofs in Portland, Oregon, USA: Greening Rooftops for Sustainable Communities ed., Chicago.

Ierodiakonou, D., Zanobetti, A., Coull, B.A., Melly, S., Postma, D.S., Boezen, H.M., et al., 2016. Ambient air pollution, lung function and airway responsiveness in children with asthma. J. Allergy Clin. Immunol. 137, 390–399. Available from: https://doi.org/10.1016/j.jaci.2015.05.028.

IPCC, 2015. Synthesis Report. Contribution of Working Groups I, II and III to the Fifth Assessment Report of the Intergovernmental Panel on Climate Change. Core Writing Team, R.K. Pachauri and L.A. Meyer. IPCC, Geneva.

Ip, K., Lam, M., Miller, A., 2010. Shading performance of a vertical deciduous climbing plant canopy. Build. Environ. 45, 81–88.

Jackson, J.I., Boutle, R., 2008. Ecological functions within a sustainable urban drainage system.: 11th International Conference on Urban Drainage, Edinburgh, Scotland, UK, pp. 1–10.

Jacob, D.J., Winner, A.A., 2008. Effect of climate change on air quality. Atmosp. Environ. 43, 51–63.

Lelieveld, J., Evans, J.S., Fnais, M., Giannadaki, D., Pozzer, A., 2015. The contribution of outdoor air pollution sources to premature mortality on a global scale. Nature 525, 367–371. Available from: https://doi.org/10.1038/nature15371.

Li, D., Bou-Zeid, E., 2013. Synergistic interactions between urban heat islands and heat waves: the impact in cities is larger than the sum of its parts, Am. Meteorol. Soc., 52. pp. 2051–2064.

Liptan, T., Strecker, E., 2003. EcoRoofs (greenroofs) - a more sustainable infrastructure. In National Conference on Urban Storm Water: Enhancing Programs at the Local Level, pp. 198–214.

Livesley, S.J., Baudinette, B., Glover, D., 2014. Rainfall interception and stem flow by eucalypt street trees – the impacts of canopy density and bark type. Urban For. Urban Gree. 13, 192–197. Available from: https://doi.org/10.1016/j.ufug.2013.09.001.

Lovett, G.M., 1994. Atmospheric deposition of nutrients and pollutants in North-America - an ecological perspective. Ecol. Applicat. 4, 629–650. Available from: https://doi.org/10.2307/1941997.

Luck, G.W., Davidson, P., Boxall, D., Smallbone, L., 2011. Relations between urban bird and plant communities and human well-being and connection to nature. Conservat. Biol. 25, 816–826. Available from: https://doi.org/10.1111/j.1523-1739.2011.01685.x.

Maher, B.A., Ahmed, I.A.M., Karloukovski, V., MacLaren, D.A., Foulds, P.G., Allsop, D., et al., 2016. Magnetite pollution nanoparticles in the human brain. PNAS, pp. 1–5 <www.pnas. org/cgi/doi/10.1073/pnas.1605941113>.

Mitchell, R., Popham, F., 2007. Greenspace, urbanity and health: relationships in England. J. Epidemiol. Commun. Health 61, 681–683. Available from: https://doi.org/10.1136/jech.2006.053553.

Nitis, T., Klaic, Z.B., Moussiopoulos, N., 2005. Effects of topography on urban heat island. In: 10th International Conference on Harmonisation within Atmospheric Dispersion Modelling for Regulatory Purposes. Crete, p. 5.

Nowak, D.J., 1994. Air pollution removal by Chicago's urban forest. In: McPherson, E.G., Nowak, D.J., Rowntree, R.A. (Eds.), Chicago's Urban Forest Ecosystem: Results of the Chicago Urban Forest Climate Project. USDA Forest Service, Radnor, PA, pp. 63–79.

Nowak, D.J., Heisler, G.M., 2010. Air Quality Effects of Urban Trees and Parks. Executive Summary. National Recreation and Parks Association, Ashburn, VA.

O'Neill, M.S., Carter, R., Kish, J.K., Gronlund, C.J., White-Newsome, J.L., Manarolla, X., et al., 2009. Preventing heat-related morbidity and mortality: new approaches in a changing climate. Maturitas 64, 98–103. Available from: https://doi.org/10.1016/j.maturitas.2009.08.005.

Oke, T.R., 1973. City size and the urban heat island. Atmos. Environ. 7, 769–779.

Oke, T.R., 1981. Canyon geometry and the nocturnal urban heat-island - comparison of scale model and field observations. J. Climatol. 1, 237–254.

Ostendorf, M., Retzlaff, W., Thompson, K., Woolbright, M., Morgan, S., Celik, S., 2011. Stormwater runoff from green retaining wall systems. Cities Alive: 9th Annual Green Roof and Wall Conference, Philadelphia, p. 15.

Parker, D.S., Barkaszi, S.F.J., Sonne, J.K., 1996. Measured Impacts of Air Conditioner Condenser Shading. Presented at The Tenth Symposium on Improving Building Systems in Hot and Humid Climates, Texas A & M University, Fort Worth, TX, May 13-14, 1996. http://www.fsec.ucf.edu/en/publications/html/FSEC-PF-302-96/.

Parker, J.H., 1983. Landscaping to reduce the energy used in cooling buildings. J. For. 81, 82–84. + 105.

Pataki, D.E., Carreiro, M.M., Cherrier, J., Grulke, N.E., Jennings, V., Pincetl, S., et al., 2011. Coupling biogeochemical cycles in urban environments: ecosystem services, green solutions, and misconceptions. Front. Ecol. Environ. 9, 27–36. Available from: https://doi.org/10.1890/090220.

Patz, J.A., Campbell-Lendrum, D., Holloway, T., Foley, J.A., 2005. Impact of regional climate change on human health. Nature 438, 310–317.

RCP, 2016. Every Breath We Take: The Lifelong Impact of Air Pollution. Royal College of Physicians, London, p. 106.

Paykel, E.S., Abbott, R., Jenkins, R., Brugha, T.S., Meltzer, H., 2000. Urban–rural mental health differences in Great Britain: findings from the National Morbidity Survey. Psychol. Med. 30, 269–280.

Rees, W.E., 1997. Urban ecosystems: the human dimension. Urban Ecosyst. 1, 63–75.

Rietveld, L.C., Siri, J.G., Chakravarty, I., Arsénio, A.M., Biswas, R., Chatterjee, A., 2016. Improving health in cities through systems approaches for urban water management. Environ. Health 15 (Suppl. 1), 171–172.

Rosenthal, J.K., Crauderueff, R., Carter, M., 2008. Urban Heat Island Mitigation Can Improve New York City's Environment: Research on the Impacts of Mitigation Strategies. Sustainable South Bronx, New York.

Rosenzweig, C., Solecki, W., Parshall L, Gaffin S, Lynn B, Goldberg R, et al. (2006) Mitigating New York city's heat island with urban forestry, living roofs, and light surfaces. Proceedings of the Sixth Symposium on the Urban Environment, January 30–Feburary 2, Atlanta, GA, p. 5. <https://ams.confex.com/ams/pdfpapers/103341.pdf>

Schmidt, M., 2009. Rainwater harvesting for mitigating local and global warming.: Fifth Urban Research Symposium 2009: Cities and Climate Change: Responding to an Urgent Agenda. Marseille, France, pp. 1–15.

Setälä, H.M., Francini, G., Allen, J.A., Hui, N., Jumpponen, A., Kotze, D.J., 2016. Vegetation type and age drive changes in soil properties, nitrogen, and carbon sequestration in urban parks under cold climate. Front. Ecol. Evolut. 4. Available from: https://doi.org/10.3389/fevo.2016.00093.

Shahmohamadi, P., Che-Ani, A.I., Etessam, I., Maulud, K.N.A., Tawil, N.M., 2011. Healthy environment: the need to mitigate urban heat island effects on human health. Proc. Eng. 20, 61–70.

Shepherd, J.M., Pierce, H., Negri, A.J., 2002. Rainfall modification by major urban areas: Observations from spaceborne rain radar on the TRMM satellite. J. Appl. Meteorol. 41, 689–701.

Simmons, M.T., Gardiner, B., Windhager, S., Tinsley, J., 2008. Green roofs are not created equal: the hydrologic and thermal performance of six different extensive green roofs and reflective and non-reflective roofs in a sub-tropical climate. Urban Ecosyst. 11, 339–348.

Sitta, V., 1983. A living epidermis for the city, Landscape Australia, 4(83), 277–286.

Solomon, P.A., Gehr, P., Bennett, D.H., Phalen, R.F., Mendez, L.B., Rothen-Rutishauser, B., et al., 2012. Macroscopic to microscopic scales of particle dosimetry: from source to fate in the body. Air Quality Atmosp. Health 5, 169–187. Available from: https://doi.org/10.1007/s11869-011-0167-y.

Stülpnagel, A., Horbert, M., Sukopp, H., 1990. The importance of vegetation for the urban climate. In: Sukopp, H. (Ed.), Urban Ecology. SPB Academic Publishing, The Hague, pp. 175–193.

Taha, H., 1997. Urban climates and heat islands: Albedo, evapotranspiration, and anthropogenic heat. Energy Build. 25, 99–103.

Tomlinson, C.J., Chapman, L., Thomes, J.E., Baker, C.J., 2011. Including the urban heat island in spatial heat health risk assessment strategies: a case study for Birmingham, UK. Int. J. Health Geographics 10, 42. Available from: https://doi.org/10.1186/1476-072X-10-42.

Ulrich, R., 1984. View through a window may influence recovery from surgery. Science 224, 420–421. Available from: https://doi.org/10.1126/science.6143402.

Watanabe, S., Nagano, K., Ishii, J., Horikoshi, T., 2014. Evaluation of outdoor thermal comfort in sunlight, building shade, and pergola shade during summer in a humid subtropical region. Build. Environ. 82, 556–565. Available from: https://doi.org/10.1016/j.buildenv.2014.10.002.

Watts, G., Pheasant, R., Horoshenkov, K., 2010. Tranquil spaces in a metropolitan area. Proceedings of the 20th International Congress on Acoustics, ICA 2010, August 23−27, 2010, Sydney Australia, pp. 1−6.

White, M.P., Alcock, I., Wheeler, B.W., Depledge, M.H., 2013. Would you be happier living in a greener urban area? A fixed-effects analysis of panel data. Psychol. Sci. 24, 920−928. Available from: https://doi.org/10.1177/0956797612464659.

Whitehead, P.G., Wade, A.J., Butterfield, D., 2009. Potential impacts of climate change on water quality and ecology in six UK rivers. Hydrol. Res. 40, 113−122. Available from: https://doi.org/10.2166/nh.2009.078.

Zölch, T., Maderspacher, J., Wamsler, C., Paulet, S., 2016. Using green infrastructure for urban climate-proofing: An evaluation of heat mitigation measures at the micro-scale. Urban For. Urban Gree. 20, 305−316. Available from: https://doi.org/10.1016/j.ufug.2016.09.011.

Chapter 1.2

Ecosystem Services in Urban Environments

Sarah Milliken

Chapter Outline

INTRODUCTION

Ecosystem services are the benefits provided by nature to society and the economy. They can be classified into four broad categories: (1) provisioning services, such as the supply of food and other raw materials; (2) regulating services, such as the mitigation of air pollution and the regulation of local and global climate; (3) cultural services, such as social relations, health and well-being; and (4) supporting services, such as water cycling and biodiversity (Millennium Ecosystem Assessment, 2005). Urban ecosystem services are generated in a diverse set of habitats, including street trees, parks, cemeteries, and private gardens. They are also generated by engineered green infrastructure, such as sustainable urban drainage systems, green roofs, and vertical greening systems.

Since the seminal article by Bolund and Hunhammar (1999), there has been a growing body of literature on the biophysical, economic, and sociocultural dimensions of ecosystem services in urban areas, and major initiatives such as the Millennium Ecosystem Assessment (2005) and The Economics of Ecosystems and Biodiversity (2011) have sparked increasing attention in public policy discourse around the world. The benefits of urban ecosystem services should be easily understood by policy makers and planners. For example, street trees contribute toward urban cooling, reducing air conditioning costs, and greenhouse gas emissions; they contribute to climate

Nature Based Strategies for Urban and Building Sustainability.
DOI: https://doi.org/10.1016/B978-0-12-812150-4.00002-1

change mitigation by sequestering and storing carbon; they improve air quality by filtering particulates and other airborne pollutants, thereby lowering health costs; they intercept stormwater, reducing the need for flood infrastructure; and they enhance the quality of place. While some benefits are directly measurable and have hard values, such as the energy savings due to the insulation provided by vertical greening systems, others are not so readily measurable and these soft values are difficult to estimate, such as the health benefits of a rooftop garden.

It is important to bear in mind that urban ecosystems do not only produce services, but also disservices, which have been defined as "functions of ecosystems that are perceived as negative for human well-being" (Lyytimäki and Sipilä, 2009: 311). For example, some species of trees and shrubs commonly planted in cities emit volatile organic compounds that can lead to secondary formation of ground-level ozone and indirectly contribute to urban smog, while some wind-pollinated plants can cause allergic reactions. The costs of disservices such as these can also be measured and valued.

Ecosystem services and disservices are coproduced by people and nature, and a social-ecological approach to planning and policies will become increasingly necessary in order to enhance human well-being in urban areas in the face of new and complex challenges such as climate change and migration (Kremer et al., 2015). Understanding and addressing resilience through urban ecosystem services should enable urban planning to become more adaptive and reflexive, but in order to improve resilience we need to understand the complex, interactive nature of urban social-ecological systems (McPhearson et al., 2015).

PROVISIONING SERVICES

Provisioning services are the material or energy outputs from ecosystems. The high proportion of impermeable surfaces in urban areas restricts the production of goods, such as food, raw materials, and medicinal resources. Nevertheless, there is some food production by community farms, domestic gardens and allotments and, in recent years, a growing number of urban farming projects have been set up in and on buildings, including open rooftop farms, rooftop greenhouses, and indoor farms, largely driven by the desire to reconnect food production and consumption. Urban farming practices range from community-based to commercial flagship projects, and present innovative opportunities for recycling resources, especially those derived from synergies between agriculture and buildings, such as the reuse of residential or industrial wastewater, waste heat, and organic waste (Ackerman et al., 2014; Buehler and Junge, 2016; Specht et al., 2014; Thomaier et al., 2014).

REGULATING SERVICES

In urban contexts regulating services include the regulation of air quality, noise, climate, and stormwater. Air pollution caused by motor vehicle exhaust emissions, such as nitrogen dioxide and particulate matter, is a major health issue in cities around the world. Urban transport infrastructure often results in the funnelling of pedestrians along major roads, where the concentration of air pollution is highest. Green corridors across cities can reduce pedestrian exposure to pollution by providing alternative routes. Vegetation removes pollutants in several ways. Plants take up gaseous pollutants through their stomata, intercept particulate matter with their leaves, and are capable of breaking down certain organic compounds such as polyaromatic hydrocarbons in their plant tissues or in the soil. In addition, they indirectly reduce air pollutants by lowering surface temperatures through transpirational cooling and by providing shade, which in turn decreases the photochemical reactions that form pollutants such as ozone in the atmosphere (Rowe, 2011).

There is a growing corpus of studies from around the world which suggest that street trees can improve air quality by trapping particulates (Baró et al., 2014; McPherson et al., 2016; Pugh et al., 2012; Russo et al., 2016; Soares et al., 2011; Tallis et al., 2011). Such studies tend to be based on modeling, leading to claims that "the removal of atmospheric pollutants by vegetation is one of the most commonly cited ecosystem services, yet it is one of the least supported empirically" (Pataki et al., 2011: 32). However, recent empirical studies have shown that roadside vegetation can indeed play an important role in the reduction of traffic induced air pollution (Islam et al., 2012; Vailshery et al., 2013). The potential of urban vegetation to remove airborne pollutants is in fact context dependent due to the high spatial variability in and among cities, and depends on multiple factors such as the weather, the pollution concentration, and the type and quality of vegetation (Setälä et al., 2013). In some instances, however, roadside planting can lead to increased concentrations of pollutants because the trees and other types of vegetation reduce the ventilation that is needed for diluting them (Vos et al., 2013; Wania et al., 2012), while landscape and tree management practices can also be polluting (Escobedo et al., 2011; Roy et al., 2012).

In contrast to trees, the filtration potential of herbaceous vegetation is comparatively understudied. Herbaceous vegetation that is diversely structured in terms of plant height, branching pattern, or leaf traits is more effective at binding particulate matter than monotonously structured vegetation (Säumel et al., 2016; Weber et al., 2014). Modeling and empirical research both suggest that green roofs and vertical greening systems can reduce air pollution both directly as well as indirectly through reduced energy consumption resulting from their insulating properties (Berardi et al., 2014; Ottelé et al., 2010; Pérez-Urrestarazu et al., 2016; Rowe, 2011; Vijayaraghavan, 2016).

One particular ecosystem service that has become a high-profile feature of climate change mitigation efforts is carbon storage. The vast majority of urban vegetation carbon stocks are attributable to trees, rather than herbaceous and woody material. Urban trees act as sinks of carbon dioxide by storing excess carbon as biomass during photosynthesis. Sequestration rates vary locally based on tree size and health, and the growth rates associated with different species and particular site conditions (Davies et al., 2013; Nowak et al., 2013; Strohbach and Haase, 2012). Green roofs and vertical greening systems can also play a small part in reducing carbon dioxide in the atmosphere, by sequestering it in the plant tissues and the soil substrate via plant litter and root exudate, and by reducing energy consumption by insulating individual buildings (Rowe, 2011).

Noise is a major pollution problem in cities and can affect human health through physiological and psychological damage. Urban soil and vegetation can attenuate noise pollution through absorption, deviation, reflection, and refraction of sound waves (Gómez-Baggethun and Barton, 2013). Green roofs can reduce noise pollution by providing increased insulation and by absorbing sound waves diffracting over roofs (Rowe, 2011; Vijayaraghavan, 2016). The vegetated surfaces of vertical greening systems can block high frequency sounds, and when constructed with a substrate or growing medium, they can also block low-frequency noises (Azkorra et al., 2015; Pérez et al., 2016).

Urban vegetation can lower air temperatures through the evaporation of water and by providing shading. Urban areas often experience elevated temperatures compared with the surrounding countryside, because of extensive heat absorbing surfaces, such as concrete and tarmac, concentrated heat production, and impeded airflow. This is known as the urban heat island effect (Taha, 1997). Heat waves during the summer pose significant health risks to urban populations either directly from the heat or from increased air pollution. The problem of the urban heat island effect is likely to get worse with climate change, as mean temperatures are predicted to rise, as are the frequencies of heat waves. Street trees can help to reduce the urban heat island effect. Their three dimensional nature means that as well as having a cool canopy, they also shade adjacent areas, which lowers the surface temperature of the shaded area and reduces the storage and convection of heat (Armson et al., 2012).

Green roofs also have the potential to reduce the urban heat island effect. Empirical and simulation studies show that green roofs increase the evapotranspiration rate through the addition of soil and plants, and reduce the proportion of infrared radiation returned to the air (Santamouris, 2014; Speak et al., 2013a; Susca et al., 2011); different types of low-growing plants have been shown to vary in their ability to cool air temperatures (Blanusa et al., 2013). While the contribution of green roofs to mitigating heat stress at the pedestrian level is negligible (Alcazar et al., 2016; Zölch et al., 2016), vertical greening systems intercept both light and heat radiation which would otherwise be largely absorbed and converted to heat by the building surfaces and then radiated back into the surrounding streetscape (Pérez et al., 2014; Perini et al., 2011). The effectiveness of this cooling effect is related primarily to the total area shaded and evapotranspiration effects of the vegetation, rather than the thickness of the vertical greening system. Other potential benefits of vertical greening systems include bioshading—reducing sunlight penetration through windows. With strategic placement, the plants in vertical greening systems can also create enough turbulence to break vertical airflow, which slows and cools down the air (Pérez-Urrestarazu et al., 2016).

Ecosystem services-based approaches can be used both to regulate the urban water cycle by reducing the amount of stormwater runoff, and to improve water quality by removing pollutants from runoff; e.g., vegetated streetscapes designed to absorb water, such as bioswales and rain gardens, have been shown to be effective (Pataki et al., 2011). Street trees intercept rainfall in their canopies and store water on their leaves and stems until it is subsequently evaporated. However, the gross interception rate varies greatly with species and tree size. Trees planted in tree pits considerably increase the infiltration rate and thereby reduce surface water runoff (Armson et al., 2013). Green roofs may also delay the timing of peak runoff, thereby alleviating stress on storm-sewer systems, by storing water in the growing medium and to a lesser extent in the vegetation canopy. A roof's ability to retain stormwater depends on factors such as the intensity and duration of the rain event as well as substrate depth, substrate moisture content at the start of the rain event, and the type, health, and density of the vegetation (Sims et al., 2016; Speak et al., 2013b; Whittinghill et al., 2015). There is still debate as to whether green roofs act as a source or sink of pollutants (Berardi et al., 2014). The specific nature of runoff quality from green roofs is highly dependent on the green roof components; and nutrient concentrations in runoff decrease over time after installation (Vijayaraghavan et al., 2012). Extensive green roofs would appear to be better than intensive systems in terms of pollutant removal, which may be related to the reduced volume of soil that can leach pollutants (Carpenter et al., 2011; Razzaghmanesh et al., 2014).

CULTURAL SERVICES

The Millennium Ecosystem Assessment defined cultural ecosystem services as "the nonmaterial benefits people obtain from ecosystems through spiritual enrichment, cognitive development, reflection, recreation, and aesthetic experiences" (Millennium Ecosystem Assessment, 2005: 40). Urban green infrastructure provides diverse social, psychological, and aesthetic benefits. Green spaces can improve mental health and the quality of community life, and researchers have observed a link between increasing urbanization and psychosis or depression (Annerstedt et al., 2012). Experimental evidence suggests that simply having views of nature can improve mood, self-esteem and concentration, increase job satisfaction, and help to treat stress and mental health disorders (Depledge et al., 2011; Douglas, 2012; Peckham et al., 2013).

While cultural ecosystem services are often neglected because they are challenging to assess, their valuation should be at the top of urban ecosystem services research priorities. Developing better understanding of urban dwellers' perceptions and the multilayered benefits and values they derive from urban cultural ecosystem services can help ensure that urban planning and decision making is grounded in and suitable for the particular social-ecological systems they serve (Kremer et al., 2015). Andersson et al. (2015) proposed that cultural ecosystem services and urban nature experiences can be used as a gateway to more informed discussions about what kind of urban green infrastructure is desirable, and can guide efforts to build support for all urban ecosystem services. Cultural ecosystem services can be especially important in cities since they are intimately known and acknowledged by most urban residents. People often notice changes in these services and can be motivated to engage in their protection or promotion. Since these are often bundled with other ecosystem services, engaging in their stewardship will implicitly include these as well (Ernstson, 2013).

SUPPORTING SERVICES

Supporting services are those that are necessary for the production of all other ecosystem services. They differ from provisioning, regulating, and cultural services in that their impacts on people are either indirect or occur over a very long time, whereas changes in the other categories can have relatively direct and short-term impacts (Millennium Ecosystem Assessment, 2005). In urban ecosystems, the most important supporting service is the provisioning of the habitat. Towns and cities are characterized by scattered habitat patches that form a network of nodes and links. The spaces between green areas are not completely blank but contribute to ecological connectivity in different ways. Theory predicts thresholds in several parameters such as patch size and connectivity below which ecological traits such as biodiversity or ecological functions decrease rapidly or even disappear completely. The planning and management of urban green areas should therefore adopt a spatially explicit approach that considers landscape connectivity, the scale of movement of different organisms, and how they use the many different habitats offered by cities. Indeed, for some species in urban environments, the configuration of the local habitat within the landscape may be as critical as the composition of the local habitat itself (Andersson and Bodin, 2008; Braaker et al., 2014).

The importance of biodiversity in underpinning the delivery of ecosystem services and the ecosystem processes that underlie them is well recognized. For example, the dynamics of soil nutrient cycles are determined by the composition of biological communities in the soil. Resilience to pests and environmental change is increased in more diverse biological communities and, in many contexts, higher biodiversity is associated with increased ecosystem functions (Mace et al., 2012). Urban biodiversity can be very high, indeed often much higher than in the surrounding agricultural landscapes, because of the high degree of heterogeneity of urban ecosystems which are characterized by a mosaic of different habitats (Andersson et al., 2014).

Increasing plant species diversity and increasing the range of vegetation types in cities can significantly increase other forms of biodiversity. Street trees enhance biodiversity by providing food, habitat, and landscape connectivity for urban fauna and, together with green roofs and vertical greening systems, should be part of an overall urban greening strategy linking different ground-level habitat patches.

THE VALUE OF ECOSYSTEM SERVICES

Presenting the economic benefits of urban ecosystem services in monetary terms allows them to be easily understood by policy and decision makers. The green infrastructure valuation toolkit (http://www.greeninfrastructurenw. co.uk) provides a flexible framework for identifying and assessing the potential economic and wider returns from investment in landscape schemes by valuing a range of ecosystem services including climate change adaptation and mitigation, water and flood management, quality of place, health, and well-being, and biodiversity (Milliken, 2013). The i-Tree software (http:// www.itreetools.org) is widely used to calculate the value of urban trees in North America and, increasingly, in Europe. The tool quantifies pollution removal, carbon storage, and stormwater reduction, in order to calculate the economic value of these ecosystem services (Baró et al., 2014; Elmqvist et al., 2015; Mullaney et al., 2015; Soares et al., 2011). Other types of modeling have been used to measure and value ozone uptake by urban trees (Manes et al., 2012), while the presence of street trees has been found to add value to the price of residential property (Donovan and Butry, 2010; Escobedo et al., 2015; Pandit et al., 2013).

CONCLUSION

Multiple benefits and both material and nonmaterial values can be produced simultaneously by the same system components. For example, community-based food production on a rooftop farm has added benefits such as regulating stormwater, enhancing biodiversity, improving human health and well-being, and fostering social cohesion. Conceiving of the multiple kinds of services—provisioning, regulating, cultural and supporting—associated with a particular place enables a more holistic understanding of the ways that humans benefit from ecosystem services and how they can be synergistically managed (Andersson et al., 2015).

Urban morphology is an important factor influencing the provision of multiple ecosystem services. Connectivity adds complexity to ecological systems, and even small patches in a fragmented urban landscape can be of disproportionally high importance in terms of the generation of ecosystem services. The widespread trend to reduce urban sprawl by developing dense, compact cities is likely to lead to a deterioration in ecosystem service provision with consequent declines in both urban biodiversity and the quality of life of the human population (Holt et al., 2015). Engineered green infrastructure, such as green roofs, vertical greening systems, and rain gardens, present opportunities for providing that all important connectivity, even in the densest of cities. Urban planners need to use them in order to create networks of green space based on key ecological processes, such as the movement patterns of pollinators or seed dispersers, in order to realize the potential of those benefits which are dependent on ecological network structures.

REFERENCES

Ackerman, K., Conard, M., Culligan, P., Plunz, R., Sutto, M.P., Whittinghill, L., 2014. Sustainable food systems for future cities: the potential of urban agriculture. Econ. Soc. Rev. 45, 189–206.

Alcazar, S.S., Olivieri, F., Neila, J., 2016. Green roofs: experimental and analytical study of its potential for urban microclimate regulation in Mediterranean-continental climates. Urban Clim. 17, 304–317.

Andersson, E., Bodin, O., 2008. Practical tool for landscape planning? An empirical investigation of network based models of habitat fragmentation. Ecography 32, 123–132.

Andersson, E., Barthel, S., Borgström, S., Colding, J., Elmqvist, T., Folke, C., et al., 2014. Reconnecting cities to the biosphere: stewardship of green infrastructure and urban ecosystem services. Ambio 43, 445–453.

Andersson, E., Tengö, M., McPhearson, T., Kremer, P., 2015. Cultural ecosystem services as a gateway for improving urban sustainability. Ecosyst. Serv. 12, 165–168.

Annerstedt, M., Östergren, P.O., Björk, J., Grahn, P., Skärbäck, E., Währborg, P., 2012. Green qualities in the neighbourhood and mental health – results from a longitudinal cohort study in Southern Sweden. BMC Public Health 12, 337. Available from: https://doi.org/10.1186/1471-2458-12-337.

Armson, D., Stringer, P., Ennos, A.R., 2012. The effect of tree shade and grass on surface and globe temperatures in an urban area. Urban For. Urban Gree. 11, 245–255.

Armson, D., Stringer, P., Ennos, A.R., 2013. The effect of street trees and amenity grass on urban surface water runoff in Manchester, UK. Urban For. Urban Gree. 12, 282–286.

Azkorra, Z., Pérez, G., Coma, J., Cabeza, L.F., Bures, S., Álvaro, J.E., 2015. Evaluation of green walls as a passive acoustic insulation system for buildings. Appl. Acoust. 89, 46–56.

Baró, F., Chaparro, L., Gómez-Baggethun, E., Langemeyer, J., Nowak, D.J., Terradas, J., 2014. Contribution of ecosystem services to air quality and climate change mitigation policies: the case of urban forests in Barcelona, Spain. Ambio 43, 466−479.

Berardi, U., GhaffarianHoseini, A.H., GhaffarianHoseini, A., 2014. State-of-the-art analysis of the environmental benefits of green roofs. Appl. Energy 115, 411−428.

Blanusa, T., Vaz Monteiro, M.M., Fantozzi, F., Vysini, E., Li, Y., Cameron, R.W.F., 2013. Alternatives to Sedum on green roofs: can broad leaf perennial plants offer better 'cooling service'? Buil. Environ. 59, 99−106.

Bolund, P., Hunhammar, S., 1999. Ecosystem services in urban areas. Ecol. Econ. 29, 293−301.

Braaker, S., Ghazoul, J., Obrist, M.M., Moretti, M., 2014. Habitat connectivity shapes urban arthropod communities: the key role of green roofs. Ecology 95, 1010−1021.

Buehler, D., Junge, R., 2016. Global trends and current status of commercial urban rooftop farming. Sustainability 8, 1108. Available from: https://doi.org/10.3390/su8111108.

Carpenter, D.D., Kaluvakolanu, P., 2011. Effect of roof surface type on stormwater run-off from full-scale roofs in a temperate climate. J. Irrigat. Drain. Eng. 137, 161−169.

Davies, Z.G., Dallimer, M., Edmondson, J.L., Leake, J.R., Gaston, K.J., 2013. Identifying potential sources of variability between vegetation carbon storage estimates for urban areas. Environ. Pollut. 183, 133−142.

Depledge, M.H., Stone, R.J., Bird, W.J., 2011. Can natural and virtual environments be used to promote improved human health and wellbeing? Environ. Sci. Technol. 45, 4660−4665.

Donovan, G.H., Butry, D.T., 2010. Trees in the city: valuing street trees in Portland, Oregon. Landscape Urban Plan. 94, 77−83.

Douglas, I., 2012. Urban ecology and urban ecosystems: understanding the links to human health and well-being. Curr. Opin. Environ. Sustain. 4, 385−392.

Elmqvist, T., Setälä, H., Handel, S.N., van der Ploeg, S., Aronson, J., Blignaut, J.N., et al., 2015. Benefits of restoring ecosystem services in urban areas. Curr. Opin. Environ. Sustain. 14, 101−108.

Ernstson, H., 2013. The social production of ecosystem services: a framework for studying environmental justice and ecological complexity in urbanized landscapes. Landscape Urban Plan. 109, 7−17.

Escobedo, F.J., Kroeger, T., Wagner, J.E., 2011. Urban forests and pollution mitigation: analyzing ecosystem services and disservices. Environ. Poll. 159, 2078−2087.

Escobedo, F.J., Adams, D.C., Timilsina, N., 2015. Urban forest structure effects on property value. Ecosyst. Serv. 12, 209−217.

Gómez-Baggethun, E., Barton, D.N., 2013. Classifying and valuing ecosystem services for urban planning. Ecol. Econ. 86, 235−245.

Holt, A.R., Mears, M., Maltby, M., Warren, P., 2015. Understanding spatial patterns in the production of multiple urban ecosystem services. Ecosyst. Serv. 16, 33−46.

Islam, M.N., Rahman, K.S., Bahar, M.M., Habib, M.A., Ando, K., Hattori, N., 2012. Pollution attenuation by roadside greenbelt in and around urban areas. Urban For. Urban Gree. 11, 460−464.

Kremer, P., Andersson, E., Elmqvist, T., McPhearson, T., 2015. Advancing the frontier of urban ecosystem services research. Ecosyst. Serv. 12, 149−151.

Lyytimäki, J., Sipilä, M., 2009. Hopping on one leg − the challenge of ecosystem disservices for urban green management. Urban For. Urban Gree. 8, 309−315.

Mace, G.M., Norris, K., Fitter, A.H., 2012. Biodiversity and ecosystem services: a multilayered relationship. Trends Ecol. Evolut. 27, 19−26.

Manes, F., Incerti, G., Salvatori, E., Vitale, M., Ricotta, C., Costanza, R., 2012. Urban ecosystem services: tree diversity and stability of tropospheric ozone removal. Ecol. Applicat. 22, 349–360.

McPhearson, T., Andersson, E., Elmqvist, T., Frantzeskaki, N., 2015. Resilience of and through urban ecosystem services. Ecosyst. Serv. 12, 152–156.

McPherson, G., van Doorn, N., de Goede, J., 2016. Structure, function and value of street trees in California, USA. Urban For. Urban Gree. 17, 104–115.

Millennium Ecosystem Assessment, 2005. Ecosystems and Human Well-Being: Synthesis. Washington, DC: Island Press.

Milliken, S., 2013. The value of green infrastructure in urban design. Urban Des. J. 126, 18–20.

Mullaney, J., Lucke, T., Trueman, S.J., 2015. A review of benefits and challenges in growing street trees in paved urban environments. Landscape Urban Plan. 134, 157–166.

Nowak, D.J., Greenfield, E.J., Hoehn, R.E., Lapoint, E., 2013. Carbon storage and sequestration by trees in urban and community areas of the United States. Environ. Poll. 178, 229–236.

Ottelé, M., van Bohemen, H.D., Fraaij, A.L.A., 2010. Quantifying the deposition of particulate matter on climber vegetation on living walls. Ecol. Eng. 36, 154–162.

Pandit, R., Polyakov, M., Tapsuwan, S., Moran, T., 2013. The effect of street trees on property value in Perth, Western Australia. Landscape Urban Plan. 110, 134–142.

Pataki, D.E., Carreiro, M.M., Cherrier, J., Grulke, N.E., Jennings, V., Pincetl, S., et al., 2011. Coupling biogeochemical cycles in urban environments: ecosystem services, green solutions, and misconceptions. Front. Ecol. Environ. 9, 27–36.

Peckham, S.C., Duinker, P.N., Ordóñez, C., 2013. Urban forest values in Canada: views of citizens in Calgary and Halifax. Urban For. Urban Gree. 12, 154–162.

Pérez, G., Coma, J., Martorell, I., Cabeza, L., 2014. Vertical Greenery Systems (VGS) for energy saving in buildings: a review. Renew. Sustain. Energy Rev. 39, 139–165.

Pérez, G., Coma, J., Barrenche, C., de Gracia, A., Urrestarazu, M., Burés, S., et al., 2016. Acoustic insulation capacity of Vertical Greenery Systems for buildings. Appl. Acoust. 110, 218–226.

Pérez-Urrestarazu, L., Fernández-Cañero, R., Franco-Salas, A., Egea, G., 2016. Vertical greening systems and sustainable cities. J. Urban Technol. 22 (4), 1–21.

Perini, K., Ottelé, M., Fraaij, A.L.A., Haas, E.M., Raiteri, R., 2011. Vertical greening systems and the effect on air flow and temperature on the building envelope. Buil. Environ. 46, 2287–2294.

Pugh, T.A.M., Mackenzie, A.R., Whyatt, J.D., Hewitt, C.N., 2012. Effectiveness of green infrastructure for improvement of air quality in urban street canyons. Environ. Sci. Technol. 46, 7692–7699.

Razzaghmanesh, M., Beecham, S., Kazemi, F., 2014. Impact of green roofs on stormwater quality in a South Australian urban environment. Sci. Total Environ. 470–471, 651–659.

Rowe, D.B., 2011. Green roofs as a means of pollution abatement. Environ. Poll. 159, 2100–2110.

Roy, S., Byrne, J., Pickering, C., 2012. A systematic quantitative review of urban tree benefits, costs, and assessment methods across cities in different climatic zones. Urban For. Urban Gree. 11, 351–363.

Russo, A., Escobedo, F.J., Zerbe, S., 2016. Quantifying the local-scale ecosystem services provided by urban treed streetscapes in Bolzano, Italy. AIMS Environ. Sci. 3, 58–76.

Santamouris, M., 2014. Cooling the cities – a review of reflective and green roof mitigation technologies to fight heat island and improve comfort in urban environments. Solar Energy 103, 682–703.

Säumel, I., Weber, F., Kowarik, I., 2016. Toward livable and healthy urban streets: roadside vegetation provides ecosystem services where people live and move. Environ. Sci. Policy 62, 24–33.

Setälä, H., Viippola, V., Rantalainen, A.L., Pennanen, A., Yli-Pelkonen, V., 2013. Does urban vegetation mitigate air pollution in northern conditions? Environ. Poll. 183, 104−112.

Sims, A.W., Robinson, C.E., Smart, C.C., Voogt, J.A., Hay, G.J., Lundholm, J.T., et al., 2016. Retention performance of green roofs in three different climate regions. J. Hydrol. 542, 115−124.

Soares, A.L., Rego, F.C., McPherson, E.G., Simpson, J.R., Peper, P.J., Xiao, Q., 2011. Benefits and costs of street trees in Lisbon, Portugal. Urban For. Urban Gree. 10, 69−78.

Speak, A., Rothwell, J., Lindley, S., Smith, C., 2013a. Reduction of the urban cooling effects of an intensive green roof due to vegetation damage. Urban Clim. 3, 40−55.

Speak, A.F., Rothwell, J.J., Lindley, S.J., Smith, C.L., 2013b. Rainwater runoff retention on an aged intensive green roof. Sci. Total Environ. 461, 28−38.

Specht, K., Siebert, R., Hartmann, I., Freisinger, U.B., Sawicka, M., Werner, A., et al., 2014. Urban agriculture of the future: an overview of sustainability aspects of food production in and on buildings. Agri. Human Values 31, 33−51.

Strohbach, M.W., Haase, D., 2012. Above-ground carbon storage by urban trees in Leipzig, Germany: analysis of patterns in a European city. Landscape Urban Plan. 104, 95−104.

Susca, T., Gaffin, S.R., Dell'Osso, G.R., 2011. Positive effects of vegetation: urban heat island and green roofs. Environ. Poll. 159, 2119−2126.

Taha, H., 1997. Urban climates and heat islands: albedo, evapotranspiration, and anthropogenic heat. Energy Buil. 25, 99−103.

Tallis, M., Taylor, G., Sinnett, D., Freer-Smith, P., 2011. Estimating the removal of atmospheric particulate pollution by the urban tree canopy of London, under current and future environments. Landscape Urban Plan. 103, 129−138.

The Economics of Ecosystems and Biodiversity, 2011. TEEB Manual for Cities: Ecosystem Services in Urban Management. <www.teebweb.org>

Thomaier, S., Specht, K., Henckel, D., Dierich, A., Siebert, S., Freisinger, U.B., et al., 2014. Farming in and on urban buildings: present practice and specific novelties of Zero-Acreage Farming (ZFarming). Renew. Agri. Food Syst. 30, 43−54.

Vailshery, L.S., Jaganmohan, M., Nagendra, H., 2013. Effect of street trees on microclimate and air pollution in a tropical city. Urban For. Urban Gree. 12, 408−415.

Vijayaraghavan, K., 2016. Greenroofs: a critical review on the role of components, benefits, limitations and trends. Renew. Sustain. Energy Rev. 57, 740−753.

Vijayaraghavan, K.U., Joshi, M., Balasubramanian, R., 2012. A field study to evaluate runoff quality from green roofs. Water Res. 46, 1337−1345.

Vos, P.E.J., Maiheu, B., Vankerkom, J., Janssen, S., 2013. Improving local air quality in cities: to tree or not to tree? Environ. Poll. 183, 113−122.

Wania, A., Bruse, M., Blond, N., Weber, C., 2012. Analysing the influence of different street vegetation on traffic-induced particle dispersion using microscale simulations. J. Environ. Manage. 94, 91−101.

Weber, F., Kowarik, I., Säumel, I., 2014. Herbaceous plants as filters: immobilization of particulates along urban street corridors. Environ. Poll. 186, 234−240.

Whittinghill, L.J., Rowe, D.B., Andresen, J.A., Cregg, B.M., 2015. Comparison of stormwater runoff from sedum, native prairie, and vegetable producing green roofs. Urban Ecosyst. 18, 13−29.

Zölch, T., Maderspacher, J., Wamsler, C., Pauleit, S., 2016. Using green infrastructure for urban climate-proofing: an evaluation of heat mitigation measures at the micro-scale. Urban For. Urban Gree. 20, 305−316.

Chapter 1.3

Incentives for Nature-Based Strategies

Gary Grant

Chapter Outline

INTRODUCTION

The main categories of green infrastructure under consideration in this book include green streets, green roofs, and vertical greening systems; these are the main focuses of this chapter, which considers incentive policies. These interventions, which are collectively described as green infrastructure, are normally created as the result of customary practice, zoning policy, regulation, or a combination of these and other factors. Where regulations on their own are insufficient as policy instruments to bring about an adequate quantum of green space and related green infrastructure elements, states, city authorities, utility companies, and non-governmental organizations (NGOs), do occasionally provide incentives to overcome resistance. There is a commonly-held perception, particularly in the private sector, that greening is an economic burden, even though there is evidence that city greening is a good investment (Greater London Authority, 2015). In the case of cities in regions with fiscal stress or stagnation, there is often a concern that, without incentives, there is the possibility that development could be stifled altogether if green infrastructure is required by regulation. Another concern is that, with insufficient vegetation and soil, low-quality environments may be created that would be more susceptible to problems (like pollution or

Nature Based Strategies for Urban and Building Sustainability.
DOI: https://doi.org/10.1016/B978-0-12-812150-4.00003-3

29

flooding) that might require expensive public-sector interventions at a later date. In these situations, incentives are considered by many towns and cities to be a good investment.

DEFINITIONS
Incentives

In the widest sense, anything that stimulates or motivates someone to take a different course of action could be considered an incentive, however in terms of planning and urban development, an incentive is taken to mean a payment, discount or reduction on fees or a concession (e.g., a rebate of local tax), which is offered by the planning authority, public utility, or occasionally an NGO, on condition that a particular green infrastructure feature of a specified quantum (and, occasionally quality) is provided.

Incentive Zoning

Incentive zoning occurs when a local government allows a developer to build in a way that would not normally be permitted in return for a public benefit that would not normally be provided. One of the earliest examples of this approach being used to create public space was in New York City in 1961 (American Planning Association, 2017). Such schemes have been controversial and have been subjected to legal challenge, but usually succeed if goals and definitions are clearly described in the regulations.

INCENTIVE SCHEMES BY CATEGORY
Tree Planting

Tree planting and the planting of avenues of trees along streets have been promoted since the Renaissance and codified in many cities for more than a century. It is now common practice in modern cities for local authorities to require the planting of street trees—usually through local ordinances. In most cases, city authorities bear the cost of planting and maintaining trees, however budgetary constraints have often limited tree planting activity, and frustration over the lack of action by the authorities has stimulated the growth of NGOs devoted to promoting urban tree planting.

These NGOs work nationally, e.g., in the United Kingdom, there is the Tree Council (2017), or more often, in individual cities, where the NGOs work closely with the authorities, businesses, and schools in partnerships. In cities all over the world, there are schemes that provide free trees, full grants for the purchase of trees, or contributions toward the cost of the purchase of trees. Although the process of incentivizing tree planting continues from year to year in most cities, individual schemes tend to be promoted as part of a campaign and are usually short lived, with funding allocated as part of a short-term local government budget or secured following a sponsorship agreement with a business.

A typical scheme involves the distribution of free trees for volunteers to plant under supervision in public spaces or on their own land. An example of this is the TreeFolk organization in Austin, Texas (TreeFolks, 2017) or the City Plants scheme in Los Angeles, where every resident is entitled to receive up to seven free trees (City Plants, 2017); however more elaborate schemes, where help-in-kind is provided, have been devised. An example of one such scheme is the Share-The-Cost program in Champaign, Illinois, in which the city spends $135 planting a tree once a written agreement is signed to ensure that newly planted trees will be maintained by local residents for at least 3 years. The program was designed to provide an incentive to both the businesses and individual residents to plant and nurture more trees (City of Champaign, 2017).

Surface Water Management

Introduction

This category of intervention includes the creation of near-natural features as part of the source-control approach to surface water management, also known as sustainable drainage systems. These interventions include swales, rain gardens, and other features designed to intercept, attenuate, and infiltrate storm water (Grant, 2016). Concern over climate change and the prediction of more extreme weather in most regions, with heavier downpours that could lead to more flooding (IPCC, 2013), has motivated local authorities and utility companies to look at ways of encouraging the installation green infrastructure features that would alleviate such problems. Schemes designed to promote near-natural drainage improvements have also been used to increase the area of green roofs (see below).

Germany

The separation of charges for draining wastewater and surface water into sewers has enabled municipalities and utility companies to provide more accurate costs for these services. This in turn has made it possible for charges to be adjusted or rebates to be made where green infrastructure interventions lead to a reduction in the volume of surface water entering drains. This approach originated in Germany. In Berlin, e.g., charges for wastewater and surface water drainage were separated in 2000 (Berlin Wasserbetriebe, 2017). This led to increases of up to 60% in charges for surface water discharge for businesses (In 2011, the €1.90 m^{-2} charge for impervious areas was the highest in Germany) and this provided a new incentive for managers operating large sites (including factories, warehouses, schools, hospitals, and offices) to introduce near-natural features as a way of reducing those charges (Nickel et al., 2014). Since that time, Germany has pioneered the development of policy instruments that promote the use of green infrastructure in stormwater management. In the basin of the River Emscher, e.g., where water quality has been a particular concern, the separation of charges for surface water discharge has led to cities and towns introducing incentive programs. Seventeen municipal authorities in the catchment have separated charges for surface water management, which promoted disconnections. These save on average, €1 m^{-2} per annum. In addition, in 2005, the Emscher authorities created a fund of €70 million to support up to 80% of the cost of green infrastructure based stormwater projects. Larger projects (e.g., streets and housing estates) are favored. By 2011 half of the money had been spent with a 4% reduction in storm water flows into public sewers achieved—a little less than the target of 5% reduction by 2010 (Nickel et al., 2014).

United States

A review of the funding of storm water programs in the United States by the Environmental Protection Agency, having noted developments in Germany, observed that charges could be adjusted to reflect the permeability of various land use types. This in turn would mean that credits or exemptions can be made for landowners who disconnect downpipes, install retention or detention basins or permeable paving (Environmental Protection Agency, 2009). Since that time, many states and cities in the United States have adopted the approach of separating charges for draining rainwater and offering incentive schemes, many of which are funded by the charges.

Indianapolis and Marion County, Indiana, working with the United Water utility company, have previously operated an annual green infrastructure grant program which disbursed up to $20,000 to nonprofit organizations for green infrastructure construction projects, including permeable pavements and rain gardens. In 2010, $100,000 was granted in total (City of Indianapolis, 2017). Now the authority operates a multi-tiered system of credits for stormwater charges to private landowners. The authority provides discounts for a range of interventions including retention and detention ponds, swales, and rain gardens. The system requires that interventions conform with the city's stormwater specification manual and are likely to perform satisfactorily during various categories of a storm event (RRSW, 2017).

In Anne Arundel County, Maryland, property owners can apply for a tax credit of up to 10% of the cost of materials and installation of an approved stormwater management practice per year, for a period of 5 years. These installations must be of a type approved by the Office of Planning and Zoning. These include infiltration planters, biofiltration installations, and green roofs. Credit is limited to a maximum of $10,000 (Anne Arundel County, 2017).

In Prince George's County, Maryland, there is a rebate scheme for various green infrastructure categories, funded by a surface water management charge (mandated across the state of Maryland), with up to $2000 available for residential projects and up to $20,000 for commercial, multifamily dwelling, nonprofit entities, or not-for-profit organizations, including housing cooperatives. The rate for individual residences is $10.00 ft^{-2} ($100 m^{-2}) for a minimum of 300 ft^2 (30 m^2) and the rate for commercial, multifamily dwelling, and nonprofit organizations is $10.00 ft^{-2} ($100 m^{-2}), increasing to $20.00 ft^{-2} ($200 m^{-2}) if the substrate is deeper than 6 in. (150 mm) (Prince George's County, 2017).

Common with many other American cities, Minneapolis, Minnesota has a scheme that discounts surface water management fees in return for property owners providing green infrastructure that attenuates rainwater. Minneapolis provides incentives for both features that improve water quality as well as those that reduce the quantity of runoff. The maximum credit (reduction in fees) for improving water quality is 50%. With stormwater quantity credit (reduction in fees) can be between 50% and 100% depending on the feature. A single feature cannot achieve more than 100% reduction in fees in total (City of Minneapolis, 2017).

United Kingdom

In the United Kingdom, the water companies collect around £1 billion per year for the cost of draining surface water. If a customer can demonstrate that surface water from the property does not drain into a public sewer, they can apply for a rebate (Ofwat, 2017). Currently, a typical annual rebate for a domestic customer is £50 (Welsh Water, 2017).

The Drain London program was operated by the Greater London Authority from 2014 to 2016 (Greater London Authority, 2017). It disbursed £3.2 million for surface water management schemes and encouraged others to provide match funding through partnerships with local authorities. The fund supported various green infrastructure projects, including the world's first vertical rain garden (Landscape Institute, 2017) (Fig. 1).

Green Roofs

Green roofs are promoted as source control mechanisms in surface water management schemes, and there may be financial incentives associated with that. However, green roofs also provide other benefits, including protecting the waterproofing from UV light and extremes of temperature, decreasing the flow of heat into and out of a building, reducing the heat island effect, reducing noise entering a building, and providing an outlook

FIGURE 1 World's first vertical rain garden funded by the Drain London program.

that improves well-being. Given these multiple benefits, a compelling economic case for roof greening can be made, see e.g., Peng and Jim (2015). However, the additional costs (in construction and maintenance) for green roofs in comparison with conventional roofs has been cited as a barrier to uptake (Zhang et al., 2012).

Netherlands

One approach to overcoming resistance to roof greening associated with cost has been the provision of grants from local authorities. One example is that of the Municipality of the Hague in the Netherlands. The city cites climate change adaptation and clean air as reasons for building green roofs and, since 2015, it has offered a subsidy of €25 m^{-2}. The subsidy is available for private individuals and small businesses. Roofs should have a minimum area of 6 m^2 and should conform to building standards and be planted "professionally" (City of The Hague, 2017). Since 2010, Amsterdam has had a similar scheme, in which €50 m^{-2} is offered to a maximum of 50% of installation costs for green roofs and living walls. Subsidies are available up to a maximum subsidy of €20,000 for each project (Amsterdam, 2017).

Germany

Although there has been a green roof industry, guided by published standards, in Germany since the 1980s (FLL, 2008), the city of Hamburg was the first German city to adopt a comprehensive green roof strategy. The intention is to plant 100 ha of green roofs in the metropolitan area in the decade beginning 2016. The Hamburg Ministry for Environment and Energy is now providing financial support for the creation of green roofs, with a total of €3 million being made available until the end of 2019. Building owners can apply for a subsidy covering up to 60% of the cost of installation. As an added incentive, because of the rainwater retention qualities of green roofs, building owners can enjoy a 50% reduction on surface water drainage fees (European Climate Adaptation Platform, 2017).

France

In France, subsidies for roof greening are still uncommon, however the Ile de France Region, which has a reputation as a leader in urban greening, provides grants for the creation of green roofs up to €45 m^{-2} (subject to various conditions). The Department of Hauts-de-Seine, is also offering subsidies for green roofs, with up to €48 m^{-2} available for eligible projects (Adivet, 2017).

United States

In the United States, there are several cities that offer incentives for roof greening, usually in the form of a subsidy, a local tax credit, as part of a general approach to improving surface water management or as a relaxation of planning rules to allow more development. An example of the latter approach is in Austin, Texas, where in a scheme known as the Green Roof Density Bonus, developers are granted an extra 8 ft^2 of extra development for every square foot of green roof provided in a new development (City of Austin, 2017). In Seattle, Washington, the planning department runs a green factor scheme as part of its incentive zoning. An extra 3 ft^2 of development is permitted for every square foot of green roof provided (City of Seattle, 2017).

Chicago has offered subsidies for green roofs in the past. When the scheme was operating, it offered up to 50% of the cost of a green roof up to a maximum of $100,000, providing the green roof covered at least 50% of the roof space. Chicago gained early recognition in the United States as an authority with a positive attitude toward roof greening. There are more than 500 green roofs in Chicago covering more than 500,000 m^2 (City of Chicago, 2017).

In Milwaukee, Wisconsin, the Milwaukee Metropolitan Sewerage District (MMSD) has offered a $5 ft^{-2} ($50 m^{-2}) for green roofs. Building owners in both the public and private sector were eligible. Although the MMSD is not promoting a specific program for green roofs in 2017, it is continuing to support green infrastructure in general (Fresh Coast, 2017).

The approach adopted in Metro Water Services in Nashville, Tennessee, is to offer property owners (within the area serviced by combined sewers) a rebate on sewerage charges if they install green roofs. The credit to be made available is $10 for every square foot of green roof installed (equivalent to $100 m^{-2}). Credits accrue for 60 months or until a maximum limit is reached, whichever comes first (Metro Government of Nashville and Davidson County, 2017).

Beginning in 2008, New York City and New York State have been providing a 1-year tax rebate of $4.50 ft^{-2} ($45 m^{-2}) for green roofs up to $100,000 or the building's tax liability (whichever is less). To be eligible for the rebate, green roof installations should cover at least 50% of the roof and have a minimum of 2 in. (50 mm) of substrate. The scheme was amended in 2013 and extended to 2018 (NYC Mayor's Office of Sustainability, 2017).

In Philadelphia, Pennsylvania, since 2007, the city has offered a tax credit of 25%, which applies to the business privilege tax. This credit applies to all costs incurred to construct a green roof up to a ceiling of $100,000. Since 2016, the credit available has been increased to 50%. To qualify for the credit, the green roof must cover 50% of the building's rooftop or 75% of what is defined as eligible rooftop space, i.e., rooftop space available to support a green roof (City of Philadelphia, 2017).

Palo Alto, California, has a Green Roof Rebate scheme, which provides $1.50 ft^{-2} ($15 m^{-2}) for the installation of a green roof to minimize storm runoff from rooftops. The rebates are limited to a maximum of $1000 per single-family residential property and $10,000 for commercial/industrial and multifamily residential properties (City of Palo Alto, 2017).

Between 2009 and 2014, Portland, Oregon, which had developed an international reputation for innovation in stormwater management by that time, in order to promote what they call ecoroofs (extensive green roofs) on both public and private property, offered building owners an incentive of $5 ft^{-2} ($50 m^{-2}) for approved projects. The program included over 330,000 ft^2 (33,000 m^2) of green roof installations with a total private investment of over $6 million (EcoMetrix Solutions, 2014).

In Syracuse, New York, the Onondaga County Green Improvement Fund has been providing grants for the installation of a range of infrastructure projects, including green roofs. Eligible expenses include design and engineering fees as well as construction costs. By 2016, GIF had provided over $9 million in funding to 79 green infrastructure projects, including green roofs (Save the Rain, 2017).

Washington, DC has a goal of installing green roofs on 20% of its buildings by 2020. Funding of $10 ft^{-2} ($100 m^{-2}) and up to $15 ft^{-2} ($150 m^{-2}) has been made available in target watersheds. There is no restriction on the size of the project eligible for the rebate and owners of properties of all types, including residential, are encouraged to apply. For buildings with a footprint of 2500 ft^2 (250 m^2) or less, funds are available to offset the cost of a structural assessment. Additional funding may be available for features that further advance environmental goals (District of Columbia, 2017).

Baltimore, Maryland, does not offer subsidies for green roofs, however it does have a separate charge for surface water discharge and installation of a green roof could result in a reduction in that fee. In Baltimore, there is an NGO, known as Blue Water Baltimore, which is dedicated to improving the quality of water in the Baltimore watershed. Blue Water Baltimore offers a payment of \$2 ft^{-2} (about \$20 m^{-2}) for green roof installations, however payments should not exceed 50% of the total cost (Blue Water Baltimore, 2017).

A departure from the usual approach of offering a tax credit, the Metropolitan Sewer District of Greater Cincinnati and the Cincinnati Office of Environment and Sustainability, working with The Ohio Environmental Protection Agency have created the Green Roof Loan Program. A \$5,000,000 fund has been made available at below-market-rates to provide loans to cover the cost of green roof installations. Green roof installation projects on residential, commercial, and industrial buildings within the Greater Cincinnati area are eligible for loans (Project Groundwork, 2017).

Canada

Toronto, Ontario, in Canada, was a pioneer in North America in roof greening, passing its green roof bylaw in 2009. The Eco Roof Incentive Program provides incentive funding to green roof on new and existing residential, commercial, and institutional buildings in Toronto, that are not subject to the Toronto's Green Roof Bylaw. The current green roof incentive is \$100 m^{-2} (City of Toronto, 2017).

Singapore

To increase greenery provision in Singapore, in 2009, the National Parks Board introduced the Skyrise Greenery Incentive Scheme (SGIS) in which they fund up to 50% of installation costs of green roofs and living walls. Since that time, the SGIS has assisted in greening more than 110 existing buildings in Singapore by retrofitting them with extensive green roofs, edible gardens, recreational rooftop gardens, and living walls. From 2015, the project has been known as SGIS 2.0 and increases the incentives available. The project expects to run until 2020, or when the funds are exhausted, whichever comes first. All types of building, both residential and commercial are eligible and must be already occupied—funding is aimed at retrofit because new builds are subject to planning rules that require greening (Skyrise Greenery, 2017).

Japan

Nagoya has established the Private Facility Greening Support Program, which provides up to 50% of the cost of green roofs and living walls. Funds are generated by a greening tax, levied by the Aichi Prefecture (Hayashi, 2010).

PLANTING FOR POLLUTION ABATEMENT

Concern over falling air quality in London and a realization that vegetation can trap and absorb air pollution, prompted the mayor to establish the London Clean Air Fund in 2011 with £5 million provided by the central government Department for Transport. Projects funded by the scheme include living walls and street tree plantings (Transport for London, 2011).

PREFERENTIAL HOME LOANS

In order to reverse a 50-year decline in open space in Nagoya, Japan, a city of 2 million people, the city government introduced an urban-greening incentive scheme in 2008. They established a voluntary certification system, in which properties may be assigned to one of three categories, namely: fair (one star), good (two stars), or excellent (three stars). Grading is calculated by considering the area of green space compared with the overall area of a site, preservation of trees, greening of roofs and walls, and suitable maintenance. The owners or prospective owners of properties that achieve two or three star ratings are eligible for a home loan at a discounted rate (0.1%–0.2%). Five banks operating in the region have agreed to support the scheme. This approach has been advocated as a way of overcoming resistance to greening by private landowners (Hayashi, 2010).

CONCLUSION

When there is a reluctance to use regulation to create the necessary quantity and quality of green infrastructure and it is perceived that education and encouragement is not enough on its own to bring about change at the pace that will be required to address declines or a lack of capacity for climate change adaptation, authorities, utility companies, and occasionally, NGOs, have introduced incentive schemes.

Tree planting subsidies and grants are the longest-established subsidies for urban greening and these schemes continue, probably because of their popularity and effectiveness. Tree planting is highly visible and there are usually opportunities for a partnership approach, which often encourage public participation.

Most of these incentive schemes that were reviewed in relation to this chapter, relate to surface water management. The separation of surface water management fees from a general charge for water supply, drainage, and treatment services, which began in Germany around the turn of the century and was copied in the United States a decade or so later, created the possibility of crediting customers for reducing rainwater run-off rates from their properties by installing green infrastructure.

Convinced by the compelling evidence of the multiple-benefits provided by green roofs and living walls, city authorities are turning their attention to ways of increasing the uptake of these features. Although it is possible for planning authorities to encourage the creation of building-integrated vegetation on new buildings through the planning system, and even though this has been common practice in German-speaking countries for some time, many administrations are reluctant to impose new regulations. Even where planning authorities require green roofs as part of the planning process (e.g., in Singapore), it is much more difficult to bring about the greening of existing buildings (a process known as retrofitting). In these situations, roof and wall greening is being encouraged through the provision of grants. Grant schemes for roof greening are particularly common in cities in the United States, but have also been established in Europe and Asia.

In addition to greening through improvements to surface water management and grants for roof greening, in a few cases, incentives are being provided for the abatement of air pollution and it is likely that this trend will continue until city transportation propulsion completes its transition from diesel to electric and hydrogen.

Finally, there is innovation in the use of financial products to promote urban greening, including loans being made at favorable rates for roof greening in the United States and, in Japan, a certification scheme for projects to green private property, which confer eligibility for home loans at favorable rates.

REFERENCES

Adivet, 2017. Incentives and subsidies [online]. Available at: <http://www.adivet.net/en/realisation/incentives-and-subsidies.html> (accessed 30.03.17.).

American Planning Association, 2017. [online] Available at: <https://www.planning.org/divisions/planningandlaw/propertytopics.htm#Incentive> (accessed 29.03.17.).

Anne Arundel County, 2017. Surface Water Management Tax Credit [pdf]. Available at: <http://www.aacounty.org/departments/public-works/highways/forms-and-publications/SWMTaxCreditApp.pdf> (accessed 30.03.17.).

Berlin Wasserbetriebe, 2017. [online]. Available at: <http://www.bwb.de/> (accessed 29.03.17.).

Blue Water Baltimore, 2017. Best Management Practices [pdf]. Available at: <https://www.bluewaterbaltimore.org/wp-content/uploads/Green-Roof-BMP-Fact-Sheet-2013.pdf> (accessed 29.03.17.).

City of Austin, 2017. Green roofs [online]. Available at: <http://www.austintexas.gov/department/green-roofs> (accessed 29.03.17.).

City of Champaign, 2017. Tree planting [online]. Available at: <http://champaignil.gov/departments/public-works/residents/trees/tree-planting/> (accessed 29.03.17.).

City of Chicago, 2017. Green roofs [online]. Available at: <https://www.cityofchicago.org/city/en/depts/dcd/supp_info/chicago_green_roofs.html> (accessed 30.03.17.).

City of Indianapolis, 2017. Green Infrastructure Grant program [online]. Available at: <http://www.indy.gov/eGov/City/DPW/SustainIndy/GreenInfra/Pages/GreenInfrastructureGrantProgram.aspx> (accessed 29.03.17.).

City of Minneapolis, 2017. Stormwater fee credits [online]. Available at: <http://www.minneapolismn. gov/publicworks/stormwater/fee/stormwater_fee_stormwater_mngmnt_feecredits> (accessed 29.03.17.).

City of Palo Alto, 2017. Green roofs [online]. Available at: <http://www.cityofpaloalto.org/gov/ depts/pwd/stormwater/rebates/greenroofs.asp> (accessed 29.03.17.).

City of Philadelphia, 2017. Green roof tax credit [online]. Available at: <https://beta.phila.gov/ services/payments-assistance-taxes/tax-credits/green-roof-tax-credit/> (accessed 29.03.17.).

City of Seattle, 2017. Green roofs in Seattle [pdf]. Available at: <http://www.seattle.gov/docu-ments/departments/ose/green-roofs-in-seattle.pdf> (accessed 30.03.17.).

City of The Hague, 2017. Subsidy for Green Roofs [online]. Available at: <https://www. denhaag.nl/en/residents/to/Subsidy-for-green-roofs.htm> (accessed 29.03.17.).

City of Toronto, 2017. Green Roof Bylaw [online]. Available at: <http://www1.toronto.ca/wps/portal/ contentonly?vgnextoid = 83520621f3161410VgnVCM10000071d60f89RCRD&vgnextchannel = 3a7a036318061410VgnVCM10000071d60f89RCRD> (accessed 29.03.17.).

City Plants, 2017 [online]. Available at: <http://www.cityplants.org/get-free-trees/our-free-trees-program> (accessed 29.03.17.).

District of Columbia, 2017. Green roofs [online]. Available at: <https://doee.dc.gov/green-roofs> (accessed 30.03.17.).

EcoMetrix Solutions, 2014. Cost Analysis for The Portland Ecoroof Incentive [pdf]. Available at: <https://www.portlandoregon.gov/bes/article/522380> (accessed 28.03.17.).

Environmental Protection Agency, 2009. Funding Stormwater [pdf]. Available at: <https:// www3.epa.gov/region1/npdes/stormwater/assets/pdfs/FundingStormwater.pdf>

European Climate Adaptation Platform, 2017. Hamburg's Green Roof Strategy [online]. Available at: <http://climate-adapt.eea.europa.eu/metadata/case-studies/four-pillars-to-hamburg2019s-green-roof-strategy-financial-incentive-dialogue-regulation-and-science> (accessed 30.03.17.).

Forschungsgesellschaft Landschaftsentwicklung Landschaftsbau (FLL), 2008. Introduction to the FLL: Guidelines for the Planning, Construction and Maintenance of Green Roofing. EV-FLL, Bonn.

Fresh Coast, 2017. Funding programs [online]. Available at: <http://www.freshcoast740. com/ funding-programs/gipp> (accessed 30.03.17.).

Grant, G., 2016. The Water Sensitive City. John Wiley & Sons, Ltd, Chichester.

Greater London Authority, 2015. Green Infrastructure Task Force Report [pdf]. Available at: <https://www.london.gov.uk/WHAT-WE-DO/environment/environment-publications/green-infrastructure-task-force-report> (accessed 29.03.17.).

Greater London Authority, 2017. DD1250 Drain London Programme 2014-2016 [online]. Available at: <https://www.london.gov.uk/mayor-assembly/gla/governing-organisation/exec-utive-team/directors-decisions/DD1250> (accessed 30.03.17.).

Hayashi, K., 2010. Economic Incentives for Green Initiatives in Nagoya City, Japan [pdf]. Available at: <http://www.eea.europa.eu/atlas/teeb/economic-incentives-for-green-initiatives-japan> (accessed 29.03.17.).

Amsterdam, I., 2017. Subsidy for green roofs and walls [online]. Available at: <http://www.iamster-dam.com/en/media-centre/city-hall/press-releases/archive-press-releases/2010-press-room-archive/ subsidy-for-green-roofs-and-walls> (accessed 29.03.17.).

Intergovernmental Panel on Climate Change, 2013. Fifth Assessment Report [online]. Available at: <http://www.ipcc.ch/report/ar5/wg1/> (accessed 29.03.17.).

Landscape Institute, 2017. London Bridge home to world's first vertical rain garden [online]. Available at:<https://www.landscapeinstitute.org/news/london-bridge-home-to-worlds-first-vertical-rain-garden/> (accessed 29.03.17.).

Metro Government of Nashville & Davidson County, 2017. Green Roof Rebate [online]. Available at: <http://www.nashville.gov/Water-Services/Developers/Low-Impact-Development/Green-Roof-Rebate.aspx> (accessed 29.03.17.).

New York City Mayor's Office of Sustainability, 2017. Roof incentives [online]. Available at: <http://www.nyc.gov/html/gbee/html/incentives/roof.shtml> (accessed 29.03.17.).

Nickel, D., Schoenfelder, W., Medearis, D., Dolowitz, D.P., Keeley, M., Shuster, W., 2014. German experience in managing stormwater with green infrastructure. J. Environ. Plan. Manage. 57 (3), 403–423.

Ofwat, 2017. Surface Water Drainage [online]. Available at: <http://www.ofwat.gov.uk/households/your-water-bill/surfacewaterdrainage/> (accessed 29.03.17.).

Peng, L.H.L., Jim, C.Y., 2015. Economic evaluation of green-roof environmental benefits in the context of climate change: the case of Hong Kong. Urban For. Urban Gree. 14, 554–561. 2015.

Prince George's County, 2017. Rebate Amounts [online]. Available at: <http://www.princegeorgescountymd.gov/327/Project-Types-Rebate-Amounts> (accessed 29.03.17.).

Project Groundwork, 2017. Green roof loan program [pdf]. Available at: <http://projectgroundwork.org/downloads/green_roof_loan_program_final.pdf> (accessed 30.03.17.).

RRSW, 2017. City of Indianapolis/Marion County Stormwater Management Regulation Compliance [online]. Available at: <http://rrstormwater.com/city-indianapolismarion-county-stormwater-management-regulation-compliance> (accessed 29.03.17.).

Save the Rain, 2017. Green improvement fund [online]. Available at: <http://savetherain.us/green-improvement-fund-gif/> (accessed 30.03.17.).

Skyrise Greenery, 2017. Incentive scheme [online]. Available at: <https://www.nparks.gov.sg/skyrisegreenery/incentive-scheme> (accessed 29.03.17.).

Transport for London, 2011. Stunning green wall unveiled at Edgware Road Tube Station to deliver cleaner air [online]. Available at: <https://tfl.gov.uk/info-for/media/press-releases/2011/november/stunning-green-wall-unveiled-at-edgware-road-tube-station-to-deliver-cleaner-air> (accessed 29.03.17.).

Tree Council, 2017. Grants [online]. Available at: <http://www.treecouncil.org.uk/grants> (accessed 29.03.17.).

TreeFolks, 2017. Neighborwoods [online]. Available at: <http://www.treefolks.org/nw/> (accessed 29.03.17.).

Welsh Water, 2017. Surface Water Rebates [pdf]. Available at: <https://www.dwrcymru.com/_library/leaflets_publications_english/surface_water_rebate.pdf>(accessed 29.03.17.).

Zhang, X., Shen, I., Tam, V.W.Y., Lee, W.Y.L., 2012. Barriers to implement extensive green roof systems: a Hong Kong study. Renew. Sustain. Energy Rev. 16 (1), 314–319.

FURTHER READING

Montgomery County, 2017. Rainscape Rebates [online]. Available at: <https://www.montgomerycountymd.gov/water/rainscapes/rebates.html> (accessed 29.03.17.).

NParks, 2017. Incentive scheme [online]. Available at: <https://www.nparks.gov.sg/skyrisegreenery/incentive-scheme> (accessed 30.03.17.).

Pincetl, S., Gillespie, T., Pataki, D.E., Saatchi, S., Saphores, J.D., 2013. Urban tree planting programs, function or fashion? Los Angeles and urban tree planting campaigns. GeoJournal 2013 (78), 475–493.

Section II

Nature Based Strategies and Technologies

Chapter 2.1

Vertical Greening Systems: Classifications, Plant Species, Substrates

Rafael Fernández-Cañero, Luis Pérez Urrestarazu and Katia Perini

Chapter Outline

INTRODUCTION

The concern of society for the environment and urban sustainability is clearly increasing. In this context, vertical gardening is a new trend, which is presented as an alternative to the traditional system of landscaping, and involves the design and construction of gardens on a vertical plane (Perini, 2012).

However, it is not something new. The growth of vegetation on buildings has been a common practice for many centuries and in different places on the planet (Köhler, 2008). Besides the well-known green roofs, it has been common to find plants growing on the facades of buildings, both planted in the ground or in pots, or hanging from balconies and windows. Undoubtedly, the famous Hanging Gardens of Babylon (600 BC), considered one of the wonders of the ancient world, still exert a powerful influence on our imagination (Polinger Foster, 1998). An example of the traditional greening of walls is found in the houses and patios from Cordoba (Spain), which manifest the ancient origins of this city—a true melting pot of cultures. The whitewashed walls support hundreds of flowering pots that become genuine orchard patios, reaching their maximum ornamental fullness in May. There is no doubt that these traditional forms have inspired the development of technologies that support the various cutting-edge types of vertical gardens.

Nature Based Strategies for Urban and Building Sustainability.
DOI: https://doi.org/10.1016/B978-0-12-812150-4.00004-5

Vertical greening systems (VGS) are different forms of vegetated wall surfaces, based on the spreading of plant species across the wall surface by using vertical structures, which may or may not be fixed to an indoor wall or to a building façade (Francis and Lorimer, 2011). VGS can be classified according to the growing method and supporting structure employed (Dunnett and Kingsbury, 2008; Köhler, 2008; Manso and Castro-Gomes, 2015; Pérez-Urrestarazu et al., 2015). Other more elaborated classifications are based on the key components and factors of green wall variants (Jim, 2015). Table 1 shows a VGS's classification including some features such as location, materials employed for their construction, typo of vegetation, etc. There are numerous typologies, which go from the simplest shape to the most sophisticated and high-tech layout. On the basis of the support structures used and the different plants selected, these systems may be split into two mayor groups: green facades and living walls (Kontoleon and Eumorfopoulou, 2010).

Green facades or green screens are traditional vertical greenery systems in which the vegetation cover is formed by climbing or hanging plants. These are mainly rooted at the base in the ground or grown in plant containers, which can be attached to the walls or integrated onto balconies at different heights of the building (Fig. 1).

Living walls or vertical gardens are contemporary VGS—generally more complex than green facades—originally invented by the French botanist Patrick Blanc, who was inspired by the epiphyte growth of several plants (Blanc, 2012). These VGS involve a supporting structure adapted to each cultivation system (Francis and Lorimer, 2011). These structures can be made out of different materials and may support a wide variety of plant species (Loh, 2008).

This chapter presents an overview of the different types of VGS, starting from simple green facades systems, i.e., extensive VGS, to then looking at living wall systems (intensive VGS), both textile and modular. This list is not exhaustive, and the examples given are only some of the living wall systems that can be found off the shelf.

EXTENSIVE SYSTEMS—GREEN FACADES

Within this category, a first group made up of systems for facade greening which use mostly climber or hanging plants can be identified. The common factor is that plants grow from the lower part of the system, where they are placed on planters that require a wire or mesh structure upon which to develop vertically, or planted directly in the ground (Fig. 1). Depending on the species used, it can take, in some cases, years for the vegetation to fully develop until the desired surface is covered (Bellomo, 2003). However, when compared to other VGS, these systems can be relatively cost effective and easy to maintain (Ottelé et al., 2011; Perini and Rosasco, 2013).

TABLE 1 VGS's Classification and Features

| | Extensive Systems | Intensive Systems | | |
	Green Façades	Living Wall—Cloth Systems	Living Wall—Modular Panel Systems	Active Living Wall Systems
Location	Outdoors	Outdoors and indoors		Mostly indoor
Structures material	Lightweight structures of stainless steel or similar	Textile or nonwoven felt with pockets	Galvanized steel, polyethylene, or recycled plastic panels	Nonwoven porous felt
Intended goals	Aesthetic, passive energy-savings, biodiversity improvement	Aesthetic, passive energy-savings, marketing, biodiversity improvement, vegetable production	Aesthetic, passive energy-savings, marketing, noise barrier, biodiversity Improvement, vegetable production	Air biofiltration, evaporative cooling, active energy-savings, marketing, aesthetic
Growing method	Organic substrate	Hydroponic cultures or organic substrate		Hydroponic cultures with porous inorganic substrates
Vegetation	Mostly Climbing and hanging plants	Wide range of species: epiphytic, lithophytic and bromeliads, ferns, succulent, herbaceous, small shrubs, climbing plants and even vegetables		
Installation Cost	Low	Medium-high		High
Maintenance	Low	Medium-high		High

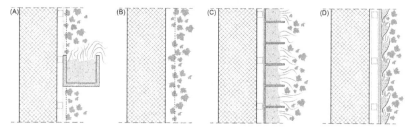

FIGURE 1 (A) Green façade with planter boxes, (B) Green façade direct greening system, (C) LWS boxes, (D) LWS felt.

In green facades, plants can be directly attached to the wall or use light-weight structures to climb and spread out over the wall. Although plants are usually cultivated in planter boxes placed on the base of buildings, it is also possible to place these planter boxes at different heights to accelerate the coverage of the surface and combine them with hanging plants. Whatever the option chosen, maintenance is relatively simple, on the one hand due to the simplicity of the system itself, but on the other hand because every task related to the irrigation or the fertilizing can be performed at the planter level.

Nowadays, green facade systems are usually based on the construction of a support structure which is placed apart from the facade, thus generating an air chamber that provides more energy benefits (Hunter et al., 2014). Additionally, this kind of solution reduces the damage usually caused to facades when climbers are attached directly to the walls. Recycling supporting structures at the end of their life also reduces the environmental impact of a system (Ottelé et al., 2011). Most options available in the market are relatively light, having a weight once planted ranging between 20 and 30 kg. Since most of the components are made of stainless steel, these systems can be installed outdoors in any condition, even in coastal environments. A variety of stainless steel wires and meshes can be employed depending on the kind of climber chosen. Stainless steel components can be combined, allowing architects to obtain a wide variety of designs. These mesh and cable systems for green facades are versatile and adaptable to curved surfaces and columns, allowing a customized design for each project.

INTENSIVE SYSTEMS—LIVING WALL SYSTEMS

Living wall systems provide different options to install VGS. According to their characteristics, two large groups can be distinguished: cloth (or felt) and panels (or boxes; Fig. 1).

Cloth or felt systems are one of the most broadly spread kinds of systems and can either consist of hydroponic cultures or have a substrate (Figs. 2—4). In turn, textile systems that contain organic substrates can use sphagnum

FIGURE 2   Living wall designed by Patrick Blanc, the Driver pub (London).

FIGURE 3   Living wall in Ho Chi Minh City, Vietnam.

FIGURE 4   Living wall at HCor—Hospital do Coração, São Paulo.

moss or topsoil. On the other hand, there can be purely hydroponic systems. The latter is one of those most widely used today. In this system, the roots receive a nutrient balanced solution with all of the essential chemical elements for the right development of the plants. They have a reduced width and are made up of a combination of layers, usually leaned against a back waterproof layer made of materials that vary depending on the commercial system. Plant species are planted by cutting the felt on-site or using modules with pockets already prepared. The irrigation system can be distributed per sector and has horizontal branches with drippers, in a way that allows the coverage of the whole surface. This system's design is a very important issue, as it needs to be adequate to guarantee the supply of the flow required for each area, depending on the kind of plants used.

As for the hydroponic felt system, Patrick Blanc became world famous with its patent back in 1988 (a recent example is given in Fig. 2). It is made up of a lining of 10 mm thick expanded PVC that is fixed to a support where alternate layers of nonwoven felt and polypropylene raffia are stapled. These layers act as the foundation for the roots of the plants, which is the reason why there is no need at all to add a substrate between the layers in these systems.

Apart from felt systems, there are different types of modular panel systems. Their classification can vary depending on the kind of growing medium used (hydroponic or substrates) and on the material of which their structure is made of—galvanized steel, polyethylene, or recycled plastic panels. The growing medium inside the panel or module can be organic substrate or inorganic compounds. The organic ones support the development of plants providing nutrients and enhancing water retention capacity. Organic fibers (coco peat) or sphagnum moss are the most common. Modular hydroponic systems enable planting without an organic medium.

They make use of inorganic substrates, such as rock wool or polyurethane foam, although many companies normally develop their own material formulas to create an optimal growth medium. Therefore, different options can be found in the market, which vary in their format and presentation. This kind of inorganic substrates do not provide the plants with any nutrients, which results in the design of the fertigation system and its operation being particularly important.

Due to the fact that these materials usually have a high capacity of absorbing and retaining water, it is indispensable to monitor the conductivity levels that are achieved as a result of a combination of irrigation and nutrients, because in many cases these could negatively affect the plants. To do so, including a monitoring system is advisable and maintenance operations must include the checking of these parameters.

Planting living wall systems can be performed on site, through perforations in the modules, into which small caliber plants are placed. The initial size of the plants affects the time it takes to reach a full coverage of the wall and therefore this aspect should be taken into account during the design process. Preplanted modules can also be used.

Another interesting system is the active living wall (ALW), which works by forcing an air flow through the living wall resulting in the biofiltration of this air (Darlington et al., 2000). At the same time, cooling of this air takes place due to the adiabatic interchange between air and plants (Franco-Salas et al., 2012; Pérez-Urrestarazu et al., 2016). This enhances the benefits linked with the environmental quality and the energy performance associated with its use. ALWs are especially effective when combined with the building's air conditioning and ventilation installations. There is a well-known example of an active biofilter in the five-floor vertical living wall in the Papadakis Integrated Sciences Building at the University of Drexel, in the United States. Far from being just an attractive interior green feature, this large wall of tropical plants acts as a living filter capturing volatile organic compounds from the air (Drexel University, n.d.).

PLANT SPECIES

Depending on the desired effect, the textures and colors offered by diverse plant species produce completely different designs and appearances. Even the period or season will affect the VGS (i.e., flowering), as it will change over the year. Plant selection is a critical aspect of their design. The plants employed in living walls are different from those used in green facades. Regardless of the system chosen, each one requires a careful selection of species that should be adapted to the specific environment in which they will grow. In addition, plant species choice can influence the performances of VGS, e.g., in terms of air quality improvement and energy saving depending on their evaporation capacity, wall coverage, etc. (Pérez et al., 2011; Cameron et al., 2014; Perini et al., 2016a). Researches show that the collecting capacity of particulate matter (PM_x) also depends on plant species characteristics (Ottelé et al., 2010; Sternberg et al., 2010; Perini et al., 2016b).

Green facades require climbing plants able to bind by themselves to facades or to structures used as supports. Fixation methods of climbing species vary greatly from one species to another and determine which species are suitable for each type of supporting structure. Some of them involve growth and tropisms of the plant such as tendrils (stem, stipule, leaf tip, prophyll, flower axis), twining or adventitious roots. In some other cases, they are just formed by arrangements of branches or leaves in a supporting design or sprawling (Bellomo, 2003). When designing a green facade without planter boxes, the plant size and growing speed have to be considered. In fact, climbing plants can grow up to 5−6, 10 or 25−30 m high, e.g., *Hedera helix* or *Wisteria* plant species (Dunnett and Kingsbury, 2008). Depending on the climate and on the performances required, evergreen or deciduous plants can be used. In the first case scenario, leaves can protect the building envelope during winter. In the second case, direct solar radiation will reach the facade.

Plants used in living walls can be epiphytic, lithophytic and bromeliads, but can also be ferns, succulent, herbaceous, small shrubs, and climbing plants. The use of native plants is possible. However, results are mixed and many native plants do not grow properly in the artificial ecosystem, which is a vertical garden.

In addition to ornamental species, there are also several experiences planting vegetables to create vertical gardens of strawberries, lettuce, and so forth.

Numerous plant species have been investigated and selected for their qualities of biofiltration, and many of them are indoor plants, such as *Dracaena godseffiana, Adiantum raddianum, Hedera helix, Spathiphyllum* sp. *'Mauna Loa', Rhododendron obtusum, Marraya sp., Vriesea splendens,* and *Dieffenbachia picta*. Being a new technology, there are no databases of plants in which species are indicated for their suitability for use in either indoor or outdoor vertical gardens. Plant selection is determined by multiple factors including the microclimatic conditions, orientation (sun exposure), system of cultivation, and stock availability in nurseries in the area of installation. Plantation design must bear in mind the gradient of moisture that occurs in the vertical direction, which can be very important in high living walls. Based on this and other microclimatic parameters such as temperature, wind or sun exposure, which can also vary on the vertical axis, the following zoning can be distinguished:

- Above: Plants with a greater need for lighting. The substrate tends to be drier between each watering. They are usually the most exposed.

- In the middle: Species with medium lighting requirements. This is a transition zone.
- Below: Saxicolous species, plants adapted to the lower level of the jungle and basins of rivers and streams are located at the bottom of the vertical garden. The plants in this area receive lower illumination and have high humidity.

CONCLUSION

As described in this chapter, different types of VGS have been developed in the last years. While green facades, i.e., the extensive VGS, require less maintenance and in most of the cases are cost effective systems, living walls have a higher aesthetic potential, allow the integration of different plant species, and can be exploited to obtain environmental benefits right after installation. Plant species, especially in the case of living walls, have to be selected considering the specific characteristics of each system, climate and environmental conditions. Many different systems are available on the market, but the several aspects involved to ensure a successful installation must be considered to make a correct choice.

REFERENCES

Bellomo, A., 2003. Pareti verdi : linee guida alla progettazione / Antonella Bellomo. Esselibri, Napoli.

Blanc, P., 2012. The Vertical Garden: From Nature to the City, Revised and Updated edition. ed. W. W. Norton & Company, New York.

Cameron, R.W.F., Taylor, J.E., Emmett, M.R., 2014. What's "cool" in the world of green façades? How plant choice influences the cooling properties of green walls. Build. Environ. 73, 198−207. Available from: https://doi.org/10.1016/j.buildenv.2013.12.005.

Darlington, A., Chan, M., Malloch, D., Pilger, C., Dixon, M.A., 2000. The Biofiltration of indoor air: implications for air quality. Indoor Air 10, 39−46. Available from: https://doi.org/10.1034/j.1600-0668.2000.010001039.x.

Drexel University, n.d. The Biowall [WWW Document]. Coll. Arts Sci. <http://drexel.edu/coas/academics/departments-centers/biology/Papadakis-Integrated-Sciences-Building/Biowall/> (accessed 10.26.16.).

Dunnett, N., Kingsbury, N., 2008. Planting Green Roofs and Living Walls. Timber Press, Portland, OR.

Francis, R.A., Lorimer, J., 2011. Urban reconciliation ecology: the potential of living roofs and walls. J. Environ. Manage. 92, 1429−1437. Available from: https://doi.org/10.1016/j.jenvman.2011.01.012.

Franco-Salas, A., Fernández-Cañero, R., Pérez-Urrestarazu, L., Valera, D.L., 2012. Wind tunnel analysis of artificial substrates used in active living walls for indoor environment conditioning in Mediterranean buildings. Build. Environ. 51, 370−378. Available from: https://doi.org/10.1016/j.buildenv.2011.12.004.

Hunter, A.M., Williams, N.S.G., Rayner, J.P., Aye, L., Hes, D., Livesley, S.J., 2014. Quantifying the thermal performance of green façades: a critical review. Ecol. Eng. 63, 102−113. Available from: https://doi.org/10.1016/j.ecoleng.2013.12.021.

Jim, C.Y., 2015. Greenwall classification and critical design-management assessments. Ecol. Eng. 77, 348−362. Available from: https://doi.org/10.1016/j.ecoleng.2015.01.021.

Köhler, M., 2008. Green facades—a view back and some visions. Urban Ecosyst. 11, 423−436. Available from: https://doi.org/10.1007/s11252-008-0063-x.

Kontoleon, K.J., Eumorfopoulou, E.A., 2010. The effect of the orientation and proportion of a plant-covered wall layer on the thermal performance of a building zone. Build. Environ. 45, 1287−1303. Available from: https://doi.org/10.1016/j.buildenv.2009.11.013.

Loh, S., 2008. Living walls - a way to green the built environment. BEDP Environ. Des. Guide 1, 1−7.

Manso, M., Castro-Gomes, J., 2015. Green wall systems: a review of their characteristics. Renew. Sustain. Energy Rev. 41, 863−871. Available from: https://doi.org/10.1016/j.rser.2014.07.203.

Ottelé, M., van Bohemen, H.D., Fraaij, A.L.A., 2010. Quantifying the deposition of particulate matter on climber vegetation on living walls. Ecol. Eng. 36, 154−162. Available from: https://doi.org/10.1016/j.ecoleng.2009.02.007.

Ottelé, M., Perini, K., Fraaij, A.L.A., Haas, E.M., Raiteri, R., 2011. Comparative life cycle analysis for green façades and living wall systems. Energy Build. 43, 3419−3429. Available from: https://doi.org/10.1016/j.enbuild.2011.09.010.

Pérez, G., Rincón, L., Vila, A., González, J.M., Cabeza, L.F., 2011. Green vertical systems for buildings as passive systems for energy savings. Appl. Energy 88, 4854−4859. Available from: https://doi.org/10.1016/j.apenergy.2011.06.032.

Pérez-Urrestarazu, L., Fernández-Cañero, R., Franco-Salas, A., Egea, G., 2015. Vertical greening systems and sustainable cities. J. Urban Technol. 22, 65−85. Available from: https://doi.org/10.1080/10630732.2015.1073900.

Pérez-Urrestarazu, L., Fernández-Cañero, R., Franco, A., Egea, G., 2016. Influence of an active living wall on indoor temperature and humidity conditions. Ecol. Eng. 90, 120−124. Available from: https://doi.org/10.1016/j.ecoleng.2016.01.050.

Perini, K., 2012. Vegetation, architecture and sustainability. Presented at the EAAE / ARCC International Conference on Architectural Research. Politecnico di Milano, Milan 480−483.

Perini, K., Rosasco, P., 2013. Cost−benefit analysis for green façades and living wall systems. Build. Environ. 70, 110−121. Available from: https://doi.org/10.1016/j.buildenv.2013.08.012.

Perini, K., Magliocco, A., Giulini, S., 2016a. Vertical greening systems evaporation measurements: does plant species influence cooling performances? Int. J. Vent. 16 (2), 152−160.

Perini, K., Ottelé, M., Giulini, S., Magliocco, A., Roccotiello, E., 2016b. Quantification of fine dust deposition on different plants of a vertical greening system. Ecol. Eng. 100, 268−276.

Polinger Foster, K., 1998. Gardens of Eden: Exotic Flora and Fauna in the Ancient Near East. ResearchGate.

Sternberg, T., Viles, H., Cathersides, A., Edwards, M., 2010. Dust particulate absorption by ivy (Hedera helix L) on historic walls in urban environments. Sci. Total Environ. 409, 162−168. Available from: https://doi.org/10.1016/j.scitotenv.2010.09.022.

Chapter 2.2

Vertical Greening Systems: Irrigation and Maintenance

Luis Pérez-Urrestarazu and Miguel Urrestarazu

Chapter Outline

INTRODUCTION

Vertical greening systems (VGS) are based on vegetation spread across a wall surface, which will require being properly watered and maintained. These systems can be divided into green facades and living walls based on the different plants and support structures used (Kontoleon and Eumorfopoulou, 2010). Depending on the type, VGS have different requirements involving irrigation and maintenance operations. In green facade systems, climber plants that cover the wall are rooted in the ground or in planter boxes. Therefore, it will only be necessary to provide water (manually or with an automated irrigation network) at the base or in each plant box. Due to their lower diversity of species and density of plants and simplicity of the structure, irrigation design is simpler compared to living walls. They also require less intensive maintenance (Ottelé et al., 2010; Perini and Rosasco, 2013).

Living walls are more complex and they are usually composed of a great variety of plant species with different water requirements (Loh, 2008). There are different methods of attachment to the supporting structure such as cloth or panels full of substrate in which the plants are rooted (Manso and Castro-gomes, 2015). Due to their characteristics, the installation of an irrigation

Nature Based Strategies for Urban and Building Sustainability.
DOI: https://doi.org/10.1016/B978-0-12-812150-4.00005-7

system capable of delivering water and nutrients at different heights is mandatory to ensure appropriate humidity conditions for plant growth and establishment all over the living wall surface.

In this chapter, the irrigation system necessary for VGS will be described. Also, some guidelines for an efficient water management will be discussed, including drainage and water recirculation. The last part will be devoted to maintenance operations.

IRRIGATION SYSTEMS

As stated previously, the design of the irrigation system will be highly dependent on the type of VGS installed. For instance, in green facades, irrigation is only required in the ground or in the intermediate plant boxes (if used). Therefore, the irrigation network is simpler and it will depend on the location of these plant boxes. Any localized irrigation system such as drip emitters, diffusers, or microsprinklers can be employed. On the other hand, the irrigation network required for a living wall is very different from conventional irrigation, due to certain distinctive features. The efficient design of the irrigation system of this new gardening concept presents the major challenge of achieving a high degree of water uniformity in the entire wall while at the same time minimizing water losses. The main difficulty in accomplishing this lies in the role that gravity plays on water distribution, as the horizontal component of water movement due to diffusion is much less significant than the vertical one. This condition makes the irrigation process of vertical surfaces more complex than in horizontal or lightly sloped surfaces, given that solutions must be provided to overcome the negative outcomes of the gravity effect; i.e., a lack of water distribution uniformity and high water losses. Furthermore, there is a difficulty for adapting to species with different water requirements as there are problems in establishing hydrozones. In this case, sectoring would be required, paying special attention to the runoff coming from the upper sectors.

The main aspects to be taken into account to design the irrigation network are the circulating flow and the working pressure. The former depends on the number of emitters installed and their flow. The working pressure is related with the type of emitter, the living wall's height, and head losses (including local head loss) throughout the network. As in living walls there is a prevalence of elevation over head losses, the main determinant in order to take into account pressure requirements will be the living wall's height. When using water directly from the network supply point, it must be verified that the pressure and flow values are adequate for the correct operation of the living wall irrigation system. If the water is pumped from a tank or reservoir, these pressure and flow values are used to choose the pump required. Other auxiliary elements may also be required such as filters (especially in recirculation systems), valves, or automation and control systems. In order to avoid pipe obstructions and improve pressure equity, closed networks (forming loops) are advised.

There are plenty of commercial systems for the application of water, but localized irrigation is advised in the case of living walls. One of the most common irrigation options for living walls is to place drip lines (Fig. 1A) at different heights so that water can move vertically and laterally in between

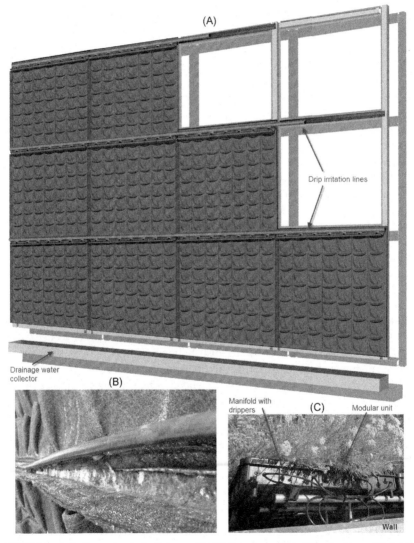

FIGURE 1   Irrigation network for living walls: (A) Location of drip irrigation lines, (B) detail of irrigation line with push-in emitters in a felt system, (C) detail of irrigation manifold for a modular unit (viewed from above).

lines by gravity and lateral diffusion. The utilization of preinstalled drip emitters simplifies the installation process. Nevertheless, they can only be used in systems with much lateral diffusion. This is because the spacing between emitters is fixed and sometimes too large for a correct water distribution. In order to avoid this problem, push-in (Fig. 1B) or inserted emitters can be a suitable choice as the designer will decide the distance between emitters. In this case, more time will be needed for the installation and problems with high pressures may appear. As an alternative, a pipe with holes or a porous pipe could also be used, but only for a living wall that is not very high.

The number of emitters to be used will depend on their flow, the living wall characteristics, and on the emitters and drip line spacing. For example, increasing the spacing between emitters or drip lines is only advisable if high flow rates are used. There is not an optimal and universal drip line spacing value but the uniformity criterion is important, as the water distribution pattern tends to improve when the irrigation lines are closer (Pérez-Urrestarazu et al., 2014). However, the decision is affected by the type of system used. For instance, in panel systems, an irrigation pipe will be usually installed for each panel as they have a higher water retention capacity, depending on the material used for its construction. In addition, as a more lateral diffusion of water is expected, the wetted area will be larger. Therefore the spacing will depend on the size of these panels. Some modular units use, for each one, a self-compensating dripper and a drainage pipe that prevents water from circulating by gravity throughout the surface of the vegetation cover to the rest of the units (Urrestarazu and Burés, 2012; Wamser et al., 2015) (Fig. 1C). Conversely, in a cloth system, the spacing between lines may not be fixed and the water flow will be more vertical, hence more emitters will be needed.

Determining the proper emitter flow is a key factor in order to decrease runoff losses and improve the water distribution uniformity. High flow emitters have shown the best water distribution uniformity but the runoff losses increase. Therefore, they are recommended when using a recirculating system. In any case, flows higher than $8 \, L \, h^{-1}$ should be avoided. It must be taken into account that when using low flow rates, irrigation times must be augmented. When the living wall is higher, water distribution will tend to be less uniform and there will be more differences between the upper part of the wall and the lower. Obviously the substrate in the lower parts of the living wall will have more moisture, hence having a higher flow rate in the lower drip lines (due to a more elevated working pressure) is not desirable. Therefore, the use of pressure compensating emitters is highly advisable given that living walls can reach considerable heights and the pressure difference between drip lines will be significant (Urrestarazu and Burés, 2012). Also, employing emitters with different flows in separate drip lines (and hence, heights) can contribute to optimize the system operation.

DRAINAGE AND WATER RECIRCULATION SYSTEMS

An excess of watering can be a problem in VGS. For that reason, mechanisms to evacuate the excess of water must be provided. When using planters for green facades, they should have draining holes. In living walls, one of the keys for their success is the drainage capability of the substrate used. It must retain water but at the same time prevent saturation in order to keep the roots zone conveniently oxygenated. This means that water will move downwards toward the bottom of the living wall. Because of this, runoff water losses are often the cause for very high water consumption. In order to prevent this effect, recirculating (closed) systems can be installed. For this purpose, runoff water must be collected at the bottom of the living wall and stored in a tank (Fig. 2), where it will be mixed with "clean" water from the network supply. It is important to decide the location and size of the tank. This will depend on economic and technical factors, such as the space available or the cost of the tank. As there will also be a need for pumping and filtering the water, the network will become more complex.

Notwithstanding, recirculating and using the water again represents a challenge, mainly due to the recirculation water quality. Hence, a close monitoring of some parameters such as the water conductivity and pH are critical during the maintenance phase. Also, water renewals must be performed periodically to improve system operation.

EFFICIENT WATER MANAGEMENT

As has been said, having a good irrigation system design is essential for the correct operation of living walls. Nevertheless, an efficient water

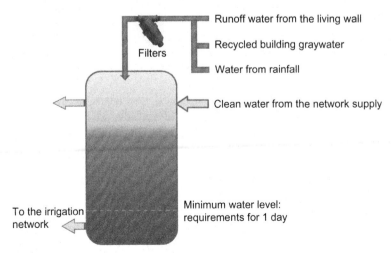

FIGURE 2   Water recirculation system.

management is equally important. In order to avoid a water shortage or an excess of humidity, especial attention must be paid to the irrigation schedule. A correct irrigation operation implies obtaining the proper water supply with low water losses.

The irrigation requirements of VGS are highly variable and depend on several factors such as location (outdoors or indoors), light exposure (direct sun light, shading, etc.), temperature and humidity conditions, type of VGS, vegetation species employed, and substrate used. For instance, Fernandez-Canero et al. (2012) measured that the water consumption of an indoor living wall using four different substrates varied from 3 to 5 L m^{-2} day^{-1} over the warm season in southern Spain. The values observed in synthetic substrates were higher than those obtained when using an organic substrate. This water consumption will be the result of the plant water uptake (irrigation requirements), direct evaporation from the substrate, and runoff and drainage water losses.

Also, based on the above mentioned factors, the operational variables—mainly irrigation duration and frequency—will have to be decided. If the system is automated, it will be easier to establish a proper irrigation programming. The best way to determine optimal irrigation times is by field observations, taking into account the resulting volume of drainage water (or runoff) for different irrigation time lengths. Irrigation events that are too short means less uniformity in water distribution,too long events, and higher water losses. It is important to take into account that the initial moisture conditions of the substrate affect the irrigation performance. If it is moderately wet, runoff is reduced during the first minutes of irrigation and distribution uniformity is slightly higher. Therefore, it is a good idea to decide on a proper irrigation frequency in order not to leave the substrate completely dry. This decision must be made according to the environmental conditions (temperature, humidity, sun exposure) (Pérez-Urrestarazu et al., 2014). Measuring the substrate's water content by means of humidity sensors may also help, though several sensors must be placed in different locations in order to take into account the lack of irrigation uniformity.

In living walls, it is possible (in large ones, even advisable) to divide the irrigation network into sectors in order to be able to provide different volumes of water in certain areas. This is especially useful when the living wall's plant species are grouped forming hydrozones according to similar water requirements. In any case, as a thumb rule, species with higher water needs should be located at the bottom part of the living wall and those that are drought-resistant in the upper region, where they will also usually receive more sun radiation.

There are other alternatives for water supply that can be explored. For instance, recovering rainwater (reducing stormwater runoff peaks) or recycling water from the evaporators of air conditioning systems can be an option. Some modern houses or buildings have separate networks for gray (wastewater generated from washbasins, showers, and baths or some household appliances) and black (originating from toilets) waters. In these cases, greywaters can also be used for the irrigation of these systems, though a previous treatment may be required. A combination of all these water origins can contribute to minimizing the volume of water required by VGS by reusing the resource in an ecological way.

MAINTENANCE OPERATIONS

Maintenance of VGS is perceived as one of the main constraints for their installation. The best maintenance program is the one that anticipates problems with appropriate prevention operations. The desired maintenance program must be taken into consideration during the design process. The maintenance required may vary and depends on the outcomes the customer wants. The main challenges of VGS maintenance are those associated with working at a height. These works require qualified personnel with a great knowledge of the systems used. This is the reason why the maintenance services are often hired along with the installation company. The most common maintenance operations are the following:

Fertilization

Green facades can be fertilized using either a solid of liquid products in the place the plants are rooted. In the case of living walls, fertigation (applying fertilizers with the irrigation water) is a common practice. Fertilization requirements depend on several factors. In living walls, it must be differentiated between the ones with lost (or run-to-waste) solution and those with recirculating systems. The latter have less fertilization requirements due to the lack of nutrient leaching. Also, it has to be taken into consideration that hydroponic systems with bare root plants will have greater fertilization requirements than systems with root balls containing growing media (i.e., source of nutrients). Recirculating systems using bare root plants require the most accurate management of fertilization due to the absence of the buffer capacity of the growing media and the fact that some ions may progressively reach toxic levels in the nutrient solution. Some recommendations for living wall fertilization are:

- continuous monitoring of the pH and Electric Conductivity of the nutrient solution, especially in recirculating systems,
- employ liquid macro and micronutrients fertigation (useful for pH corrections),
- alternate the latter with organic fertilizers (slow nutrients release),

- perform periodic foliar and water analyses to identify nutrient deficiencies or toxicities,
- application of aminoacid-based foliar fertilizers may stimulate vegetation after periods of biotic or abiotic stress.

Living Wall Pruning

Pruning frequency will depend on the VGS type and vegetative vigor. There are always plant species with different pruning requirements, or which require some special care, so pruning operations must be always scheduled.

Pest and Disease Management

This maintenance operation depends on many factors such as local climate, plant species, location (indoor or outdoor), or the owner's environmental concern. Pests frequently observed in temperate regions are aphids, spider mites, whiteflies, mealybugs, or leaf miners. They can be controlled by a foliar spray or dissolution in irrigation water containing commercial pesticides. A biological control of pests with other beneficial organisms is an alternative to chemical control.

Weed Control

Weeds may represent a problem, more in living walls than in green facades, especially in outdoor locations where weed seeds may be transported by air. Weed control is usually done by manually removing weeds during the pruning and cleaning operations. It is not recommended to perform chemical weed control, due to, among other reasons, the pollution problems that residues of herbicides cause.

Additionally, a malfunction in the irrigation system would have the worst impact on the VGS. Therefore, all the elements (emitters and pipes, hydraulic devices, filtering and control systems, pumps and water tanks) must be regularly checked.

CONCLUSION

In order to have a healthy and good looking VGS, irrigation and proper maintenance operations are crucial. Establishing maintenance plans particularized for each installation will ensure a correct performance of the VGS and reduce complications. Though irrigating green facades is in most cases simple, living walls present peculiarities such as mixture of species with different requirements, or diverse environmental conditions that make an efficient water management challenging. Hence, more studies on living wall irrigation are required in order to improve the systems used for an optimal use of resources.

REFERENCES

Fernandez-Canero, R., Pérez-Urrestarazu, L., Franco Salas, A., 2012. Assessment of the cooling potential of an indoor living wall using different substrates in a warm climate. Indoor Buil. Environ. 21, 642–650. Available from: https://doi.org/10.1177/1420326X11420457.

Kontoleon, K.J., Eumorfopoulou, E.A., 2010. The effect of the orientation and proportion of a plant-covered wall layer on the thermal performance of a building zone. Build. Environ. 45, 1287–1303. Available from: https://doi.org/10.1016/j.buildenv.2009.11.013.

Loh, S., 2008. Living walls – a way to green the built environment. Actions Towards ds Sustainable Outcomes Living walls.

Manso, M., Castro-gomes, J., 2015. Green wall systems: a review of their characteristics. Renew. Sustain. Energy Rev. 41, 863–871. Available from: https://doi.org/10.1016/j.rser.2014.07.203.

Ottelé, M., van Bohemen, H.D., Fraaij, A.L.A., 2010. Quantifying the deposition of particulate matter on climber vegetation on living walls. Ecol. Eng. 36, 154–162. Available from: https://doi.org/10.1016/j.ecoleng.2009.02.007.

Pérez-Urrestarazu, L., Egea, G., Franco-Salas, A., Fernández-Cañero, R., 2014. Irrigation systems evaluation for living walls. J. Irrig. Drain. Eng. 140, 4013024. Available from: https://doi.org/10.1061/(ASCE)IR.1943-4774.0000702.

Perini, K., Rosasco, P., 2013. Cost-benefit analysis for green façades and living wall systems. Build. Environ. 70, 110–121. Available from: https://doi.org/10.1016/j.buildenv.2013.08.012.

Urrestarazu, M., Burés, S., 2012. Sustainable green walls in architecture. J. Food Agric. Environ. 10, 792–794.

Wamser, A.F., Morales, I., Álvaro, J.E., Urrestarazu, M., 2015. Effect of the drip flow rate with multiple manifolds on the homogeneity of the delivered volume. J. Irrig. Drain. Eng. 141, 4014048. Available from: https://doi.org/10.1061/(ASCE)IR.1943-4774.0000780.

Chapter 2.3

Green Roofs Classifications, Plant Species, Substrates

Gabriel Pérez and Julià Coma

Chapter Outline

INTRODUCTION

Green roof systems have been established all over the world as important construction nature-based strategies that offer interesting benefits over conventional gray solutions not only from the environmental point of view but also as cost-effective, aesthetic, and social valuable solutions (European Commission. Expert group on Nature-Based Solutions and Re-Naturing Cities, 2015).

There is convincing evidence that green roofs along with other urban infrastructure features can provide several ecosystem services at both building and urban scales. At building scale, green roofs contribute to the indoor thermal regulation, with the consequent energy savings, providing acoustic insulation, increasing the durability of waterproofing membranes, recovering open urban spaces for landscaping, gardening, food production, and social activities. At urban scale, green roofs contribute to the urban heat island effect mitigation, storm water management, noise and air pollution reduction, support to biodiversity, offering spaces for recreation and for human well-being and health improvement (Vijayaraghavan, 2016).

On the other hand, some challenges were overcome in order to deploy these systems in the built environment, such as the use of friendly materials in the designs instead of traditional unsustainable construction materials, the

Nature Based Strategies for Urban and Building Sustainability.
DOI: https://doi.org/10.1016/B978-0-12-812150-4.00006-9

associated expenses due to the initial investment and maintenance, the possible damage of vegetation to the building, among others (Zhang et al., 2012).

Both aspects, to maximize the benefits and to minimize the drawbacks, have influenced in the current green roof designs so that a wide range of commercial solutions can be found in the market to fit the specific requirements of a building project.

This chapter presents the classification of green roofs, which mainly depends on its morphological composition, i.e., layers and materials. In addition, the main features of each of these layers and materials are addressed. Although green roofs are in the current high level of technification in its design, the fact that consolidates these systems as a reliable and well established construction system, there is still room for improvements.

CLASSIFICATION

At its most elemental, a green roof consists of introducing plants or seeds that grow in a medium on a rooftop (Snodgrass and Snodgrass, 2006). But this simple concept has become a high-tech construction solution that allows answering to all the technical requirements demanded by the building sector. As result, the current state of the art for greening a roof recognizes two main approaches, extensive and intensive systems (Dunnet and Kingsbury, 2008), although intermediate solutions may be found.

This grouping derives from the multilayer composition of a green roof and the materials that make up the system, which generically are, from top to bottom above the structure of the roof, the vegetation layer, substrate layer, filter layer, drainage layer, protection and water retention layer, and finally the root barrier and waterproofing layer (FLL guidelines, 2008) (Figs. 1 and 2).

The main characteristics of these green roof typologies are the following:

- *Extensive green roofs* are not designed for public use and are mainly developed for aesthetic and ecological benefits. They are distinguished by being low cost, lightweight, and with thin mineral substrates. Minimal maintenance is required and inspection is performed 1−2 times per year. Plants selected tend to be of the low maintenance and self-generative type (Wong et al., 2007).
- *Intensive green roofs* are usually called roof gardens. They involve high loads and have thick substrate layers with a higher amount of organic material than extensive systems. They are developed so as to be accessible to people and are used as parks or building amenities. Hence, they usually incorporate areas of paving and seating. The added weight, higher capital cost, intensive planting, and higher maintenance requirements characterize intensive green roofs. Plant selection ranges from ornamental lawns, to shrubs, bushes and trees; consequently it affects the weight,

FIGURE 1   Left: Intensive green roof, London 2009. Right: Extensive green roof, Bern 2016.

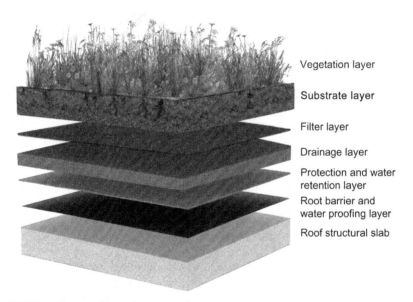

Vegetation layer

Substrate layer

Filter layer

Drainage layer

Protection and water retention layer

Root barrier and water proofing layer

Roof structural slab

FIGURE 2   Green roof layered structure scheme.

build-up heights, and costs of the roof garden. Moreover, further regular garden maintenance such as mowing, fertilizing, watering, and weeding is required for intensive landscapes (Wong et al., 2007).

- In addition to the two extreme systems, it is possible to define an intermediate typology of green roof, named *semiintensive green roofs*, in which a deeper substrate allows more possibilities for landscaping than with the extensive approach by assuming the increase of weight, maintenance, and costs.

TABLE 1 Green Roofs Typologies and Main Features

	Extensive	Semiintensive	Intensive
Weight at maximum water capacity	$50-150$ kg m^{-2}	$120-350$ kg m^{-2}	>350 kg m^{-2}
Substrate layer thickness	$6-20$ cm	$10-25$ cm	>25 cm
Plant typologies	Succulent, herbaceous, and grasses	Herbaceous, grasses, and shrubs	Grasses, shrubs, and trees
Slope	$<100\%$	$<20\%$	$<5\%$
Irrigation	Never or periodically	Periodically	Regularly
Maintenance	Low	Moderate	High
Costs	Low	Middle	High
Use	Only accessible for maintenance	Pedestrian areas but with a moderate use	Pedestrian/recreation areas

Although a wide range of values can be found in both the standards of different countries and in the information supplied by the companies, Table 1 summarizes the main features of these three typologies of green roofs.

Along the existence of contemporary green roofs, some national or local guidelines have been established. These regulations provide valuable information about the different typologies and main characteristics of green roofs as well as referring to the requirements for their implementation and maintenance.

However, they are mainly addressed to manufacturers and installers more than to the end-users. In addition, these guidelines are mainly focused on the technical aspects to guarantee the green roof survival, without taking into consideration the possibility to maximize the several ecosystem services provided by green roofs, such as thermal and acoustic benefits, runoff control, support to biodiversity, etc.

VEGETATION LAYER

The type of vegetation that can be used in a green roof was directly related to the construction typology (Table 1), so that in extensive green roofs, due to the limited thickness of the substrate, the type of vegetation will always be mainly herbaceous, succulents, or grasses. By increasing the substrate thickness, as well as the support systems (irrigation and maintenance), from the semiintensive up to the intensive approaches, shrubs and tree species can be incorporated into the landscaping palette.

Even small variations in the thickness of the substrate can influence the growth of plant species. According to Heim and Lundholm (2014) in a study in which the substrate thickness ranged from 5 to 15 cm, it was observed that *Sedum* sp. tend to dominate in lower substrate depths, while greater depths are required for the persistence of grasses.

In addition to the substrate thickness, the possibility to incorporate water and nutrients as well as the periodic maintenance, implies that in semiintensive and intensive systems any plant species can be used being the basic choice criteria, the aesthetic and compositional.

Thus, the real challenge remains in extensive green roofs, without irrigation and with minimum maintenance, in which the vegetation choice has multiple variables to consider. According to Vijayaraghavan (2016), considering the extreme environment on roof tops, the favorable characteristics of vegetation for extensive green roofs must be; the ability to withstand drought conditions, to survive under minimal nutrient conditions, to achieve good ground coverage, to require less maintenance, to have rapid multiplication, with short and soft roots, and finally, plants that can provide phytoremediation.

Generally speaking, to ensure a good and stable development of vegetation the main factors that determine the growth of plants must be identified. In reference to the climatic requirements, the general climate and the specific microclimatic conditions of the roof must be taken into account.

Among the most important general climatic parameters pluviometry, drought periods, periods of frost and winds prevailing should be considered. Regarding the specific conditions of each green roof project, the plant selection, the sunny and shaded areas, the effect of emissions from flues and other facilities of the building, the specific conditions of wind circulation among others, must be considered from the beginning.

Finally, it can set aside the own plant species characteristics and needs, such as competition between species (invasive species), sensitivity to gases or pollution, plants with aggressive rhizomes, sensitivity to light reflections and thermal accumulation, stability against wind, etc.

Bevilacqua et al. (2015) highlighted these effects in an extensive green roof case study. Both, temporal and spatial changes were found on floristic composition implying that plant cover in that extensive green roof was not homogeneous, not only spatially on the roof surface but also seasonally over time. The spatial effect was related basically to the solar radiation exposure and wind effect, which directly influences not only the substrate temperatures performance but also the floristic composition. The seasonal changes were consequence mainly to the invasion of autochthonous foreign species from the surrounding environment, which after 4 years from the roof installation were perfectly established on the roof.

According to Benvenuti and Bacci (2010), succulent plants are generally considered the most appropriate plants to apply on extensive green roofs, due to their shallow root systems, their crassulacean acid metabolism, the efficient water use, and tolerance to extreme drought conditions. Within succulent, *Sedum* species are the most used worldwide. Butler and Orians (2011) highlighted the potential of *Sedum*s by facilitating the performance of neighboring plants by reducing soil temperature during dry weather conditions, thereby decreasing abiotic stress for other life forms.

Although succulent species are very interesting for extensive green roofs, in order not to fall into monocultures simplicity the large range of native species perfectly adapted to the climatic requirements must be considered because they offer great and interesting possibilities of design.

According to Ondoño et al. (2015) the use of autochthonous species of wild flora in xerogardening, landscaping, and revegetation is of increasing interest because of their capacity to adapt to adverse environmental conditions. While green roof industries rely mainly on the use of succulent and some other tried-and-true plants (i.e., plants tested and found capable for use on green roofs), often applied in low species numbers or limited to only one life form, a range of unexplored herbaceous perennial and annual plants exist with the necessary drought adaptations. The use of different life forms has been shown to provide better ecosystem functioning and resistance to environmental stress, mainly due to niche complementarity and facilitation.

SUBSTRATE LAYER

The growing medium is the main element for the vegetation development in green roofs since it is not only the physical support but it makes available water and nutrients for plants.

Substrate should allow roots and underground organs of plants good penetration and development. Green roof substrate must be structurally stable, able to retain water and make it available to the plants, which will also allow the surplus of water to seep completely in the drainage layer. The substrate must be able to contain enough volume of air for the implanted type of vegetation, even if reaching water saturation.

Again, the main challenge is to design substrates for extensive green roofs (and some semiintensive), because there are not limitations in the case of intensive green roofs in which the same criteria, than in any landscaping project, can be applied.

According to Vijayaraghavan (2016), the desirable substrate characteristics will be high stability, good anchorage, and will support a wide variety of plants, minimal organic content, low bulk density, high water holding capacity, high air-filled porosity, high hydraulic conductivity, high sorption capacity, and less leaching. Taking all factors into consideration, it is a difficult task to identify or prepare a green roof substrate that possesses all favorable characteristics. Some of the characteristics may be toned down to improve others. Low bulk density achieved by using lightweight minerals may compromise the stability of substrate and plant anchorage. By decreasing the particle size and improving organic matter content in an effort to increase water holding, capacity may affect air-filled porosity and hydraulic conductivity. However, a good substrate design should preserve the main benefits provided by the green roof.

Achieving all these requirements has resulted in the design of multiple compositions of specific substrates for green roofs. The general practice is to mix different components that provide necessary and desirable characteristics to the substrate, usually separated into inorganic fraction and organic fraction.

For the inorganic fraction, multiple kinds of materials can be used, such as scoria, ash, pumice, zeolite, vermiculite, sand, coir, pine bark, and even recycled materials such as crushed bricks, porcelain, and tiles, etc.

Relating to the organic fraction, that is peat, the recommendation is to minimize it because, although organic matter improves plant growth and the substrate water content, it usually has a relatively short lifespan, degrading and slumping over time, and may become water repellent and difficult to rewet if it dries out. Generally guidelines recommend less than 6% of organic matter for extensive green roofs.

The green roof substrate depth influences the plant species that can be used and establishes the different typologies of green roof: extensive, semiintensive, and intensive (Table 1). Typically the substrate layer depth tends to be constant across the surface of the green roof, which was often linked to the homogeneity of the selected plant species palette, especially in extensive green roofs.

In a more creative and closest to the natural environment conception of green roofs, the incorporation of heterogeneity in the depth of the substrate becomes a key factor. According to Gedge et al. (2013), the most common way to establish substrate depth heterogeneity on green roofs is to create areas with less depth (hollows), contrasting with areas where the substrate is mounded. This creates heterogeneity not only in substrate depth, but possibly also in exposure, where mounds would be more prone to wind and desiccation (being above the main grade of the roof), and in hydrology, where water could drain off the mounded areas into the hollows. This is expected to contribute to greater diversity of invertebrates as well as plants.

FILTER LAYER

This layer allows separating the substrate layer from the drainage layer. Water can go down through the filter layer but not the small substrate particles that can obstruct the drainage layer. The usual, main materials are polypropylene or polyester geotextile felts.

Despite the good performance provided by these types of fabrics, new alternatives for future materials must be found because due to its synthetic origin they are not the best choice from the point of view of sustainable construction approach that inspires the use of green roofs. Previous life cycle assessment-based studies highlighted this idea (Rincón et al., 2014).

DRAINAGE LAYER

Its aim is to ensure a good balance between water and air in the green roof system. The drainage layer must be able to drain water excess during rain events but to retain a timely quantity for plants later use. In addition space for air must be guaranteed in order to ensure suitable oxygenation of roots.

To achieve all these contradictory requirements is a complex challenge that has been overcome following different strategies, which can basically be grouped into two main approaches:

- Alveolar panels: Made of polyethylene or polystyrene allows evacuating excess of water during rain events while some is stored in the alveolus, and space for air is also available between alveolus.
- Granular porous materials: Made of porous stone materials with some water retention capacity, such as expanded clay, expanded shale, pumice, natural pozzolana, etc. They have a good performance because of the macroporous between particles and the microporous within the particles that allows evacuating the excess of water during rain events through the macroporous, while retains water in the microporous.

Again, the current materials used are in contradiction with the desirable sustainability for these systems, both by using plastic derivate or limited petrous materials that must be removed from nature with the consequent environment impacts.

In this regard some actions have been conducted to improve this issue, such as the possibility to use rubber crumbs from out of use tires as drainage material, with good performance both as drainage as well as insulation material (Pérez et al., 2012), or the use of bioplastics instead of the traditional oil-based ones.

BOTTOM LAYERS

The protection and water retention layer provides mechanical protection to the below root barrier and waterproofing layer. This is very important during the installation phase. The main usual materials are polypropylene or polyester geotextile felts.

The function of the root barrier and waterproofing layer is to protect the building from the action of roots and water; usually bitumen or polyvinyl chloride (PVC) plastic membranes, reinforced with polyester, fiberglass, plastics, and mineral granules. There are also some made with synthetic rubber or polyethylene. The material resistant to the roots' penetration can be embedded in the waterproofing membrane, or may be a separate membrane placed above the waterproofing membrane, which protects it from the roots' action. This material can be a physical or chemical barrier.

CONCLUSION

In a context in which the architecture and building sector are tending toward a sustainable and eco-friendly approach, green roofs have become one of the most promising construction systems in this direction.

The fact that current commercial solutions allow solving any complex architectural situation, highlights the technical maturity reached by these systems. The classification is well established, the structural composition by layers is clear, several national and local guidelines and standards guarantee the uniformity and quality of these solutions.

Further research and designs must deal with the sustainability of all the materials used in green roofs solutions as well as in taking into account the different ecosystem services provided by these solutions, in order to maximize the results.

ACKNOWLEDGMENTS

This study has received funding from European Union's Horizon 2020 research and innovation program under grant agreement N°723596 (Innova MicroSolar) and N°657466 (INPATH-TES), from the European Commission Seventh Framework Program (FP/2007-2013) under grant agreement No PIRSES-GA-2013-610692 (INNOSTORAGE). The work is partially funded by the Spanish government (ENE2015-64117-C5-1-R (MINECO/FEDER) and ENE2015-64117-C5-3-R (MINECO/FEDER)). GREA is certified agent TECNIO in the category of technology developers from the government of Catalonia. The authors would like to thank the Catalan Government for the quality accreditation given to their research group (2014 SGR 123). Finally, Julià Coma wants to thank the Departament d'Universitats, Recerca i Societat de la Informació de la Generalitat de Catalunya for his research fellowship.

REFERENCES

Benvenuti, S., Bacci, D., 2010. Initial agronomic performances of Mediterranean xerophytes in simulated dry green roofs. Urban Ecosyst. 13, 349–363.

Bevilacqua, P., et al., 2015. Plant cover and floristic composition effect on thermal behavior of extensive green roofs. Building Environ. 92, 305–316.

Butler, C., Orians, C.M., 2011. Sedum cools soil and can improve neighboring plant performance during water deficit on a green roof. Ecol. Eng. 37, 1796–1803.

European Commission, 2015. Directorate-General for Research and Innovation. Directorate I—Climate Action and Resource Efficiency. Unit I.3—Sustainable Management of Natural Resources. Nature-Based Solutions & Re-Naturing Cities. Final Report of the Horizon 2020 Expert Group on 'Nature-Based Solutions and Re-Naturing Cities' 2015.

Dunnett, N., Kingsbury, N., 2008. Planting Green Roofs and Living Walls. Timber Press Inc., Portland, Oregon.

FLL-Guidelines for the Planning, Construction and Maintenance of Green Roofing. Green Roofing Guideline. The Landscape Development and Landscaping Research Society e. V. (FLL), 2008.

Gedge, D., et al., 2013. Creating Green Roofs forInvertebrates A Best Practice Guide. Buglife — The Invertebrate ConservationTrust.

Heim, A., Lundholm, J., 2014. The effects of substrate depth heterogeneity on plant species coexistence on an extensive green roof. Ecol. Eng. 68, 184—188.

Ondoño, S., et al., 2015. Evaluating the growth of several Mediterranean endemic species in artificial substrates: are these species suitable for their future use in green roofs? Ecol. Eng. 81, 405—417.

Pérez, et al., 2012. Use of rubber crumbs as drainage layer in green roofs as potential energy improvement material. Appl. Energy 97, 347—354.

Rincón, et al., 2014. Environmental performance of recycled rubber as drainage layer in extensive green roofs. A comparative Life Cycle Assessment. Building Environ. 74, 22—30.

Snodgrass, C., Snodgrass, L.L., 2006. Green Roof Plants: A Resource and Planting Guide. Timber Press Inc., Portland, Oregon.

Vijayaraghavan, K., 2016. Green roofs: a critical review on the role of components, benefits, limitations and trends. Renew. Sustain. Energy Rev. 57, 740—752.

Wong, N.H., et al., 2007. Study of thermal performance of extensive rooftop greenery systems in the tropical climate. Building Environ. 42, 25—54.

Zhang, X.L., et al., 2012. Barriers to implement extensive green roof systems: a Hong Kong study. Renew. Sustain. Energy Rev. 16, 314—319.

Chapter 2.4

Green Roofs: Irrigation and Maintenance

Panayiotis A. Nektarios

Chapter Outline

INTRODUCTION

Several plants do possess the capacity to withstand drought or a water deficit stress by altering their metabolism while others utilize a drought avoidance mechanism. One of the most common drought avoidance mechanisms is the development of deeper root systems in order to utilize water quantities that are found in deeper soil depths. However, in most green roof systems the expression of such a mechanism is hindered due to the shallow substrate depth especially in the cases of extensive or semi-intensive green roof systems.

The guidelines that are in existence permit the use of irrigation whenever it is demanded and especially during the establishment of plant material. In

Nature Based Strategies for Urban and Building Sustainability.
DOI: https://doi.org/10.1016/B978-0-12-812150-4.00007-0

shallow green roof systems, it seems that the guidelines would prefer to utilize succulent plant species, such as *Sedum* or *Delosperma,* that do not require irrigation. However, these guidelines are mainly applicable to northern climates and a significant adjustment is demanded for arid and semi-arid climates. Even though they are reports for several plant species that could withstand extreme drought conditions on extensive green roofs without any water application under Mediterranean climatic conditions (*Dianthus fruticosus, Sedum sendiforme, Chrithmum maritimum,* Nektarios et al., 2011, 2015, 2016a, b), irrigation should be considered as a necessary practice that would permit the increase of the utilized plant palette. This necessity is supported further in arid and semi-arid climates due to the poor performance of *Sedum* plant species that are commonly utilized in the northern climatic zones.

Furthermore, it is substantiated by several researchers that green roofs could contribute to urban environment amelioration only under the provision that they will be implemented in adequately large city surfaces (Getter and Rowe, 2006) and would transpire in order to contribute to urban heat island mitigation and microclimate amelioration (Akbari et al., 2001). Thus, it is obvious that irrigation should be considered as a necessity especially in the arid and semi-arid regions.

Due to the importance of irrigation to green roof sustainability, the aim of the current chapter is to analyze the different contemporary irrigation systems which are mostly utilized in conjunction with other maintenance procedures. The knowledge of the characteristics and applicability of each irrigation system combined with the necessary maintenance procedures is expected to facilitate the decision-making processes that would lead to sustainable and aesthetically pleasing green roof systems having minimum requirements for water resources inputs.

IRRIGATION EQUIPMENT

In most cases, green roof irrigation has similar hardware and software with the common land-based systems. The irrigation initiates with a controller that controls several valves which deliver water sequentially in different parts of the roof. Water delivery is performed either through sprinklers or drip irrigation pipes placed on top of the growth substrate surface or below it. However, there are several modifications of the irrigation application systems adapted to the particularities of green roofs. The controller is usually equipped with rainfall sensors in order to prevent irrigation during rain events or it can be connected to a substrate moisture sensor in order to apply irrigation based on the substrate moisture content. Finally, there are controllers connected to small meteorological stations capable of determining the reference evapotranspiration and apply irrigation according to the corresponding needs of the green roof plants.

SAFETY AND ALARM SYSTEMS IN GREEN ROOF IRRIGATION INSTALLATIONS

Due to the lack of frequent maintenance visits on green roof systems, especially for the extensive and semi-intensive types, it is necessary to install alarms in case of irrigation system failures such as broken pipes or malfunctioning valves. There are alarm systems that calculate the amount of water that should normally be applied on the green roof and compare it to the actual water volume using a flow meter. In cases that the actual flow is greater than the calculated one, the alarm is turned on indicating an irrigation system failure. Alternatively, other alarm systems depend on substrate moisture sensors, which are installed in different parts of the growing substrate and send an alarm signal whenever substrate moisture content rises to unexpectedly high levels excluding cases of intense rainfall.

SURFACE IRRIGATION SYSTEMS

Sprinkler Irrigation

Surface irrigation could be applied to intensive and extensive green roof systems either through sprinkler systems or drip irrigation. Even though sprinklers are mostly utilized for ground covers such as lawns, succulent plants can also be irrigated through sprinklers. Rowe et al. (2014) reported that for the *Sedum* plants characterized by short root systems, sprinkler irrigation was more efficient compared to drip irrigation due to the fact that green roof substrates are coarse textured and gravity easily overcomes the capillary forces within the substrates particles. In cases that a sprinkler system is decided to be utilized, several parameters should be taken into consideration for maximizing uniformity of irrigation application. More specifically, it is recommended to utilize sprinklers with low flow trajectory in order to confront the problem of wind gusts which are more severe and frequent on roof tops compared to the ground level. Alternatively, sprinklers that do throw stream lines instead of droplets are also preferably used on green roofs (Precision, The TORO Co; MP Rotators, Hunter Industries; R-VAN, Rainbird and others).

Drip Irrigation

Drip irrigation can accommodate organic and informal shapes of plantings much easier compared to the sprinkler systems, and with increased uniformity. In cases that drip irrigation is selected, either drippers or surface drip pipes can be utilized depending on the selected plants and the uniformity of the landscape design. In horizontal or slightly inclined green roofs the drip lines are placed at 40−50 cm spacing depending on the plant species selection. In contrast, in pitched green roofs the spacing of the drip lines is closer toward the top of the roof and gradually becomes sparser in order to compensate for the increased moisture accumulation toward the lowest part of the roof.

SUBSURFACE IRRIGATION SYSTEMS

In the case of extensive or semi-intensive green roofs subirrigation is also an option for green roof irrigation. In those cases, there are several systems and different methodologies that could be utilized.

Subsurface Drip Irrigation

Subsurface drip irrigation pipes are placed within the substrate at a depth that should be determined by the type of the substrate. Usually spacers are utilized to retain the drip pipes at the determined depth before the application of the substrate layer. Alternatively, the substrate is placed on the green roof in a two-step procedure. At the first step, the substrate is placed and leveled at a height that coincides with the depth that subsurface irrigation pipes should be placed. The drip pipes are laid on top of the temporary substrate surface. Then the remaining substrate is added as a second step in order to bury the drip pipes in the appropriate depth. Apart from the usual hardware that is utilized for automated systems, subsurface drip irrigation is complemented by a specialized filter comprised of successive disks impregnated with a chemical substance (trifluralin). The filter disks are constantly and slowly delivering the chemical during each irrigation event in order to prevent small roots from intruding in the dripper hole and meander. However, there is a lack of information about the leaching capacity of the specific chemical through the green roof systems.

Subirrigation From Drainage Troughs

In case that specialized drainage boards are utilized, irrigation is partially fulfilled through the infiltrating water that accumulates into the drainage layer troughs in conjunction with the water stored within the underlying protection mat. It is a common phenomenon that thin roots penetrate the geotextile used as a separation filter between the substrate and the drainage layer. Roots rotate within the drainage layer troughs and pump the accumulated water as soon as it becomes available.

Irrigation Using a Perched Water Table

In cases of leveled green roofs, irrigation can be performed by creating an artificial water table. This approach utilizes roof dam elements with adjustable height that are higher than the roof surface (Fig. 1). Thus, the infiltrating water is accumulated on the roof top and its height is determined

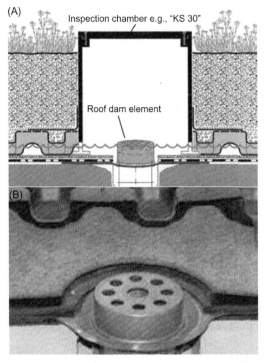

FIGURE 1 Irrigation for leveled green roofs using dam elements (A. Sketch; B. Photograph) with adjustable height to create an artificial water table (Copyright: ZinCo GmbH).

by screwing and unscrewing the body of the dam element. The depth of the water table is adjusted for different roofs as well as in-between different seasons of the year. This type of overflow systems are placed within inspection chambers in order to be easily accessible and to permit the adjustment of the water table. The use of a constant water table has the advantage of uniform water application through the roof. However, this approach also possesses disadvantages that include the exertion of a continuous water pressure toward the waterproofing membrane and equipment and the increased weight of the green roof construction toward the building framework.

Apart from the systems that depend on the accumulation of the infiltrating water, there are others that replenish water whenever is demanded through a floater system (Fig. 2) in order to keep the predetermined height of the artificial water table (Diadem, Optigreen).

Wicking Ropes and Mats

Modifications of the described subirrigation systems are also in existence. One of them utilizes wicking ropes that absorb water from a lower water

FIGURE 2 Green roof irrigation using an artificial water table using a floater replenishing system (Copyright: Optigreen).

depot found in the drainage troughs (Fig. 3) and pumps it closer to the plants' root system (Wicking mat, ZinCo). Another system utilizes a wicking mat that has embedded drip lines. The mat is laid within the substrate at depths that depend on the planting type, with shallower depths corresponding in plants with smaller root systems (Eco Mat, Hunter Industries) or above the drainage layer (Aquafleece, ZinCo).

Irrigation of Pitched Green Roofs

In cases of pitched green roofs, the inclined substrate tends to accumulate moisture in the lowest parts of the green roof. In such cases the increased difference between the upper and lower portions of the green roof has a direct effect on plant growth. In cases of phytocommunities, a distinctive separation is observed between the plants that are drought tolerant and occupy the top and middle part of the roof, and those that are not which tend to accumulate at the lowest part of the roof. In such cases obstacles that are placed below the substrate surface in a perpendicular direction compared to the inclination create a dam effect permitting the accumulation of substrate moisture in smaller green roof segments compared to the whole roof length, and thus reducing substrate moisture fluctuation along the roof length.

Irrigation of Reinforced Pitched Green Roofs

A similar effect can be created on inclined green roof systems where geogrids are utilized to stabilize the substrate by preventing its erosion (DiaDomino, Diadem; Cable and net, Optigreen; Georaster, Zinco). Apart from their substrate stabilizing effects, geogrids and geocells also act as temporary water barriers, contributing to the amelioration of substrate moisture differentiation along the inclined roof length.

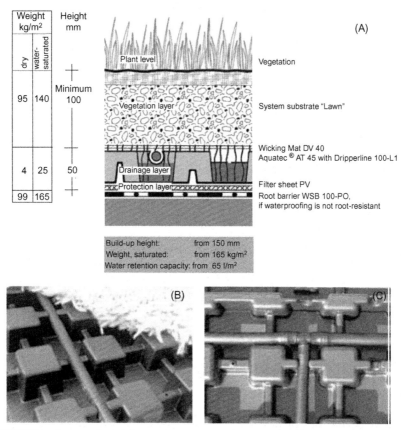

Weight kg/m²		Height mm		
dry	water-saturated			
95	140	Minimum 100		
4	25	50		
99	165			

(A)

Plant level — Vegetation

Vegetation layer — System substrate "Lawn"

Wicking Mat DV 40
Aquatec ® AT 45 with Dripperline 100-L1

Drainage layer

Protection layer — Filter sheet PV
Root barrier WSB 100-PO,
if waterproofing is not root-resistant

Build-up height:	from 150 mm
Weight, saturated:	from 165 kg/m²
Water retention capacity:	from 65 l/m²

(B)

(C)

FIGURE 3 Green roof sub-irrigation facilitated with wicking ropes (A. Sketch; B. Photograph of wicking ropes on top of the drainage layer; C. Drip pipes placed in the drainage layer) (Copyright: ZinCo GmbH).

Irrigation of Modular Green Roof Systems

There are several modular green roof systems in the market using different approaches in order to facilitate the installation procedure, the partial replacement of the green roof, and to secure the integrity of the waterproofing and thermal insulation layers. Several of those systems are complemented with the corresponding irrigation system. In most cases, modular systems are designed with specialized spaces to clip the drip irrigation pipes either from the side or from the middle of each module.

PLANT WATER DEMANDS

Irrigation Quantity

Irrigation demands can vary significantly depending on the utilized systems (extensive, semi-intensive, intensive) as well as within each category based on the utilized plant species, substrate type, and depth characteristics. Ntoulas et al. (2013) have provided a thorough flow chart to determine irrigation demands based on the abovementioned parameters. Irrigation during the periods of maximum demand can reach $7 \, mm \, day^{-1}$ for the most demanding plant species established in intensive green roof types. In contrast, there are plant species in extensive green roofs that might not require any irrigation at all after the first year of establishment. These species are plants possessing obligatory or facultative crassulacean acid metabolism characterized by minimal evapotranspirational demands. Most xerophytic plants can adequately grow in shallow extensive green roofs irrigated at 30% of the daily evaporation (Benvenuti and Bacci, 2010; Kokkinou et al., 2016; Nektarios et al., 2011, 2015, 2016a, b).

Irrigation Frequency

Irrigation frequency is of great importance due to the small substrate volume that results in speedy depletion of the substrate moisture. Most resistant plants can withstand drought for 10−15 days while succulents can withstand more prolonged drought periods. However, it is advisable to split the irrigation water quantity in smaller batches applied in more frequent time intervals.

GREEN ROOF IRRIGATION AUTOMATIONS

Apart from the safety requirements, green roofs are equipped with innovations that will contribute to water savings and prudent irrigation applications. Such innovations involve sensors that will stop irrigation application whenever rainfall occurs, while more elaborate systems include the control of the irrigation program with the use of substrate moisture sensors. Such sensors initiate irrigation whenever substrate moisture content drops below a predetermined level (18−20% v/v for most plants and 15% v/v for drought tolerant plants) while they prevent irrigation events whenever the substrate moisture content is above a maximum level (25−28% v/v for most plants).

GREEN ROOF IRRIGATION MAINTENANCE

Green roof irrigation system requires maintenance that comprises of cleaning the filters and fine-tuning irrigation schedules depending on the plant types and seasonal changes. Filters can be found ahead of the valves as well as within the sprinkler bodies. In case of subsurface drip irrigation, the filter containing the impregnated disks with trifluralin should be changed once a year. In addition, in areas that frost and low temperatures prevail during winter time, the irrigation system should be drained in order to prevent pipe breakage and valve failures from ice formation.

GREEN ROOF MAINTENANCE

Apart from the irrigation system, the whole green roof structure requires maintenance. The intensity of the maintenance will be related to the green roof type. Thus, in intensive green roof types, maintenance will be increased compared to the semi-intensive or extensive types. In intensive green roofs, the maintenance procedures will be identical to a private garden or a park, including pruning, disease, insect and weed prevention or eradication, mowing, fertilizing, etc. However, the personnel must be trained to the particularities of the green roof systems to prevent harming the integrity of the thermal and water insulation layers. While in most cases the intensive and semi-intensive green roofs are equipped with safety railings, in extensive green roofs the workers must be hooked through belts to safety points that can be either rods or anchors. In extensive green roofs, the maintenance is rare and might occur only once or twice a year. In those cases, weeding is performed, cleaning of the drainage system (peripheral gravel, surface drainage outlets, and inspection chambers) from debris, checking, fine-tuning and maintaining the irrigation system, and replacing plants that might have failed. In these cases, the eradication of invasive weeds, that might have been established on the extensive green roofs, is of major importance. Such examples in the southern climates could be *Ailanthus altissima*, *Cynodon dactylon*, and other weed species that might overtake the roof or destabilize its integrity. During the biannual maintenance visits, it might be appropriate to fertilize, especially during the first couple years after establishment to promote plant growth. Fertilization must be avoided during drought conditions since a lush plant growth has been correlated with reduced drought tolerance of the extensive green roof plants (Nagase and Dunnett, 2011; Ntoulas et al., 2013). If the roof is planted with native plants, cutting and removing seasonal growth might also be appropriate.

CONCLUSION

Green roof systems possess particularities that must be taken into account when designing their irrigation system and organizing their maintenance. Establishment of an irrigation system seems to be a necessity even in northern climates, taking into consideration the global climatic changes. Irrigation has the capacity to secure the sustainability of the green roof system. Decisions in concern to the irrigation type and maintenance should be based on site specific requirements, taking into account the type and character of the green roof as well as the regional climatic conditions.

REFERENCES

Akbari, H., Pomerrantz, M., Taha, H., 2001. Cool surfaces and shade trees to reduce energy use and improve air quality in urban areas. Solar Energy 70, 295–310.

Benvenuti, S., Bacci, D., 2010. Initial agronomic performances of Mediterranean xerophytes in simulated dry green roofs. Urban Ecosyst. 13 (3), 349–363.

Getter, K.L., Rowe, D.B., 2006. The role of extensive green roofs in sustainable development. HortScience 41, 1276–1285.

Kokkinou, I, Ntoulas, N, Nektarios, P.A., Varela, D., 2016. Response of native aromatic and medicinal plant species to water stress when grown on extensive green roof systems. Hort Sci. 51 (5), 608–614.

Nagase, A., Dunnett, N., 2011. The relationship between percentage of organic matter in substrate and plant growth in extensive green roofs. Landscape Urban Plan. 103 (2), 230–236.

Nektarios, P.A., Amountzias, I., Kokkinou, I., Ntoulas, N., 2011. Green roof substrate type and depth affect the growth of the native species *Dianthus fruticosus* under reduced irrigation regimens. HortScience 46 (8), 1208–1216.

Nektarios, P.A., Ntoulas, N., Nydrioti, E., Kokkinou, I., Bali, E.-M., Amountzias, I., 2015. Drought stress response of *Sedum sediforme* grown in extensive green roof systems with different substrate types and depths. Sci. Horticul. 181, 52–61.

Nektarios, P.A., Nydrioti, E., Kapsali, T., Ntoulas, N., 2016a. *Crithmum maritimum* growth in extensive green roof systems having different substrate type and depth and irrigation regime. Acta Horticul. 1108, 303–308.

Nektarios, P.A., Nydrioti, E., Kapsali, T., Ntoulas, N., 2016b. Substrate type, depth and irrigation regime effects on *Ebenus cretica* growth in extensive green roof. Acta Horticul. 1108, 297–302.

Ntoulas, N., Nektarios, P.A., Charalabous, E., Psaroulis, A., 2013. *Zoysia matrella* cover rate and drought tolerance in adaptive extensive green roof systems. Urban For. Urban Green. 12 (4), 522–531.

Rowe, D.B., Kolp, M.R., Greer, S.E., Getter, K.L., 2014. Comparison of irrigation efficiency and plant health of overhead, drip, and sub-irrigation for extensive green roofs. Ecol. Eng. 64, 306–313.

Chapter 2.5

Green Streets: Classifications, Plant Species, Substrates, Irrigation, and Maintenance

Tijana Blanusa and Madalena Vaz Monteiro

Chapter Outline

INTRODUCTION

In addition to vertical greening and green roofs, which were discussed in Sections 2.1 and 2.2, there is a range of other green infrastructure forms and interventions that can be found in cities. The list includes, but is not limited to, street trees, domestic gardens, public parks, sports fields and (school) playgrounds, rain gardens/bioswales, community gardens and allotments, urban orchards or woodlands, and many more.

This chapter focuses on street trees, domestic gardens, and rain gardens/bioswales. Recent research suggests that urban residents value and understand the environmental benefit of various types of street-level greening (Weber et al., 2014). Trees and orderly-maintained vegetation were generally preferred, but the contribution of more wildlife type vegetation was also valued in a study carried out in Cologne and Berlin (Germany) (Weber et al., 2014). Depending on the location and cultural context of the city, the proportion and composition of various forms of street greening will differ between different cities and parts of the world.

Nature Based Strategies for Urban and Building Sustainability.
DOI: https://doi.org/10.1016/B978-0-12-812150-4.00008-2

STREET TREES

Street tree diversity is limited in many cities across the word. In central and northern European cities, 3–5 genera, typically including *Tilia, Acer, Platanus, Aesculus, Quercus,* or *Fraxinus*, may account for 50%–70% of all street trees and often only a few species dominate (Pauleit et al., 2002). This finding was corroborated by a study using data from ten Nordic cities, including Oslo, e.g., where 67% of the surveyed street trees belonged to *Tilia* sp., *Acer platanoides,* and *Aesculus hippocastanum* (Sjöman et al., 2012). Likewise in south European cities, the number of dominant tree species in streets tend to be small, with *Platanus xhispanica, Celtis australis* or *Tilia* sp. being among the most common (frequently accounting for around 30% of the street tree population—Baró et al., 2014; Soares et al., 2011). A similar trend of low street tree diversity is also found in North America, as shown by a study evaluating street trees inventories from 12 cities, including New York, Chicago, and Toronto, where 15%–57% of all street trees considered within each city were *Acer* sp. (Raupp et al., 2006). This was true in Asia, as exemplified by studies performed in Taipei, Taiwan, where three species (*Cinnamomum camphora, Ficus microcarpa,* and *Koelreuteria elegans*) accounted for 49% of the inventoried street trees (Jim and Chen, 2009) and in Bangkok, Thailand, where 42% of the surveyed street trees belonged to a single species: *Pterocarpus indicus* (Thaiutsa et al., 2008). Although some citywide street tree inventories have revealed high species diversities (e.g., Melbourne, Australia—Frank et al., 2006; Bangalore, India—Nagendra and Gopal, 2010), most point toward an overwhelming presence of a small number of species.

This widespread lack of tree species diversity in streetscapes appears to be linked to a traditional fashion for uniformity at the street level, but also to the challenging conditions imposed by such sites, which reduce the number of adaptable species. In cities, trees typically grow under higher temperatures and higher levels of air pollution than in rural environments (Akbari et al., 2001). In addition, street trees are frequently planted in small pits surrounded by impervious materials (Fig. 1). In these areas, water infiltration is restricted by the pavement which seals the surface and channels water runoff primarily to drains (Mullaney et al., 2015). Furthermore, urban soils are frequently compacted, of low quality and contaminated with man-made waste, pollutants, or heavy metals (Craul, 1985). These surface and soil properties reduce the soil volume available for root development, together with the amount of water and nutrients necessary for proper tree establishment and growth. Consequently, mortality of newly planted street trees is often high, commonly in the region of 20%–25% in the first 3 years (Gilbertson and Bradshaw, 1990; Roman et al., 2014). Even when trees survive beyond the first few years, they may still not reach their full growth potential (Grabosky and Gilman, 2004).

FIGURE 1   Tree dying in a public car park in a small tree pit surrounded by impervious materials. *Image copyright: authors.*

The adverse effects created by the site may be partially offset by a tree management strategy providing guidelines for street tree establishment and maintenance; however the completeness of such plans, if existent, differs considerably among regions (Pauleit et al., 2002). For example, common practices during planting and establishment used within many, but not all, European cities surveyed by Pauleit et al. (2002) included sourcing of good plant material, site preparation (with specially formulated substrates and fertilization), and aftercare consisting of irrigation up to 3 years, weeding, mulching, pruning, staking, and protection from mechanical damage. But tree management strategies should also plan for the future, with the tree selection process aiming at planting "the right trees in the right places" and considering imminent pressures such as climate change and the risk of pests and diseases. Climate change is already shifting the geographical distribution of native plants, and this may force a change in the historical tree composition in many cities, where air temperatures are elevated compared to rural regions (Lanza and Stone, 2016). Likewise, the arrival of new pests and diseases can lead to high numbers of tree loss if preventive measures are not put in place (Raupp et al., 2006). The resilience of the street tree population to future threats could be improved if the diversity of tree species (including native and nonnative) was increased at a city level (Sjöman et al., 2012; Raupp et al., 2006). However, care has to be taken not to replace well-adapted species with poorly adapted ones (Kendal et al., 2014) and to ensure that nonnative species do not become invasive (Sjöman et al., 2012; Raupp et al., 2006).

DOMESTIC GARDENS

Domestic gardens (defined as outdoor spaces adjacent to a domestic dwelling) can constitute a significant proportion of the urban green space: more than 50% in Dunedin, New Zealand (Mathieu et al., 2007) and 35%–47% in the United Kingdom (Loram et al., 2007). In the UK context, anything between 16% and 25% of total area in major towns and cities is occupied by domestic gardens (Gaston et al., 2005) and, often a large proportion of those is vegetated. Urbanization, however, is decreasing the proportion of area dedicated to gardens (Mathieu et al., 2007; Smith, 2010), through existing gardens being used for housing developments and new housing stock having smaller gardens.

Sizes of urban gardens, naturally, vary greatly. There are indications that housing type and density influences the proportion of green space available in its surroundings (Whitford et al., 2001), with older built stock correlating with more extensive areas of green cover, and modern housing stock having the lowest proportion of green space around it. Composition of domestic gardens is also very versatile. Lawns are ubiquitous and constitute in the region of 60% (United Kingdom, Gaston et al., 2005) or 55% (New Zealand, Mathieu et al., 2007) of the area. In the United States, lawns (private, public, and sports turf) cover between 8 and 16 million ha, far surpassing land coverage of major crops (e.g., cotton, Robbins et al., 2001) and making it the largest irrigated monoculture plant system in the country (Marshall et al., 2015). Larger domestic gardens (Lin et al., 2017), those associated with older properties (Hope et al., 2003), or with higher income or tertiary-educated residents tend to have proportionally more vegetation, greater diversity of plants, and more complex garden styles (Daniels and Kirkpatrick, 2006). Domestic gardens are diverse mixtures of both planted and volunteer species (Smith et al., 2006). A study in Sheffield, UK showed that within almost 1200 plant species identified in 61 studied gardens, 70% were alien; but at the family level, 72% were native. Their records also showed that alien species generally had lower occupancy than natives, with overwhelming majority being recorded only once (Smith et al., 2006). Subsequent work also showed benefits of having mixes of native and near-native plant species, with a selection of exotics in domestic gardens to extend flowering season, provide resources to specialist groups of insect pollinators, and increase pollinator abundance (Salisbury et al., 2015).

Front and back gardens often differ in character, with visual impact traditionally being more important in front gardens, with the back used for functional purposes. A recent survey for the Royal Horticultural Society, UK's largest gardening charity, shows a rapid decline in vegetated cover, particularly in the UK's front gardens, with almost 40% of front gardens having only 25% or less green cover (Anon, 2016a) (Fig. 2).

FIGURE 2   (A) Many front gardens of domestic dwellings are paved over to create parking spaces. (B) However, thoughtful garden design can incorporate both the planting and car parking spaces. *Images copyright: RHS.*

Unsurprisingly, a whole range of garden soil types is found in domestic gardens. Compared to agricultural land, garden soils are generally less managed, but this depends on garden composition (% of lawn and annual plants vs more perennial and woody vegetation, the soil management typically being less intensive in the latter) (Cameron et al., 2012). There is also often quite a difference between the soil health/fertility between new build vs. mature gardens, the latter harboring more organic matter and better soil structure (Anon, 2016b). Additionally in domestic gardens, a range of growing media are used, sometimes as a soil amendment (Barrett et al., 2016). Proximity of the household to the road infrastructure or to the sites of current or industrial activity is linked to higher concentrations of various metals in soils (e.g., Antisari et al., 2015). Soil metal concentrations, however, are not always correlated with metal concentrations in plant parts (which also vary in different plant organs of the same plant); this lack of correlation can be attributed to factors such as the effect that soil pH, organic matter content, and other soil properties have on metal solubility and bioavailability (McBride et al., 2014).

The extent of garden maintenance will be largely influenced by the type of planting in it, but also the owner and community's perception of what is required (Loram et al., 2011). Lawns are management-intensive and consume more resources (energy, water, fertilizer) than, e.g., some forms of perennial woody vegetation which may require just an annual pruning (Cameron et al., 2012).

The extent of garden management significantly influences the garden's environmental benefit. As a general rule less intensively managed gardens, with less soil disturbance, less application of pesticides and fertilizers, smaller main water inputs will have a greater environmental benefit (Cameron and Blanuša, 2016). There is however also an argument that sustainable increase in management inputs (e.g., in using additional irrigation—including from collected rainwater or household graywater—to maintain high plant transpiration rates and provide localized cooling) can be environmentally justified (Cameron and Blanuša, 2016). Additional water and management input in maintaining trees and shrubs located in close proximity to buildings or climbing plants up a building wall, might be offset by savings in cooling and heating use in a building due to the better summertime cooling and wintertime insulation (Cameron et al., 2014).

RAIN GARDENS AND BIOSWALES

Vegetative strips incorporated in the urban landscape, with the main purpose of retaining water, filtering pollutants, and slowing the rate of water flow into the conventional drainage pipes and sewers, are becoming a feature in many cities. Definitions and their nomenclature vary depending on where they are located, what specific function they perform, what planting arrangements they encompass, and from which disciplinary angles they are studied. In the United Kingdom they were categorized by Dunnett and Clayden (2007) as either: (1) true rain gardens (planted, slightly sunken areas in the landscape to temporarily collect, store, and infiltrate excess rainwater runoff) (Fig. 3); (2) infiltration basins and strips (sunken planted features in predominantly hard surrounds); and (3) storm-water planters (raised planted beds at the base of a building that take runoff directly from a building's roofs or adjacent pavement areas) (Fig. 3).

In the United States, rain gardens are defined more generally as installations for domestic/small scale use, using a natural, existing soil type, installed primarily to reduce water runoff (American Society for Soil Science, https://www.soils.org/discover-soils/soils-in-the-city/green-infrastructure/important-terms/rain-gardens-bioswales). The additional concept of bioswales in the United States is defined as more engineered, larger structures, which are reducing run off from specific impervious surfaces (parking lots, roads, or other large sealed surfaces). Bioswales often use engineered soils, are deeper, and with more extensive vegetation than rain gardens. The key aim with

FIGURE 3 An example of a rain garden that contains planting adapted to intermittent wet and dry weather periods. *Images copyright: RHS.*

bioswales is the regulation of water quality (primarily to lower concentrations of N and P in order to reduce/prevent eutrophication) by choosing the combinations of plant and soil types which would maximize the uptake and immobilization of these elements (Anon, 2017).

Plants for rain gardens have to be able to withstand cyclic periods of excess water and prolonged soil drying. The choice of vegetation in rain gardens can include a mixture of forbs, shrubs, and trees (Read et al., 2008). Genera such as *Betula* and *Magnolia* (trees), *Viburnum* and *Ithea* (shrubs), *Helianthus* (herbaceous perennials), and various grasses have been shown to be able to grow well in a field study in North Carolina (US) under conditions of typical local annual rainfall (Turk et al., 2014). Larger plants were linked in some studies to improved water retention (on canopies and in soil) and filtration (Bratieres et al., 2008; Read et al., 2008). Choice and density of vegetation can influence the efficiency of water filtration (Mazer et al., 2001; Scharenbroch et al., 2016; Read et al., 2008). However, external factors such as light levels (which influence vegetation emergence and growth) and duration of inundation (which in turn is linked to the extent of vegetation and litter) will also be influencing capacity of vegetation and soil to retain and filter runoff (Mazer et al., 2001). Although not used as much in rain gardens, trees may be well-suited to improving the performance of biofilters because of their large above-ground surface area onto which water is retained and below-ground biomass, to maximize soil water uptake. The role of tree roots, however, goes beyond direct absorption of water and pollutants; by altering and improving soil structure with its deep-penetrating roots, trees in a rain garden can limit runoff by increasing water infiltration rates (Bartens et al., 2008).

Choice of rain garden substrate will directly impact the garden's ability to retain and filter stormwater. Soil-based substrate (composed of 50% loam soil and 50% pine bark) conducts storm water more quickly than sand-based (80% sand/15% silt/5% pine bark) and slate-based (80% slate/20% pine bark); however, slate-based substrate retains most N and P (Turk et al., 2014). Additionally, soil fauna has the potential to directly influence removal of nutrients and heavy metals from the water in the substrate, but also indirectly—through its impact on plant growth and soil structure (Mehring and Levin, 2015).

Long-term, the life span and performance of rain gardens may be compromised by clogging due to deposition of fine-grained sediments (Virahsawmy et al., 2014). Follow-on maintenance of rain gardens is therefore essential and involves the activities that support the establishment of healthy vegetation (e.g., weed removal and pest control, annual removal of dead material from herbaceous plants) and the maintenance of substrate permeability (e.g., light cultivation of soil surface and removal of dead organic matter) (Cameron and Hitchmough, 2016). In order to identify the level of maintenance required to ensure continuous functionality, rain gardens should be assessed regularly (Asleson et al., 2009).

CONCLUSION

This chapter summarized information on the most common street trees and planting compositions of urban gardens and rain gardens primarily in the temperate and continental climate context. The choice of vegetation and how it is managed, can, however, have a dramatic influence of the extent of the benefits that are reaped (Cameron and Blanuša, 2016). From the point of view of most practitioners, the plant choice within urban green infrastructure is still determined primarily by cost, aesthetic preferences, and chances of plant survival in a given location. New research findings suggest strongly that plant selection based on the plant's ability to deliver a greater range of ecosystem services to a higher degree, and use of the diversity of forms and species/cultivars has to be an important component of enhancing the delivery of ecosystem services by urban vegetation. It is also clear that vegetation contributes multiple environmental services to the urban environment, and that the positive contribution of urban vegetation other than trees had been underestimated and can therefore be improved (Säumel et al., 2016).

REFERENCES

Akbari, H., Pomerantz, M., Taha, H., 2001. Cool surfaces and shade trees to reduce energy use and improve air quality in urban areas. Solar Energy 70 (3), 295–310.

Anon, 2016a. Ipsos MORI. <https://www.ipsos-mori.com/researchpublications/researcharchive/3738/How-green-are-British-front-gardens.aspx> (accessed 29.11.16.).

Anon, 2016b. Royal Horticultural Society. <https://www.rhs.org.uk/advice/profile?PID = 632> (accessed 29.11.16.).

Anon, 2017. Rain Gardens and Bioswales. Soil Science Society of America. <https://www.soils.org/discover-soils/soils-in-the-city/green-infrastructure/important-terms/rain-gardens-bioswales> (accessed 05.01.17.).

Antisari, L.V., Orsini, F., Marchetti, L., Vianello, G., Gianquinto, G., 2015. Heavy metal accumulation in vegetables grown in urban gardens. Agron. Sustain. Develop. 35 (3), 1139–1147.

Asleson, B.C., Nestingen, R.S., Gulliver, J.S., Hozalski, R.M., Nieber, J.L., 2009. Performance assessment of rain gardens. J. Am. Water Resour. Assoc. 45, 1019–1031.

Baró, F., Chaparro, L., Gómez-Baggethun, E., Langemeyer, J., Nowak, D.J., Terradas, J., 2014. Contribution of ecosystem services to air quality and climate change mitigation policies: the case of urban forests in Barcelona, Spain. AMBIO 43 (4), 466–479.

Barrett, G.E., Alexander, P.D., Robinson, J.S., Bragg, N.C., 2016. Achieving environmentally sustainable growing media for soilless plant cultivation systems – a review. Sci. Horticul. 212, 220–234.

Bartens, J., Day, S.D., Harris, J.R., Dove, J.E., Wynn, T.M., 2008. Can urban tree roots improve infiltration through compacted subsoils for stormwater management? J. Environ. Quality 37 (6), 2048–2057.

Bratieres, K., Fletcher, T.D., Deletic, A., Zinger, Y., 2008. Nutrient and sediment removal by stormwater biofilters: a large-scale design optimisation study. Water Res. 42 (14), 3930–3940.

Cameron, R., Hitchmough, J., 2016. New green space interventions-green walls, green roofs and rain gardens. Environmental Horticulture: Science and Management of Green Landscapes. CABI, Boston, MA, pp. 260–283.

Cameron, R.W.F., Blanuša, T., 2016. Green infrastructure and ecosystem services – is the devil in the detail? Ann. Botany 118, 377–391.

Cameron, R.W.F., Blanusa, T., Taylor, J.E., Salisbury, A., Halstead, A.J., Henricot, B., et al., 2012. The domestic garden - its contribution to urban green infrastructure. Urban For. Urban Gree. 11 (2), 129–137.

Cameron, R.W.F., Taylor, J.E., Emmett, M.R., 2014. What's 'cool' in the world of green façades? How plant choice influences the cooling properties of green walls. Buil. Environ. 73 (0), 198–207.

Craul, P.J., 1985. A description of urban soils and their desired characteristics. J. Arboricul. 11 (11), 330–339.

Daniels, G.D., Kirkpatrick, J.B., 2006. Comparing the characteristics of front and back domestic gardens in Hobart, Tasmania, Australia. Landscape Urban Plan. 78 (4), 344–352.

Dunnett, N., Clayden, A., 2007. Rain gardens: managing water sustainably in the garden and designed landscape. Timber Press, OR.

Frank, S., Waters, G., Beer, R., May, P., 2006. An analysis of the street tree population of greater Melbourne at the beginning of the 21st century. Arboricul. Urban For. 32 (4), 155–163.

Gaston, K.J., Warren, P.H., Thompson, K., Smith, R.M., 2005. Urban domestic gardens (IV): the extent of the resource and its associated features. Biodiv. Conserv. 14 (14), 3327–3349.

Gilbertson, P., Bradshaw, A.D., 1990. The survival of newly planted trees in inner cities. Arboricul. J. 14 (4), 287–309.

Grabosky, J., Gilman, E., 2004. Measurement and prediction of tree growth reduction from tree planting space design in established parking lots. J. Arboricul. 30, 154–164.

Hope, D., Gries, C., Zhu, W., Fagan, W.F., Redman, C.L., Grimm, N.B., et al., 2003. Socioeconomics drive urban plant diversity. Proc. Natl. Acad. Sci. USA 100 (15), 8788–8792.

Jim, C.Y., Chen, W.Y., 2009. Diversity and distribution of landscape trees in the compact Asian city of Taipei. Appl. Geogr. 29 (4), 577–587.

Kendal, D., Dobbs, C., Lohr, V.I., 2014. Global patterns of diversity in the urban forest: is there evidence to support the 10/20/30 rule? Urban For. Urban Gree. 13 (3), 411–417.

Lanza, K., Stone Jr., B., 2016. Climate adaptation in cities: what trees are suitable for urban heat management? Landscape Urban Plan. 153, 74–82.

Lin, B., Gaston, K., Fuller, R., Wu, D., Bush, R., Shanahan, D., 2017. How green is your garden?: Urban form and socio-demographic factors influence yard vegetation, visitation, and ecosystem service benefits. Landscape Urban Plan. 157, 239–246.

Loram, A., Tratalos, J., Warren, P.H., Gaston, K.J., 2007. Urban domestic gardens (X): the extent & structure of the resource in five major cities. Landscape Ecol. 22 (4), 601–615.

Loram, A., Warren, P., Thompson, K., Gaston, K., 2011. Urban domestic gardens: the effects of human interventions on garden composition. Environ. Manage. 48 (4), 808–824.

Marshall, S., Orr, D., Bradley, L., Moorman, C., 2015. A review of organic lawn care practices and policies in North America and the implications of lawn plant diversity and insect pest management. HortTechnology 25 (4), 437–443.

Mathieu, R., Freeman, C., Aryal, J., 2007. Mapping private gardens in urban areas using object-oriented techniques and very high-resolution satellite imagery. Landscape Urban Plan. 81 (3), 179–192.

Mazer, G., Booth, D., Ewing, K., 2001. Limitations to vegetation establishment and growth in biofiltration swales. Ecol. Eng. 17 (4), 429–443.

McBride, M.B., Shayler, H.A., Spliethoff, H.M., Mitchell, R.G., Marquez-Bravo, L.G., Ferenz, G.S., et al., 2014. Concentrations of lead, cadmium and barium in urban garden-grown vegetables: the impact of soil variables. Environ. Pollut. 194, 254–261.

Mehring, A.S., Levin, L.A., 2015. Review: potential roles of soil fauna in improving the efficiency of rain gardens used as natural stormwater treatment systems. J. Appl. Ecol. 52 (6), 1445–1454.

Mullaney, J., Lucke, T., Trueman, S.J., 2015. A review of benefits and challenges in growing street trees in paved urban environments. Landscape Urban Plan. 134, 157–166.

Nagendra, H., Gopal, D., 2010. Street trees in Bangalore: density, diversity, composition and distribution. Urban For. Urban Gree. 9 (2), 129–137.

Pauleit, S., Jones, N., Garcia-Martin, G., Garcia-Valdecantos, J.L., Rivière, L.M., Vidal-Beaudet, L., et al., 2002. Tree establishment practice in towns and cities – results from a European survey. Urban For. Urban Gree. 1 (2), 83–96.

Raupp, M.J., Cumming, A.B., Raupp, E.C., 2006. Street tree diversity in eastern North America and its potential for tree loss to exotic borers. Arboricul. Urban For. 32 (6), 297–304.

Read, J., Wevill, T., Fletcher, T., Deletic, A., 2008. Variation among plant species in pollutant removal from stormwater in biofiltration systems. Water Res. 42 (4–5), 893–902.

Robbins, P., Polderman, A., Birkenholtz, T., 2001. Lawns and toxins - an ecology of the city. Cities 18 (6), 369–380.

Roman, L.A., Battles, J.J., McBride, J.R., 2014. The balance of planting and mortality in a street tree population. Urban Ecosyst. 17 (2), 387–404.

Salisbury, A., Armitage, J., Bostock, H., Perry, J., Tatchell, M., Thompson, K., 2015. Enhancing gardens as habitats for flower-visiting aerial insects (pollinators): should we plant native or exotic species?. J. Appl. Ecol. 52 (5), 1156–1164.

Säumel, I., Weber, F., Kowarik, I., 2016. Toward livable and healthy urban streets: roadside vegetation provides ecosystem services where people live and move. Environ. Sci. Policy 62, 24–33.

Scharenbroch, B.C., Morgenroth, J., Maule, B., 2016. Tree species suitability to bioswales and impact on the urban water budget. J. Environ. Quality 45 (1), 199–206.

Sjöman, H., Östberg, J., Bühler, O., 2012. Diversity and distribution of the urban tree population in ten major Nordic cities. Urban For. Urban Gree. 11 (1), 31–39.

Smith, C., 2010. London: Garden city? London Wildlife Trust, Greenspace Information for Greater London, Greater London Authority, London.

Smith, R.M., Thompson, K., Hodgson, J.G., Warren, P.H., Gaston, K.J., 2006. Urban domestic gardens (IX): composition and richness of the vascular plant flora, and implications for native biodiversity. Biol. Conserv. 129, 312–322.

Soares, A.L., Rego, F.C., McPherson, E.G., Simpson, J.R., Peper, P.J., Xiao, Q., 2011. Benefits and costs of street trees in Lisbon, Portugal. Urban For. Urban Gree. 10 (2), 69–78.

Thaiutsa, B., Puangchit, L., Kjelgren, R., Arunpraparut, W., 2008. Urban green space, street tree and heritage large tree assessment in Bangkok, Thailand. Urban For. Urban Gree. 7 (3), 219–229.

Turk, R.L., Kraus, H.T., Bilderback, T.E., Hunt, W.F., Fonteno, W.C., 2014. Rain garden filter bed substrates affect stormwater nutrient remediation. HortScience 49 (5), 645–652.

Virahsawmy, H.K., Stewardson, M.J., Vietz, G., Fletcher, T.D., 2014. Factors that affect the hydraulic performance of raingardens: implications for design and maintenance. Water Sci. Technol. 69, 982–988.

Weber, F., Kowarik, I., Säumel, I., 2014. A walk on the wild side: perceptions of roadside vegetation beyond trees. Urban For. Urban Gree. 13 (2), 205–212.

Whitford, V., Ennos, A.R., Handley, J.F., 2001. "City form and natural process"—indicators for the ecological performance of urban areas and their application to Merseyside. UK Landscape Urban Plann. 57, 91–103.

FURTHER READING

Dietz, M., Clausen, J., 2005. A field evaluation of rain garden flow and pollutant treatment. Water Air Soil Pollut. 167 (1), 123–138.

Gibbons, S., Mourato, S., Resende, G., 2011. The amenity value of English nature: a hedonic price approach. Discussion Paper, SERCDP0074. Spatial Economics Research Centre (SERC), London School of Economics and Political Sciences London, UK.

Mehring, A.S., Hatt, B.E., Kraikittikun, D., Orelo, B.D., Rippy, M.A., Grant, S.B., et al., 2016. Soil invertebrates in Australian rain gardens and their potential roles in storage and processing of nitrogen. Ecol. Eng. 97, 138–143.

Yang, H., Dick, W.A., McCoy, E.L., Phelan, P.L., Grewal, P.S., 2013. Field evaluation of a new biphasic rain garden for stormwater flow management and pollutant removal. Ecol. Eng. 54, 22–31.

Section III

Nature Based Strategies: Benefits and Challenges

Chapter 3.1

Vertical Greening Systems to Enhance the Thermal Performance of Buildings and Outdoor Comfort

Gabriel Pérez, Julià Coma and Luisa F. Cabeza

Chapter Outline

INTRODUCTION

Recently, the concept of urban green infrastructure has been defined by the European Commission-Environment (2012) as a set of man-made elements that provide multiple ecosystem services both at building and urban scale. Among these functions, the building energy savings as well as the reduction of ambient temperatures and urban heat island (UHI) effect stand out. Some of the most innovative and interesting construction systems for this purpose are the vertical greening systems (VGS).

Since the vertical surface available in a building is much larger than the horizontal, especially in high rise buildings, it seems obvious that the potential of VGS in providing these services will also be proportionately higher. In view of these opportunities in the last years, a set of rigorous research studies have reviewed the use of VGS as a passive tool for energy savings, at building scale, as well as for the reduction of the UHI effect at city scale (Pérez et al., 2014).

Nature Based Strategies for Urban and Building Sustainability.
DOI: https://doi.org/10.1016/B978-0-12-812150-4.00009-4

In this chapter, the main influencing aspects for this purpose, which are the differences between the current construction systems, the climate and plant species influence, and the operating methods, will be analyzed. Moreover, a short review throughout the main research findings on the topic will be presented to finish with some suggestions for further research in the conclusions.

MAIN INFLUENCING ASPECTS

Differences Between Construction Systems

Over the years, different VGS construction systems have been developed, in terms of materials and layers that make up the system, in the used support elements, the plants species as well as the maintenance requirements (extensive or intensive), etc. (Fig. 1). According to Pérez et al. (2011), these differences can have a big influence on the ecosystem services provided by each vertical greening system, e.g., energy savings or UHI reduction that must be taken into consideration already from the design phase. Table 1 summarizes the main VGS constructive systems, their layers and the level of maintenance (water and nutrients) (Pérez et al., 2011), and the great differences between systems, which results on different thermal behavior can be observed.

In order to deal with the dispersion on designs, and with the aim to compare research studies relating to the thermal performance of these systems; in Pérez et al. (2014) the building external wall surface temperature reduction (°C) due to the effect of the VGS was established as the most relevant parameter for the research studies comparison, because it is the first and most direct effect arising from the presence of sunscreen. In addition, neither

FIGURE 1 Examples of different vertical greening systems. Green facades (traditional and double-skin) and green walls.

TABLE 1 Main Vertical Greening Constructive Systems and Their Layers Composition

VGS System		Layer				Plants	Level of Maintenance
		Support Structure	Air Gap	Substrate			
Green façade	Traditional	No	No	No		Climber plants	Extensive
	Double-skin	Very light, steel wires or mesh	Yes, usually open	No		Climber plants	Extensive
	Perimeter flowerpots	Flowerpots	Yes, usually open	Only in the pots		Climber and hanging shrubs	Intensive
Green wall	Geotextile	Geotextile felts	No	No		Shrubs and hanging shrubs	Intensive
	Modular panels	Heavy and strong, anchored to the facade wall	Yes, open or closed	Yes, inside the modules		Shrubs	Intensive

the heat fluxes through the wall nor the interior surface wall temperature are comparable due to the differences between constructive systems of the facade building wall between case studies or research studies.

The Climate Influence

When the passive energy saving potential of vertical vegetation systems for buildings is the aim, the big influence of climate conditions over their operation must be considered. Weather not only affects directly on the thermal performance of the building but also on specific aspects relating to plants such as their growth (foliage density, plant height, etc.) and their physiological responses (transpiration, position of leaves, etc.), and consequently over the thermal behavior of the whole system. In this regard, the most influential climatic parameters will be solar radiation, temperatures and relative humidity, rainfall, and finally the wind.

This fact is very important for two main reasons. On one hand, it must take it into account in the most suitable system choice during the design phase so that it fulfills the local climatic constraints. On the other hand, from research's point of view, these aspects should be taken into account when comparing research results. Pérez et al. (2014) suggested the use of Köppen classification in order to unify criteria and to properly compare research results relating to the VGS as a passive tool for energy savings, because it is based on the annual and monthly temperature and precipitation averages which are the most influential parameters for vegetation development.

In this study it was found that there is still a lack of research regarding the use of VGS as a tool for passive energy saving in many climatic zones of the world. Specifically, a lack of studies in areas of the world that receive more radiation is observed and therefore where these systems could be more effective is due to the shade effect.

Plant Species Influence

Another aspect to consider in VGS (when used as passive energy savings systems) is the plant species used because each constructive system uses different types of plants. Thus, green facade climbing plants are usually used whereas in green wall shrubs and herbaceous plants are more common. Consequently, plants used for green facades can be deciduous or evergreen species, but in green wall plants are usually evergreen. This fact has a strong influence in each system's thermal performance because when using perennial plants both cooling and heating periods are influenced by plant coverage, and when using deciduous plants only the cooling period is affected since the solar radiation will pass during the heating period (leafless period).

An important issue to consider is the solar gains on the building during the periods of transition, i.e., spring and autumn, when deciduous plants are used for VGS. Since leaves of different species grow at different moments with different speeds during spring, and not all species lose their leaves at the same moment or according to the same speed during autumn, all must be taken into account in the yearly thermal behavior of the whole system (Pérez et al., 2010).

For research purposes the number of species used during the last years for green facades is very limited, being the most used one perennial species, Ivy (*Hereda* sp.) and two deciduous species, Boston Ivy (*Parthenocissus tricuspidata*) and Wisteria (*Wisteria sinensis*). In the case of green walls, herbaceous and shrub species are the most common, usually well-adapted to local climatic conditions, and always evergreen. The possibility to use a higher number of species is bigger than in green facades, though this heterogeneity will have strong consequences in the final thermal behavior.

In general, there is a lack of studies relating the possibility to use a larger number of species in different climates in order to create catalogs by climate, to facilitate the decision for each particular project. In this sense, the risk of excessive assumptions relating the species behavior (thermal conductivity, shading coefficient, etc.), especially during simulation studies, is also preoccupying.

OPERATING METHODS

Since VGS are constructively different (basically green walls or green facades), the mechanisms that regulate their thermal behavior could also differ. Four main effects for this purpose were established by Pérez et al. (2011): the shade effect, the cooling effect, the insulation effect, and the wind barrier effect. Table 2 summarizes the operating methods and parameters that could influence green vertical systems as passive systems for energy savings in buildings.

MAIN RESEARCH FINDINGS

According to the considerations made in previous sections, the study of the VGS contribution on saving energy at building scale can be clearly differentiated by types of constructive systems.

Thus, studying about traditional systems, reductions in the outer surface temperature of the building wall ranging from 1.2°C (Perini et al., 2011) to 13°C (Hoyano, 1988) in warm temperate climate (C) and between 7.9°C and 16°C (Di and Wang, 1999) in snow climate (D), during the summer period have been measured. In previous studies relating to traditional VGS the great influence of the foliage thickness layer and the building facade orientation on the final results can be observed.

From the previous studies referring to double-skin green facades, the reduction on the exterior surface temperature of the building facade wall ranged

TABLE 2 Operating Methods and Main Parameters That Could Influence of Green Vertical Systems as Passive Systems for Energy Savings in Buildings

Effect	Method	Parameters That Could Influence
Shade	Solar radiation interception provided by plants	Density of the foliage (leaf area index (LAI))
Cooling	Evapotranspiration from the plants and substrates	Density of the foliage (LAI)Type of plant (transpiration coefficient) Climate conditions (dry/wet)Wind speedSubstrate moisture[a]
Insulation	Insulation capacity of the different construction system layers: plants, air, substrates, felts, panels, etc.	Density of the foliage (LAI)Air gap thicknessSubstrate thickness and composition[a]
Wind barrier	Wind effect modification by plants and support structures	Density of the foliage (LAI)Facade orientationWind speed

[a]*Only in certain types of vertical greening systems, such as living walls.*

from 1°C (Hoyano, 1988) to 15.18°C (Pérez et al., 2011) for the studies located in warm temperate climate (C). In this case, unlike traditional VGS, often an air gap of variable thickness between the plant layer and the building facade wall is generated, which despite being open in most cases may influence the final results. In addition, again the orientation of the facade and the foliage layer thickness are factors to consider for the thermal behavior of the whole system. Unfortunately, there are still very few data on these two factors.

With green walls, reductions in external surface temperatures of the building facade wall were considerable in warm temperate climate (C), ranging from 12°C, according to Mazzali et al. (2013) to 20.8°C (Chen et al., 2013) in the summer period. In this typology the most interesting parameters to consider in their analysis and consequently in the subsequent design, are the period of study, the species used, the facade orientation, and the foliage thickness (or the coverage percentage), the substrate typology and thickness, and the air gap thickness between the plant layer and the building facade wall.

Recent studies conducted by Coma et al. (2014) under Mediterranean continental climate provided a real scale analysis of the thermal performance for a double-screen green facade made with a deciduous species as a passive system for energy savings in buildings. As a further step, the same authors compared the thermal performance between green walls and green facades in which large energy savings for both typologies during the cooling period, and promising behavior during heating periods were obtained. Fig. 2 shows the experimental set-up and the main results from this investigation.

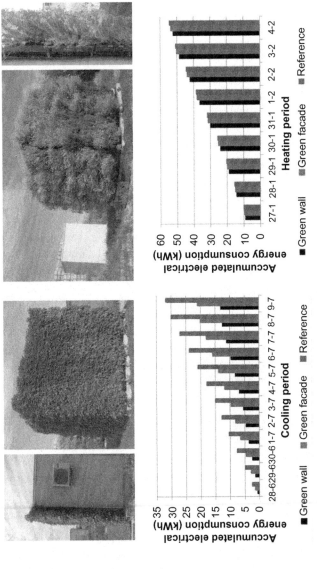

FIGURE 2 Experimental set-up in *Puigverd de Lleida* and the main results on energy savings achieved for both summer and winter.

From the simulation studies on VGS, it can be generally stated that the VGS are an effective tool for energy savings during the cooling period in warm temperate (C) and arid (B) climates, with reductions between 5% and 50%, being the most frequent between 20% and 30%, taking into special consideration the influence of west facade orientation. The studies conducted by Wong et al. (2009) and Kontoleon and Eumorfopoulou (2010) provided interesting data in this regard. The main weaknesses observed in simulation studies regarding to VGS refer to the difficulty to characterize the different plant species used, the lack of real scale experimental tests to validate the theoretical results obtained from the simulation with high quality real data from enough length experiments.

From the research conducted to the date, it could be stated that VGS has good potential as a passive tool for energy savings in buildings, providing good results during cooling periods (summer), variable depending on the constructive systems used. In general, a lack of studies relating to the performance during heating periods (winter), the contribution not only from shade effect but from insulation, cooling and wind barrier effects, as well as the influence of leaf area index (LAI) and facade orientation, can be highlighted.

At urban scale, vegetation contributes to the reduction of the UHI effect. The UHI effect is caused by a variety of factors, such as anthropogenic activity (combustion heat, people, etc.), less evaporative cooling due to the lack of vegetation, less cold wind in the streets, the configuration of streets, solar heat stored in the urban fabric, etc. In addition, cities have large areas of asphalt and other dark materials that have low albedo (reflectivity) resulting in the absorption of radiant heat from the sun and reradiation at night. Since vegetation not only has higher albedo than most of the commonly building materials used but also provides cooling through evapotranspiration, plants contribute to the reduction of the heat island effect (Taha, 1997).

Only a few authors have studied specifically the global effect of VGS on the whole urban environment. Alexandri and Jones (2008) simulated the thermal effect of covering building envelopes with vegetation on the built environment microclimate for a set of climates and urban canyon geometries. The main conclusion was that there is an important potential of lowering urban temperatures when the building envelope is covered with vegetation. On the contrary, Wong et al. (2010) conducted a large experiment in which data on the thermal behavior of eight VGS in Singapore (tropical climate) were recorded. Although in the experiment the average experimental wall surface temperature reduction ranges from 6°C to more than 10°C, the VGS influence on the surrounding built environment temperature was not significant. Wilmers (1990) who highlighted the importance of green areas on the improvement of the urban climate pointed out the LAI, the evapotranspiration, and the wind as the most important factors.

Recently, Price et al. (2015) point out the use of vegetation as an optimal mitigation strategy for UHIs and VGS as one of the most promising ways to add this vegetation to the city due to their significantly larger area in comparison with green roofs, which therefore implies a potentially larger overall contribution in heat reduction.

CONCLUSION

VGS are a part of urban green infrastructure that contribute to the provision of several ecosystem services in the built environment. Among others, VGS highlights its ability as a passive tool for saving energy at building scale and reducing the heat island effect in large cities.

In recent years it has developed a set of constructive systems (green facades and green walls in a general approximation) that should be differentiated when considering their potential for this purpose.

Among the different operational methods, the shadow effect during cooling periods is perhaps the most important, and therefore the most studied.

The future challenges should face issues such as the contribution of evapotranspiration cooling effect, insulation effect, and wind barrier effect. Other important future research should be related to measurement of the LAI in VGS relating to the energy savings provided, the orientation influence, and the thermal behavior of VGS during the spring and autumn periods when deciduous leaves are growing or falling. Finally, several studies must address the local behavior of different plant species in order to create local catalogs that help the future VGS designs.

ACKNOWLEDGMENTS

The work partially was funded by the Spanish government (ENE2015-64117-C5-1-R (MINECO/FEDER) and ULLE10-4E-1305). The authors would like to thank the Catalan Government for the quality accreditation given to their research group (2014 SGR 123) and to the city hall of Puigverd de Lleida. These projects have received funding from the European Commission Seventh Framework Program (FP/2007-2013) under grant agreement No PIRSES-GA-2013-610692 (INNOSTORAGE) and from European Union's Horizon 2020 research and innovation program under grant agreement No 657466 (INPATH-TES). Julià Coma would like to thank the Departament d'Universitats, Recerca i Societat de la Informació de la Generalitat de Catalunya for his research fellowship.

REFERENCES

European Commission-Environment, 2012. The Multifunctionality of Green Infrastructure. Science for Environment Policy. In-depth Reports.

Alexandri, E., Jones, Ph, 2008. Temperature decreases in an urban canyon due to green walls and green roofs in diverse climates. Build. Environ. 43, 480–493.

Chen, Q., et al., 2013. An experimental evaluation of the living wall system in hot humid climate. Energy Build. 61, 298–307.

Coma, J., et al., 2014. New green facades as passive systems for energy savings on buildings. ISES Solar World Congress. Energy Proc. 57, 1851–1859.

Di, H.F., Wang, D.N., 1999. Cooling effect of ivy on a wall. Exp. Heat Trans. 12 (3), 235–245.

Hoyano, A., 1988. Climatological uses of plants for solar control and the effects on the thermal environment of a building. Energy Build. 11, 181–199.

Kontoleon, K.J., Eumorfopoulou, E.A., 2010. The effect of the orientation and proportion of a plant-covered wall layer on the thermal performance of a building zone. Build. Environ. 45, 1287–1303.

Mazzali, U., et al., 2013. Experimental investigation on the energy performance of living walls in a temperate climate. Build. Environ. 64, 57–66.

Pérez, G., 2010. Façanes vegetades. Estudi del seu potencial com a sistema passiu d'estalvi d'energia, en clima mediterrani continental. PhD thesis. Universitat Politècnica de Catalunya.

Pérez, G., et al., 2011. Green vertical systems for buildings as passive systems for energy savings. Appl. Energy 88, 4854–4859.

Pérez, G., et al., 2014. Vertical Greenery Systems (VGS) for energy saving in buildings: a review. Renew. Sustain. Energy Rev. 39, 139–165.

Perini, K., et al., 2011. Vertical greening systems and the effect on air flow and temperature on the building envelope. Build. Environ. 46, 2287–2294.

Price, A., 2015. Vertical greenery systems as a strategy in Urban Heat Island Mitigation. Water Air Soil Pollut. 226, 247. Available from: https://doi.org/10.1007/s11270-015-2464-9.

Taha, H., 1997. Urban climates and heat islands: albedo, evapotranspiration, and anthropogenic heat. Energy Build. 25, 99–103.

Wilmers, F., 1990. Effects of vegetation on urban climate and buildings. Energy Build. 15-16, 507–514.

Wong, N.H., et al., 2009. Energy simulation of greenery systems. Energy Build. 41, 1401–1408.

Wong, N.H., et al., 2010. Thermal evaluation of vertical greenery systems for building walls. Build. Environ. 45, 663–672.

Chapter 3.2

Green Roofs to Enhance the Thermal Performance of Buildings and Outdoor Comfort

Julià Coma, Gabriel Pérez and Luisa F. Cabeza

Chapter Outline

INTRODUCTION

Since the building sector is responsible for about 33% of the global final energy consumption and one-third of total direct and indirect CO_2 emissions, their significant reduction are key targets for all countries for the next decades (International Energy Agency, 2013). Thus, many countries have agreed on implementing new and more restrictive policies in increasing energy efficiency and reducing the energy demand of buildings to meet the long-term greenhouse gas reduction target by 2050 (European Commission 2014; A policy framework for climate and energy in the period from 2020 to 2030).

Within this context, the use of green infrastructure solutions such as green roofs in urban spaces have become more popular during the last few decades, with promising contributions in reducing greenhouse gas (GHG) emissions because they contribute in reducing the energy for heating and

Nature Based Strategies for Urban and Building Sustainability.
DOI: https://doi.org/10.1016/B978-0-12-812150-4.00010-0
109

cooling in buildings. At the same time these systems provide multiple other benefits in comparison to traditional gray solutions (ecosystem services such as rainwater runoff reduction, air purification, support to biodiversity, positive influence on the health and well-being of the city's inhabitants, etc.). This idea takes more relevance as more than 54% of the world population lives in densely urban areas, a percentage that is expected to increase up to 66% by 2050 (United Nations, 2017).

Nowadays, green infrastructures such as green roofs are well known and globally implemented. This evidence forces architects, engineers, and stakeholders to have some clear ideas before designing a green roof system. The first one is about the function of this system when implemented on a building envelope (to generate recreation areas on rooftops, to have better aesthetics, to increase the energy efficiency of a building, to enlarge the biodiversity in urban areas, to mitigate urban heat island (UHI), etc.), and the second one concerns to the specific climatic conditions of the place (continental, Mediterranean, Atlantic, tropical, etc.).

Northern EU countries such as Switzerland, United Kingdom, and Germany, are examples of countries that have implemented these systems for many years as passive systems to reduce the energy demand in buildings. They are leaders in implementing green roofs because they have been (and they currently are) promoting these systems through two main measures: financial incentives and building regulations (Green infrastructure implementation and efficiency, 2011).

With the aim to disseminate the important role of the thermal performance of green roofs when implemented on building envelopes, this chapter presents the main influencing parameters that affect the thermal performance of these systems (typology, climatic conditions, plant species, growing media, and the drainage layer) and the most relevant findings of experimental research studies.

MAIN INFLUENCING ASPECTS

Differences Between Construction Systems

Generally, green roofs are classified in different categories according to three main aspects: the final use, their physical properties (thickness and composition of the different layers), and the maintenance required during their operational lives (International Green Roof Association, 2017). After a literature review, three typologies of green roofs were found: extensive, semi-intensive, and intensive systems (for specific technical info go to Chapter 2.3: Green Roofs Classifications, Plant Species, Substrates).

Among these systems the extensive ones are, under a sustainable approach, the most promising because they require a lower initial investment, less maintenance during their lifetime, and have high potential to be used for refurbishment of old buildings because they do not usually require extra reinforcement of the building structure. However, the extensive green roof systems require a comprehensive technical analysis when they are implemented on building envelopes for thermal purposes.

On the other hand, the advantages of intensive green roofs in comparison to the extensive types are in terms of biomass quality and complexity, potential to support biodiversity, the microclimatic effect, and landscape and aesthetic values (Jim and Tsang, 2011A).

Climatic Conditions and Plant Species

Green roof systems provide passive energy savings in buildings, which are generally reflected in reducing the energy demand for cooling or heating purposes. This passive potential varies according to the climatic conditions, but also other important variables such as the variety of plant species, their intrinsic properties (foliage density, plant height, etc.) as well as their physiological responses (transpiration, position of leaves, etc.) can also affect the final thermal response of these systems. After studying six different plant species (two succulent and four broad-leaved) in temperate maritime climatic conditions, Monteiro et al. (2017) concluded that not all vegetated components of green roofs provide the same benefits, so each plant species can strongly determine the type and level of thermal benefits provided.

From the scientific research approach, it is interesting to highlight the necessity of a standardized classification to establish adequate and comprehensive comparisons between performed studies worldwide. Within this topic, Pérez et al. (2014) suggested the use of Köppen climate classification provided by Kottek et al. (2006), to unify criteria and properly compare the thermal performance, because it considers the monthly and annual temperature and precipitation averages, which are the most influential parameters for plant development.

In addition to the climatic conditions, the typology of plant species also varies depending on the green roof typology. Extensive green roof solutions with thin substrate layers (6−20 cm) generally allow implementing moss, sedums, herbs, and lawns, while plant communities in semi-intensives (10−25 cm) and intensive systems (>25 cm up to 100 cm) usually are shrubs, large bushes, and trees.

Regarding the thermal performance of aforementioned green roofs, intensive and semi-intensive systems are the ones that provide higher thermal performances not only for their thick soils, which isolate the internal ceiling temperatures from the external temperature fluctuation as a large heat sink, but for the typology of plants (bushes and trees) with high shade values, which can be estimated by means of the leaf area index (LAI). Therefore, this type of plant provides a higher cooling effect by reducing the solar irradiance into the soil surface in comparison to sedums or lawns.

Growing Media

Besides providing nutrients and water for plant development, the substrate layer is crucial for the overall thermal performance of a green roof, especially in extensive systems during the plant growth period, which can last up to 2 years after the installation as it can be seen in Fig. 1. However, as the growth of vegetation is variable depending on external factors such as weather conditions, diseases, etc.; the coverage of plants cannot ensure uniformity and consequently the "shade effect" cannot be considered as a constant parameter.

Substrates are generally made by a mix of materials grouped in two main fractions: the organic fraction to ensure the adequate living conditions for plants, and the inorganic fraction such as scoria, pumice, vermiculite, sand, crushed bricks, tiles, etc. Currently, there is still a lack of valuable information about the specific thermo-physical properties of these complex aggregations of materials. It is quite common that the composition of substrates indeed depends on the local availability of materials and varies according to national recommendations. A different composition is connected to different thermal responses of a substrate and, consequently, of the whole green roof system. Therefore, it is important to have accurate information about the growing media intended to be used, especially in the design phase, in which heat transfer numerical models often require such information. Although there is detailed information for natural soils in the literature (Nidal and Randall, 2000; Zhang et al., 2017) it is difficult to deduce thermal properties of green roof substrates from these data.

In addition to all of these variables, the moisture content capacity of each mixed substrate and consequently the thermal response, can vary considerably. A study conducted by Sailor and Hagos (2011), after experimentally evaluating the thermal conductivity of 12 common substrates found in the western United States where composition and moisture content were varied from dry to saturated conditions, it was observed that thermal conductivity of dry green roof substrates doubles when the moisture is increased to about 35% saturation, and triples when fully saturated. Thus, the water content

FIGURE 1 Two years evolution of an extensive green roof in an experimental set-up under Mediterranean continental climate conditions. Puigverd de Lleida, Spain (Coma et al., 2016).

plays the most important role over the final thermal response of the substrate and consequently of the whole green roof system.

Moreover, Jim and Tsang (2011A) evaluated the thermal performance of an intensive green roof of a 100 cm thickness on a rooftop in Hong Kong by measuring the heat flux penetration into the building. In summer, the results showed that a thin soil layer of 10 cm provides enough heat penetration reduction into the building in a humid subtropical climate (Cfa, according to Kottek et al., 2006). However, in winter they concluded that green roofs cause notable heat losses from the substrate to the ambient air during this period, thus increasing the energy consumption to heat up the indoor air.

Finally, due to the lack of experimental studies and because there are many variations in the composition of growing media used in different geographical locations, gathering data regarding the thermal properties of different substrate mixtures in different climatic conditions has become a current key research topic. This is crucial information to be implemented in simulation tools to provide more accurate results relating to green roofs energy performance.

Drainage Layer

Besides providing the water storage capacity to ensure survival in dry periods and to drain the excess water during rain events, the drainage layer plays an important role in the thermal performance of a green roof, particularly in extensive types.

A study conducted by Coma et al. (2016) shows that there were significant differences in energy savings when two identical extensive green roofs with different drainage layer materials (one with 4 cm of pozzolana and the other one with 4 cm of recycled rubber crumbs) were experimentally evaluated in both cooling and heating periods (Fig. 2). The green roof with rubber crumbs

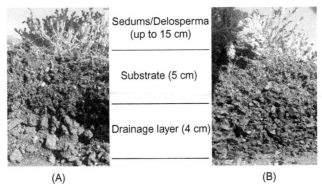

Sedums/Delosperma
(up to 15 cm)

Substrate (5 cm)

Drainage layer (4 cm)

(A)   (B)

FIGURE 2 Detailed section of the two extensive green roofs with different drainage layer materials; (A) with pozzolana, (B) with rubber crumbs (Coma et al., 2016).

TABLE 1 The Main Parameters and Effects That Have Influence on the Thermal Performance of a Green Roof When Implanted as Passive Energy Saving Systems in Buildings

Effect	Method	Influencing Parameter
Shade	Solar irradiation interception provided by different plant species	Density of the foliage (leaf area index (LAI))
Cooling	Evapotranspiration from the plants and substrates	Density of the foliage (LAI)
		Type of plant (transpiration coefficient)
		Climate conditions (dry/wet)
		Substrate moisture content
		Wind speed
Insulation	Insulation capacity of the different green roof layers: plants, air, substrates, geotextile felts, drainage layer, etc.	Density of the foliage (LAI)
		Substrate thickness and composition
Wind barrier	Wind effect modification by plants	Density of the foliage (LAI)
		Wind speed

reduced in 14%, the electrical energy consumption of HVAC for cooling, and 5% for heating purposes in comparison to green roof with pozzolana.

OPERATING METHODS

Green roofs provide energy improvements to the building basically through four main effects: shade, cooling, insulation, and wind barrier. Table 1 summarizes the methods by the means of these effects take place, as well as the main influencing parameters.

MAIN RESEARCH FINDINGS

There is wide selection of literature in relation with the energy savings provided by green roofs on buildings, thus some of the most interesting findings are summarized below.

Related to the seasonal thermal performance, Silva et al. (2016) developed a numerical model, which was experimentally validated, to quantify the energy savings of the different green roof types in the Mediterranean continental climate conditions of Lisbon (Csa). The main outcomes after evaluating the three green roofs (extensive, semi-intensive, and intensive) lead to similar heating energy needs, but the extensive green roof solution shows higher cooling energy needs than semi-intensive and intensive ones, of 2.8 and 5.9 times more, respectively.

Under similar climatic conditions, Pérez et al. (2012) and Coma et al. (2016) studied the seasonal passive potential of noninsulated extensive green roofs in a Mediterranean continental climate (Csa), in which the energy savings due to the evolution of the area covered by the vegetation (from 20% to >85%), were evaluated. The results showed representative energy savings for cooling (16.7%), while an increment of the energy consumption for heating (11.1%) was observed when compared to a traditional insulated flat roof. Moreover, in the experimental study conducted by Spolek (2008) under the mild climate (Csb) of Portland, Oregon similar results were obtained for summer. However, heat transfer reductions of around 13% were observed in winter.

As an example, for a humid subtropical region (Cfa) of Austin, Texas with high temperatures and intense rain events, Simmons et al. (2008) evaluated six different green roof platforms. The authors concluded that all the studied systems showed significantly lower internal temperatures on warm days, while in cold days no differences were observed when compared to traditional and cool roofs.

Concerning the shade effect provided by plants, Rahman et al. (2015) compared the rate of growth, morphology, cooling effectiveness, and stress tolerance of five different tree species under the same temperate maritime climatic conditions (Cfb). The results showed that some varieties of trees grow three times faster, providing higher LAI and higher stomatal conductivity, thus providing a more cooling effect. Within the same topic, Monteiro et al. (2017) concluded that different types of plants significantly differ in their cooling and insulation benefits during hot periods, when it is most needed. The study concludes that nonsucculent canopies, in particular light-colored ones, with high leaf stomatal conductance and high LAI provided maximum potential for substrate insulation and environmental cooling in hot periods.

According to Saiz et al. (2016), the cooling effect of extensive green roofs in summer is driven by shade and the water evaporation from substrates. In their study, differences up to 5°C at 1 m above the ground comparing areas with moist soil and gray surfaces were obtained. The preliminary conclusion extracted from this study is that the effect of green roofs is limited and it is largely affected by wind speed and solar radiation.

On the other hand, according to Jim and Tsang (2011B), for an intensive green roof located in the humid subtropical climate conditions (Cfb) of Hong Kong, the wind effect does not play a major role in facilitating transpiration rate in the four seasons, as indicated by a rather low correlation coefficient between the wind above the canopy and transpiration. On sunny days, the sensible heat loss through air convection is not significant because of low wind speed. Although the wind speed is higher on rainy days, the effect of evaporative cooling is partly offset due to suppression by high relative humidity.

CONCLUSION

It can be stated that each green roof needs to be designed according to unique requirements of a project under specific circumstances. There is not a green roof that offers constant thermal properties (U_{value}) when it is compared with traditional insulation materials implemented on building envelopes (rock wool, XPS, mineral wool, etc.) due to their variable parameters such as specific climate conditions, plant species, moisture content, thickness and substrate composition, wind effect, etc.

Moreover, after reviewing related literature it could be stated that the passive cooling potential of green roofs is widely studied and experimentally validated for many climatic conditions. However, there are still controversial results regarding the heating performance of these systems depending on the climate region and type of system.

Finally, as a general overview, green roofs can contribute significantly to achieving many of the EU's key policy objectives, while at the same time they contribute to mitigate the GHG emissions and promoting many ecosystems services at both, building and urban scales.

ACKNOWLEDGMENTS

This study has received funding from European Union's Horizon 2020 research and innovation program under grant agreement No 657466 (INPATH-TES), from the European Commission Seventh Framework Program (FP/2007-2013) under grant agreement No PIRSES-GA-2013-610692 (INNOSTORAGE). The work is partially funded by the Spanish Government (ENE2015-64117-C5-1-R (MINECO/FEDER)). GREA is certified agent TECNIO in the category of technology developers from the government of Catalonia. The authors would like to thank the Catalan Government for the quality accreditation given to their research group (2014 SGR 123).

REFERENCES

A policy framework for climate and energy in the period from 2020 to 2030. European Commission 2014. Available at: <https://ec.europa.eu/energy/en/topics/energy-strategy/2030-energy-strategy> (last access January 2017).

Coma, J., Pérez, G., Solé, C., Castell, A., Cabeza, L.F., 2016. Thermal assessment of extensive green roofs as passive tool for energy savings in buildings. Renew. Energy 85, 1106–1115.

Green infrastructure implementation and efficiency, 2011. Institute for European Environmental Policy. Final report; December 2011. Available at: <http://ec.europa.eu/environment/nature/ecosystems/studies.htm> (last access January 2017).

IGRA, 2017 .Green roof types. In: International Green Roof Association. Available at: <http://www.igra-world.com/types_of_green_roofs/index.php> (last access January 2017).

Jim, C.Y., Tsang, S.W., 2011a. Biophysical properties and thermal performance of an intensive green roof. Build. Environ. 46 (6), 1263–1274.

Jim, C.Y., Tsang, S.W., 2011b. Ecological energetics of tropical intensive green roof. Energy Build. 43 (10), 2696–2704.

Kottek, M., Grieser, J., Beck, C., Rudolf, B., Rubel, F., 2006. World map of the Köppen-Geiger climate classification updated. Meteorol. Zeitschrift 15 (3), 259–263.

Monteiro, M.V., Blanuša, T., Verhoef, A., Richardson, M., Hadley, P., Cameron, R.W.F., 2017. Functional green roofs: importance of plant choice in maximising summertime environmental cooling and substrate insulation potential. Energy Build. Available from: http://dx.doi.org/10.1016/j.enbuild.2017.02.011.

Nidal, H., Randall, C., 2000. Soil thermal conductivity, Effects of Density, Moisture, Salt Concentration and Organic Matter, 64. Soil Science Society of America, pp. 1285–1290.

Pérez, G., Coma, J., Martorell, I., Cabeza, L.F., 2014. Vertical Greenery Systems (VGS) for energy saving in buildings: a review. Renew. Sustain. Energy Rev. 39, 139–165.

Pérez, G., Vila, A., Rincónn, L., Solé, C., Cabeza, L.F., 2012. Use of rubber crumbs as drainage layer in green roofs as potential energy improvement material. Appl. Energy 97, 347–354.

Rahman, M.A., Armson, D., Ennos, A.R., 2015. A comparison of the growth and cooling effectiveness of five commonly planted urban tree species. Urban Ecosyst 18, 371. Available from: https://doi.org/10.1007/s11252-014-0407-7.

Sailor, D.J., Hagos, M., 2011. An updated and expanded set of thermal property data for green roof growing media. Energy Build. 43, 2298–2303.

Saiz, S., Olivieri, F., Neila, J., 2016. Green roofs: experimental and analytical study of its potential for urban microclimate regulation in Mediterranean–continental climates. Urban Clim. 17, 304–317.

Silva, C.M., Gomes, M.G., Silva, M., 2016. Green roofs energy performance in Mediterranean climate. Energy Build. 116 (15), 318–325.

Simmons, M.T., Gardiner, B., Windhager, S., Tinsley, J., 2008. Green roofs are not created equal: the hydrologic and thermal performance of six different extensive green roofs and reflective and non-reflective roofs in a sub-tropical climate. Urban Ecosyst 11 (4), 339–348.

Spolek, G., 2008. Performance monitoring of three ecoroofs in Portland, Oregon. Urban Ecosyst 11, 349–359.

Transition to Sustainable Buildings; Strategies and opportunities to 2050. International Energy Agency, 2013. Available at: <www.iea.org/etp/buildings> (last access February 2016).

United Nations. Available at: www.un.org (last access March 2017).

Zhang, T., Cai, G., Liu, S., Puppala, A.J., 2017. Investigation on thermal characteristics and prediction models of soils. Int. J. Heat Mass Transfer 106, 1074–1086.

FURTHER READING

Energy Technology Perspectives, 2016. Towards Sustainable Urban Energy Systems. Available at: <www.iea.org/publications/> (last access January 2017).

Chapter 3.3

Green Streets to Enhance Outdoor Comfort

Katia Perini, Ata Chokhachian and Thomas Auer

Chapter Outline

INTRODUCTION

Environmental conditions in cities affect human health and quality of life. However, dense urban settlements could reduce the impact of anthropogenic activities, lowering the emissions deriving from transportation and building energy use (Ewing et al., 2008; Hamin and Gurran, 2009) and, at the same time, allow protecting of underdeveloped lands and habitats (Farr, 2008).

Thus, environmental issues in cities should be mitigated for human well-being. As an example, the urban heat island (UHI) phenomenon causes higher temperatures in cities (up to $2-5°C$) compared to the surrounding rural areas (Taha, 1997). This phenomenon is mainly related to (Grimmond, 2007; Oke, 1982):

- the amount of manmade surfaces (paved areas) in cities with low albedo, which store and later radiate solar radiations,
- anthropogenic heat emissions (e.g., air conditioning, transportation, industrial plants),
- low evaporation due to the amount of impermeable surfaces,
- dense buildings which affect wind speed and the release of heat.

Outdoor thermal comfort strongly influences human well-being, especially during summer due to the UHI phenomenon temperatures remaining high

Nature Based Strategies for Urban and Building Sustainability.
DOI: https://doi.org/10.1016/B978-0-12-812150-4.00011-2

during night time (Taha, 1997). As demonstrated by several studies (Onishi et al., 2010; Perini and Magliocco, 2014), greenery highly influences microclimatic conditions, mainly due to evapotranspiration and shadow effects.

UHI can be mitigated by large amount of high albedo surfaces (Rizwan et al., 2008)—e.g., white painted, grass, etc.—but urban parks and other large green areas may be more effective (Petralli et al., 2006).

According to Givoni (2009), during summer, higher temperatures caused by the UHI phenomenon, lower wind speeds, and high mean radiant temperatures due to solar access are drivers of thermal discomfort in cities. However, in winter cases the scenario is inversed. For example, during winter, solar radiation increases outdoor comfort, while during summer it can be a major cause of discomfort.

Thermal comfort depends on the thermal balance between the human body and the environment (Givoni, 2009). Since this is a subjective sensation, measurement methods are very important. Among these, the most common indices are the predicted mean vote (PMV), the physiological equivalent temperature (PET), the standard effective temperature (SET), and the universal thermal climate index (UTCI). The PMV, developed by Fanger (Fanger, 1973), considers comfort as the thermal equilibrium in the environment and boundary conditions. SET reflects the heat stress or cold felt by the occupants (Gagge et al., 1986). PET (Höppe, 1999) compares complex outdoor conditions to a typical steady-state indoor setting. A recent developed index is UTCI, which is based on a dynamic physiological response model (Bröde et al., 2010).

In this chapter, the effects of green streets on the most important parameters influencing outdoor thermal comfort are analyzed. Studies regarding the positive effects of greenery on outdoor thermal comfort are described. Finally, some research relating to planting trees and their influence on the outdoor comfort in urban canyons is presented.

OUTDOOR COMFORT AND VEGETATION

Green streets can have climatic and environmental functions if correctly designed; shading effects during summer periods guarantee a partial scattered solar radiation transmission, the reduction or increase of air flow, etc., mitigate microclimatic conditions, and play an important role with respect to outdoor well-being (Scudo and Ochoa de la Torre, 2003).

As summarized in Table 1, the characteristics of plant species determine performances, and effects outdoor comfort parameters, depending on the season, climate conditions, latitude, etc.

According to Morakinyo et al. (2017) the characteristics of trees, specifically in terms of morphological properties, differently regulate thermal comfort due to solar attenuation capacity. A study regarding the effects of trees on outdoor comfort in Hong Kong shows that leaf area index (LAI, $[m^2m^{-2}]$),

TABLE 1 Main Characteristics of Vegetation Affecting Outdoor Comfort Parameters (based on Grosso, 2012; Perini, 2013; Scudo and Ochoa de la Torre, 2003).

Foliage shape and dimensions
- Regarding mean radiant temperature, foliage determines shadow area, depending on the site latitude.
- Row/group of trees can create a barrier or increase air flow.
- Foliage affects plants evapotranspiration, which results in reduced air temperatures and increased air humidity.

Height of trunk
- Regarding mean radiant temperature, the trunk's height determines shadow area, depending on the site latitude.
- In order to protect itself from winter wind, the trunk's height should be reduced.

Leaf area density (LAD)
- High values reduce the solar radiation transmitted during summer.
- LAD determines the air flow through the foliage (low or high).
- LAD affects plants evapotranspiration, which results in reduced air temperatures and increased air humidity.

Seasonal cycle
- Deciduous plant species avoid winter shading.
- Evergreen species are required for winter air flow control.

Daily transpiration
- High levels of daily transpiration cool the air flow passing through trees.
- Transpiration implies a thermal energy absorption able to decrease summer overheating and increase air humidity.

tree height, and trunk height are the most influential parameters in improving microclimate. Leaf area density (LAD, $[m^2m^{-3}]$) is another important foliage modeling parameter for trees and shrubs (Fahmy et al., 2010).

Lee et al. (2016) show that trees, compared to grasslands, have higher performances in mitigating human heat stress. Results show, during the day, a mitigation up to 3.4°C and 2.7°C for air temperature, 7.5°C and 39.1°C for mean radiant temperature, and 4.9 and 17.4 for PET, respectively in the presence of grasslands, and of trees over grasslands.

Microclimate studies address different levels: on one side, urban features dealing with the form of the built environment; on the other hand, the scale of the study which could be defined in canyon, street, and district scale. According to Nyuk-Hien Wong and Yu Chen (2009) while plants can reduce high ambient temperatures in a built environment in Singapore, with regard to the UHI, the effects of urban greening are evident only when a large surface is considered (e.g., urban parks, road trees). Morakinyo et al. (2017) show that with higher urban densities, the reduced effects of trees on outdoor thermal comfort are recorded. Depending on canyon measures, more suitable trees can be selected, e.g., tall trees of low canopy density with high trunks in deeper canyons, and vice versa for open areas.

In order to quantify the effects of greenery on microclimate and outdoor comfort, environmental modeling can be used. For example ENVI-met models allow for evaluating the effects of urban forms and vegetation on microclimate (Krüger et al., 2011; Wang et al., 2016).

Figs. 1 and 2 show UTCI values for an urban configuration which includes two urban canyons (NS and WE oriented) and a courtyard without

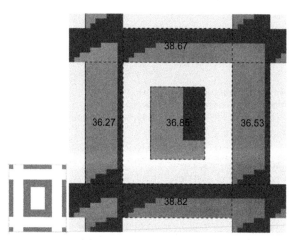

43.5% Very strong heat stress** 56.4% Strong heat stress

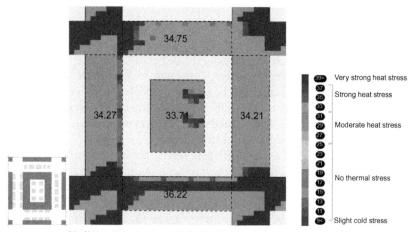

28.4% Very strong heat stress** 71.5% Strong heat stress

FIGURE 1 UTCI map at 3:00 p.m. for canyons (NS and EW oriented) and courtyard with (right) and without (left) vegetation, at 1.5 m from the ground, with average UTCI value recorded for each canyon and for the courtyard.

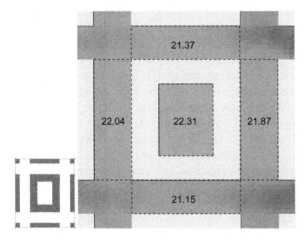

100.% No thermal stress

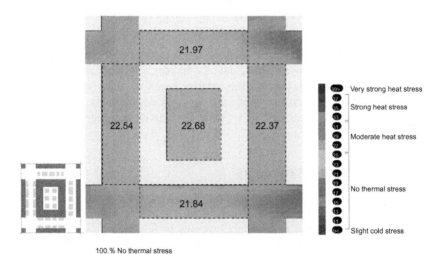

100.% No thermal stress

FIGURE 2   UTCI map at 10:00 p.m. for canyons (NS and EW oriented) and courtyard with (right) and without (left) vegetation, at 1.5 m from the ground, with average UTCI value recorded for each canyon and for the courtyard.

vegetation and with vegetation. The latter is based on 15 m high dense trees placed in the middle of the canyon and roadside. The simulations are set during a hot summer day in the city of Munich (12th August) to monitor the maximum effect of the parameters analyzed. Climate data are collected from a weather station in the city center (minimum and maximum temperatures respectively 15.3°C and 33.9°C, wind speed 1 m s^{-1}, relative humidity in a range of 27%–97%). The simulations' outputs (outdoor comfort parameters) are imported to Grasshopper (http://www.grasshopper3d.com/) and combined to result UTCI values.

Results show that, during the day (Fig. 1), 15 m high trees having LAD from 0.15 to 2.18 $m^2 m^{-3}$ influence outdoor comfort with a UTCI reduction of 15% of areas with "very strong heat stress." UTCI values are lower near building facades, as a result of shadow effect. Lower values are observed in the case with trees, varying between 1.5 and 8 m high, depending on solar exposure. Regarding orientation of the canyons, UTCI values for east-west canyons are about 3.25°C higher in the case without vegetation in average. In the case of canyon oriented north-south the delta is 2.13°C. This effect could be related to wider ground surface of the canyon exposed to direct radiation for east-west oriented canyons.

During the night (Fig. 2) the difference is less evident, which is about 1°C of UTCI between cases with and without trees. Higher values are recorded in the canyon with greenery.

EFFECTS OF GREENERY ON MICROCLIMATE

In order to investigate influential factors determining UTCI values, a better understanding of each parameter is necessary. The parameters, depending on the urban environment, used to calculate the indices and determining outdoor comfort are mean radiant temperature, relative humidity, wind speed, and air temperature. However, the effect of each factor is not the same and remains different on overall calculation of UTCI depending on time period, climate, and location of simulations.

Air Temperature and Humidity

Inside green areas, humidity increases due to evapotranspiration, a phenomenon connected to photosynthesis which implies a thermal energy absorption is able to mitigate urban temperatures; thanks to this phenomenon most of solar radiation is transformed into latent heat (Nyuk-Hien Wong and Yu Chen, 2009).

Reflection, transmission, and absorption processes can sensibly modify the solar radiation, with effects varying during time (day/night, summer/winter). The photosynthetic process absorbs only a part of the total solar radiation that falls on the leaves, whereas the major part of it is reflected, transmitted, and reemitted in the form of sensible and latent heat. In general, we can affirm that broad-leaved vegetation during foliation period reflects almost 20% of the total solar radiation, absorbs less than 5% thanks to photosynthesis, absorbs and reemits as 65% latent and sensible heat, and transmits less than 10% (Scudo and Ochoa de la Torre, 2003). Furthermore, leaves dispose themselves following the solar radiation for photosynthesis processes, improving their shading capacity (Bellomo, 2003).

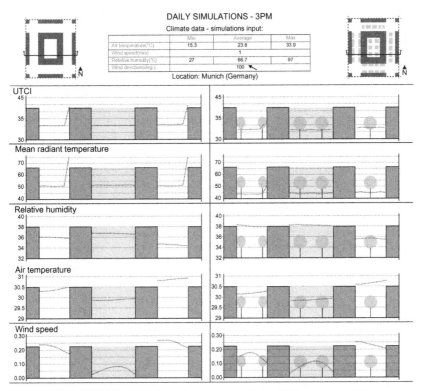

FIGURE 3 Cross section of ENVI-met simulations results for UTCI, mean radiant temperature, wind speed, air quality, and relative humidity values at 1.5 m from the ground at 3:00 p.m.

As shown in Fig. 3, which represents the results (cross section) of an ENVI-met simulation (the same reported in Figs. 1 and 2), daily temperatures are lower in greening urban canyons, with a similar trend for both roadside trees and trees in the middle of the canyon. This highlights that green areas reduce the daily thermal excursion. In the courtyard area fewer effects on air temperatures are observed. Looking at relative humidity, an opposite trend is noticed, with higher values recorded for the configuration with trees. In this case, the effect of the higher number of trees (i.e., canyon with roadside trees and courtyard) is more evident. The night time simulation shows a similar behavior of trees with slightly lower temperatures (in the case with trees). Differently, relative humidity values are very similar. This is explained by the absence of photosynthesis during the night.

The results of the simulations show that the effects of green streets on air temperature and humidity depend on the relation between vegetation and built space. As shown in Table 1, specific characteristics of vegetation play a key role as well. In fact, dense (high LAD) and wide foliage will affect evapotranspiration, i.e., the most influencing phenomenon for temperature mitigation.

Air Movement

The presence of plants (trees or shrubs) can control air flow direction and intensity, creating screens to stop, conduct, divert, or filter. This effect regulates microclimate more or less depending on temperatures, humidity, and wind speed. Vegetation can be used as a wind barrier, thanks to a certain permeability that enhances less turbulences around it (Scudo and Ochoa de la Torre, 2003). The wind speed reduction provided by a vegetable barrier (disposition of trees and shrubs) is affected by the form, height, and porosity of the barrier itself (Grosso, 2012).

Air flow during summer is an important parameter for thermal outdoor comfort. Trees can reduce air movement, resulting in a negative effect. Figs. 3 and 4, clearly show that trees, especially in the roadside position both during the day and night. Therefore, in order to exploit the effects of green areas on thermal outdoor comfort, this aspect should be considered. Using different types of plant species, shrubs, e.g., can be a possible design strategy, although, in this case, a less effect on mean radiant temperature would be recorded.

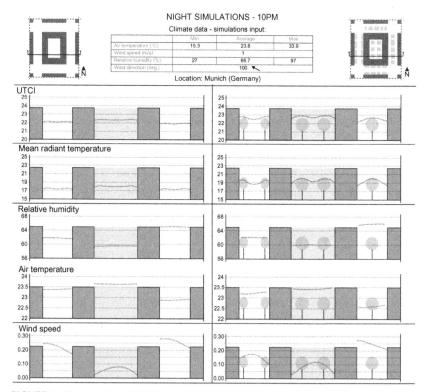

FIGURE 4 Cross section of ENVI-met simulations results for UTCI, mean radiant temperature, wind speed, air quality, and relative humidity values at 1.5 m from the ground at 10:00 p.m.

Mean Radiant Temperature

The effects of green streets on the mean radiant temperature are mainly related to plants' shading. Leaves can limit incoming solar radiation to reach building and street artificial materials, therefore greening paved surfaces with vegetation to intercept the radiation can reduce the amount of heat absorbed by hard surfaces. As shown by Chokhachian et al. (2017) mean radiant temperature is mainly affected by material properties of built environment, therefore increasing the amount of high albedo surfaces (as grasslands) can highly increase outdoor comfort (highly affected by mean radiant temperature).

Nyuk-Hien Wong and Yu Chen (2009) explored the shading effect of plants on a building envelope, conducting field measurements on low rise buildings with trees planted nearby, comparing it to a no-trees scenario. Results show lower surface temperatures on the external walls with differences of $1-2°C$ at night and around $4-8°C$ during the daytime. The tree shading effect is more evident during the morning, if trees are planted along the eastern side.

In order to reduce mean radiant temperature values, greenery shading effect should be optimized choosing the most effective plants (especially in terms of shape, dimensions, LAD). For the same woody plant, the shadow area is nearer to the tree base as much as the solar inclination is high, or rather it is closer to the equator. In low latitudes the sun is high for a good part of the day, therefore wide and umbrella shaped foliages can have an effective cooling function; in medium latitudes, wherein the hottest time is reached halfway through the afternoon, wide foliages have a significant influence on the solar radiation, which is lower (Grosso, 2012).

This phenomenon is evident in Figs. 3 and 4. During the day, at 3:00 p.m., mean radiant temperatures are lower (about $10°C$) below trees, in all the cases analyzed. In particular, thanks to trees shading effect, facades are much cooler. During the night, vegetation works the other way around, resulting in slightly higher mean radiant temperature values.

Looking at UTCI values, a similar trend is noticed. This demonstrates the effects of trees on thermal outdoor comfort and the important role played by mean radiant temperature reductions.

CONCLUSION

Green streets allow for improving thermal outdoor comfort in cities, with effects evident at a range of scales (city, district, canyon scale). In order to exploit the use of trees and shrubs for microclimate regulation, plant species characteristics (e.g., foliage shape and dimension, LAD, and seasonal cycle) have to be considered in relation to the built environment. Simulations results of summer performances of trees planted in the middle of a canyon, roadside and in a courtyard, show variable effects: roadside trees decrease wind speed—which has a negative effect on outdoor comfort during summer—and have higher effects on relative humidity and air temperatures. Mean radiant temperature is mitigated both during the day and night by trees, resulting in improved thermal comfort (lower UTCI values).

ACKNOWLEDGMENTS

The authors would like to thank the University of Genoa for providing the funding which allowed the collaboration between the Technische Universität München, Chair of Building Technology and Climate Responsive Design and the Architecture and Design Department (University of Genoa).

REFERENCES

Bellomo, A., 2003. Pareti verdi: linee guida alla progettazione / Antonella Bellomo. Esselibri, Napoli.

Bröde, P., Jendritzky, G., Fiala, D., Havenith, G., Weihs, P., Batchvarova, E., et al., 2010. The universal thermal climate index UTCI. Berichte Meteorol. Inst. Albert-Ludwigs-Univ. Freibg. 20, 184–188.

Chokhachian, A., Perini, K., Dong, S., Auer, T., 2017. How Material Performance of Building Façade Affect Urban Microclimate, in: Powerskin Conference | Proceedings. Presented at the Powerskin Conference, Tu Delft, Delft, The Netherlands, pp. 83–96.

Ewing, R.H., Anderson, G., Winkelman, S., Walters, J., Chen, D., 2008. Growing Cooler the Evidence on Urban Development and Climate Change. ULI, Washington, DC.

Fahmy, M., Sharples, S., Yahiya, M., 2010. LAI based trees selection for mid latitude urban developments: a microclimatic study in Cairo. Egypt. Build. Environ. 45 (2), 345–357. Available from: https://doi.org/10.1016/j.buildenv.2009.06.014.

Fanger, P.O., 1973. Thermal Comfort. McGraw-Hill Inc, New York.

Farr, D., 2008. Sustainable Urbanism: Urban Design With Nature. Wiley, Hoboken, NJ.

Gagge, A.P., Fobelets, A.P., Berglund, L.G., 1986. A standard predictive index of human response to the thermal environment. ASHRAE Trans. US 92, 2B.

Givoni, B., 2009. Thermal comfort issue and implications in high-density cities, Designing High-Density Cities: For Social and Environmental Sustainability. Edited by Edward Ng, London. Routledge, 87–106.

Grimmond, S., 2007. Urbanization and global environmental change: local effects of urban warming. Geogr. J. 173. Available from: https://doi.org/10.1111/j.1475-4959.2007.232_3.x.

Grosso, M., 2012. La Ventilazione Naturale Controllata e il Raffrescamento Passivo Ventilativo degli edifici [WWW Document]. <http://porto.polito.it/2579961/> (accessed 02.02.17.).

Hamin, E.M., Gurran, N., 2009. Urban form and climate change: balancing adaptation and mitigation in the U.S. and Australia. Habitat. Int. 33, 238–245. Available from: https://doi.org/10.1016/j.habitatint.2008.10.005.

Höppe, P., 1999. The physiological equivalent temperature – a universal index for the biometeorological assessment of the thermal environment. Int. J. Biometeorol. 43, 71–75. Available from: https://doi.org/10.1007/s004840050118.

Krüger, E.L., Minella, F.O., Rasia, F., 2011. Impact of urban geometry on outdoor thermal comfort and air quality from field measurements in Curitiba, Brazil. Build. Environ. 46, 621–634. Available from: https://doi.org/10.1016/j.buildenv.2010.09.006.

Lee, H., Mayer, H., Chen, L., 2016. Contribution of trees and grasslands to the mitigation of human heat stress in a residential district of Freiburg, Southwest Germany. Landsc. Urban Plan. 148, 37–50. Available from: https://doi.org/10.1016/j.landurbplan.2015.12.004.

Morakinyo, T.E., Kong, L., Lau, K.K.-L., Yuan, C., Ng, E., 2017. A study on the impact of shadow-cast and tree species on in-canyon and neighborhood's thermal comfort. Build. Environ. 115, 1–17. Available from: https://doi.org/10.1016/j.buildenv.2017.01.005.

Oke, T.R., 1982. The energetic basis of the urban heat island. Quart. J. Royal Meteorol. Soc. 108, 1–24. References - Scientific Research Publish. Q. J. R. Meteorol. Soc. vol. 108, pp1–24.

Onishi, A., Cao, X., Ito, T., Shi, F., Imura, H., 2010. Evaluating the potential for urban heat-island mitigation by greening parking lots. Urban For. Urban Green. 9, 323–332. Available from: https://doi.org/10.1016/j.ufug.2010.06.002.

Perini, K., 2013. Progettare il verde in città: una strategia per l'architettura sostenibile. F. Angeli, Milano.

Perini, K., Magliocco, A., 2014. Effects of vegetation, urban density, building height, and atmospheric conditions on local temperatures and thermal comfort. Urban For. Urban Green. Available from: https://doi.org/10.1016/j.ufug.2014.03.003.

Petralli, M., Prokopp, A., Morabito, M., Bartolini, G., Torrigiani, T., Orlandini, S., 2006. Ruolo delle aree verdi nella mitigazione dell'isola di calore urbana: uno studio nella città di Firenze. Riv. Ital. Agrometeorol. 1, 51–58.

Rizwan, A.M., Dennis, L.Y., Liu, C., 2008. A review on the generation, determination and mitigation of Urban Heat Island. J. Environ. Sci. 20, 120–128.

Scudo, G., Ochoa de la Torre, J.M., 2003. Spazi verdi urbani: la vegetazione come strumento di progetto per il comfort ambientale negli spazi abitati. Sistemi editoriali: Esselibri-Simone, [Napoli].

Taha, H., 1997. Urban climates and heat islands: albedo, evapotranspiration, and anthropogenic heat. Energy Build. 25, 99–103. Available from: https://doi.org/10.1016/S0378-7788(96)00999-1.

Wang, Y., Berardi, U., Akbari, H., 2016. Comparing the effects of urban heat island mitigation strategies for Toronto, Canada. Energy Build. 114, 2–19. Available from: https://doi.org/10.1016/j.enbuild.2015.06.046.

Wong, N.-H., Chen, Y., 2009. The role of urban greenery in high-density cities. Designing High-Density Cities: For Social and Environmental Sustainability. Edited by Edward Ng, London. Routledge, pp. 87–106.

Chapter 3.4

Vertical Greening Systems for Pollutants Reduction

Katia Perini and Enrica Roccotiello

Chapter Outline

INTRODUCTION

By 2020 it is estimated that almost 80% of EU citizens will be living in cities (European Commission, 2016a), and their quality of life will be directly influenced by the state of the urban environment. Vertical greening systems (VGS) can significantly improve the environmental quality of dense urban areas by providing several ecosystem services, i.e., reducing the urban heat island (UHI) effect, improving air quality and energy performance of buildings, fostering biodiversity, etc. As also highlighted by the European Commission, nature-based solutions can provide "environmental, social, and economic benefits and help build resilience" (European Commission, 2016b).

Air quality is a major issue in many cities around the world. Epidemiological studies have proven a strong correlation between increased air pollution and adverse health effects leading to an increase in mortality and disease incidences with a decrease in life expectancy (Merbitz et al., 2012; WHO, 2013).

In this challenge, vertical surfaces could be exploited to improve air quality, especially in specific urban structures that represent major pollution hotspots, such as constricted street canyons formed by tall buildings on either side of the road (Pugh et al., 2012; Tallis et al., 2011). Focusing on

Nature Based Strategies for Urban and Building Sustainability.
DOI: https://doi.org/10.1016/B978-0-12-812150-4.00012-4

TABLE 1 Main Influencing Parameters for Pollutants Reduction

	VGS Type	Plant Species	Site, Season, Climate	Pollutants' Mitigation
Leaf macro- and micromorphology		x		PM_x
Leaf surface (LAI)	x	x		PM_x, gases
Foliage density (LAD)	x	x		PM_x, gases
Plant's health	x	x	x	PM_x, gases
VGS and urban canyon's structure and size			x	PM_x, gases

(greening) street canyons instead of urban parks can potentially be more effective due to the difficulty in finding empty spaces for large green areas inside dense cities.

Applying VGS on both new and existing buildings can offer multiple environmental benefits and a sustainable approach in terms of energy saving, nutrients and water management, and efficient preservation of buildings (Pérez-Urrestarazu et al., 2015). Although VGS are mainly studied in order to improve the energetic performance of building (Pérez et al., 2014), they can also provide an interesting opportunity to mitigate air pollutants improving air quality (Perini et al., 2017; Pugh et al., 2012).

In this chapter, the effects of VGS on air quality are analyzed and the main influencing parameters are described. Researches regarding the effects of plants on the major pollutants influencing urban air quality, extensively investigated in epidemiological research for their adverse health effects (European Environment Agency, 2015; WHO, 2005), are considered; in particular particulate matter (PM_x) and gaseous pollutants, i.e., nitrogen oxides (NO_x), and ozone (O_3). Finally, indirect effects of VGS are discussed (Table 1).

MAIN INFLUENCING PARAMETERS

The air quality improvement through plants is mainly related to the absorption of fine dust particles (PM_x) and the uptake of gaseous pollutants, such as carbon dioxide CO_2, NO_2, and sulfur dioxide (SO_2) (Baik et al., 2012). Fine dust particles adhere to the plant surfaces, therefore plants are a perfect anchor for airborne particles at different heights (Ottelé et al., 2010).

Unlike the other greening strategies for urban areas, depending on the technology/system used, the performances of VGS can be different. In fact, VGS can be based on living wall modules, panels planted with different species (also not growing vertically in nature as small shrubs) or on climbing plants planted in front of a facade or in planter boxes. In addition, VGS provide a potentially different obstacle to PM deposition and gaseous air pollutant absorption compared to trees and shrubs used for other urban greening solutions (Thönessen et al., 2008).

Several parameters determine the performances of VGS in improving air quality. First, is the foliage density (Janhäll, 2015). This parameter can be measured by means of the leaf area index (LAI, i.e., leaf area per m^2 of wall surface, in the case of VGS). Considering the case of green facades based on a climbing plant, this index depends on the specific characteristic of the plant used. When considering living wall systems based on several plant species, all the plants should be analyzed, by cutting leaves or using specific tools, when literature about LAI values (e.g., Wolter et al., 2009) is not available. In general, high values of LAI identify dense plants with wide surfaces for the collection of pollutants. Evergreen plants will provide similar benefits both in summer and winter.

Another important parameter which should be considered is the ability of plants to live in a certain environment/location, which determines the plants' health and growth, especially in the case of living wall systems (Perini et al., 2011).

Plant species have different performances in terms of PM_x collecting capacity, i.e., the sink capacity of leaves depends on their macro- and micro-morphology. Leaves with different micromorphology (i.e., thick cuticle, cuticular waxes, hairs, etc.) are proven to be effective in collecting $PM_{2.5}$ (smaller than 2.5 μm) (Perini et al., 2017; Roccotiello et al., 2016), among the most dangerous pollutants for human health (Powe and Willis, 2004). In addition, specific leaf characteristics (i.e., morphology, leaf area index, porosity, leaf area density, etc. (Lin et al., 2016) can influence and interact with PM_x deposition and dispersion, playing a key role in potential air quality improvements (Janhäll, 2015).

The choice of plant species should also consider that air pollutants can impact on plant morphology, physiology, and biochemistry. Depending on the specific characteristics of plants, PM_x may be adsorbed to leaf surface and decrease plant photosynthetic performance, or absorbed damaging the photosynthetic apparatus (Nowak et al., 2014). In addition, gaseous pollutants may enter the leaf through stomatal fluxes (Manes et al., 2016) and disrupt leaf gas exchanges.

When speaking about VGS and air quality, air dynamics have to be considered as well. Due to their important role in forecasting plants' behavior, there is an increasing interest in studies—typically with computational fluid dynamic simulations—within urban street canyons and on the effect of plants on air quality. Some studies show that nature-based solutions could not have the desired impact on air quality, e.g., roadside urban plants could increase pollutant concentrations locally, reducing air flows which would allow diluting air pollutants from traffic source (Vos et al., 2013). Differently, VGS, as shown by Thönessen et al. (2008), do not create this issue due to their typical geometry. VGS can thus contribute to mitigate both air pollution and climate change.

On the other hand, some papers have shown that, for some climate and environmental conditions, the usage of plants may not have the desired impact on air quality in cities (Wania et al., 2012), causing ecosystem disservices (e.g., allergens and volatile organic compounds−VOCs emission) (von Döhren and Haase, 2015). Therefore, caution should be exercised to avoid such ecosystem disservices, while making urban greening plans (Escobedo et al., 2011).

VERTICAL GREENING SYSTEMS AND PARTICULATE MATTER

PM_x represents the seventh major cause of death across the world, responsible for cardiopulmonary disease, cancer, and respiratory infections in urban areas (Cohen et al., 2005). As mentioned above, the effects of VGS on air quality, and PM_x collecting capacity, depend on site and plant species specific characteristics (Fig. 1).

Considering the site/location, the most influencing features are canyon (building) geometry—which influences PM_x deposition—weather conditions (specifically wind speed and directions), and distance from traffic source (Janhäll, 2015). In addition, in order to select the most appropriate plants,

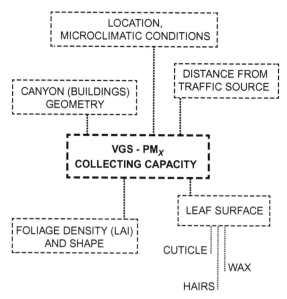

FIGURE 1 PM_x collecting capacity, main influencing factors.

foliage density and shapes have to be considered as well as leaf surface and structure (cuticle, wax, and hairs). These specific micromorphological characteristics of leaves can play a key role, as demonstrated by Perini et al. (2017). Analyzing different species planted on INPS Green Facade (a pilot project in the city of Genoa, Italy—Fig. 2), the performances of *Trachelospermum jasminoides* (Lindl.) Lem., *Hedera helix* L., *Cistus* "Jessamy Beauty," and *Phlomis fruticosa* L. in terms of PM comparing collecting capacity were investigated. The study used the counting method, developed by Ottelé et al. (2010) and based on PM counting on $100 \times$, $250 \times$, $500 \times$, $2500 \times$ magnifications, taken with an ESEM microscope. An example of ESEM micrograph is given in Fig. 3, where different sizes of PM_x are visible thanks to $100 \times$, $500 \times$, $1000 \times$, and $5000 \times$ magnifications.

Fig. 4 shows a comparison of the four plants analyzed, in terms of average number of particles in 1 mm^2 area, under the same conditions (height/location, pollution exposition, weather). The results of the study highlight a different behavior among the plant species analyzed. This is mainly related to the plant species and structure (leaf shape, epidermis, roughness, etc.): the waxy leaves of *T. jasminoides* collect a higher number of particles, while the *P. fruticosa* hairy leaves collect the smallest amount. This experiment demonstrates that the effects of VGS on PM_x concentration depend on the specific characteristics of plants and on the level of air pollution in the area.

Particle size distribution in Fig. 4 shows that mainly bigger particles PM_{10} and smaller particles $PM_{2.5}$ adhere to the leaves' surface; it was also demonstrated that such small particles are not washed away by rain water (Ottelé et al., 2010; Perini et al., 2017). Also according to other authors

FIGURE 2 INPS Green Facade, Genoa (Italy). Photo by Anna Positano.

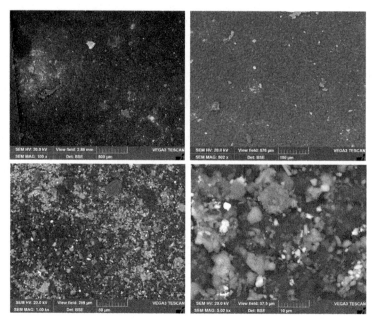

FIGURE 3 PM$_x$ specifically binds to leaf epidermis in ivy, *Hedera helix*. ESEM micrographs (100 ×, 500 ×, 1000 ×, and 5000 × from left to right, respectively).

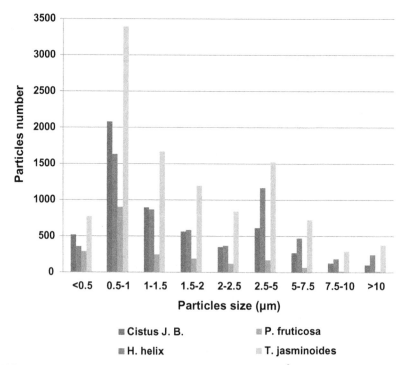

FIGURE 4 Average number and size (μm) of particles in 1 mm^2 for *H. helix*, *C.* "Jessamy Beauty," *P. fruticosa, T. jasminoides*, based on 100 ×, 250 ×, 500 × magnifications ($n = 96$), based on Perini et al. (2017).

(Terzaghi et al., 2013) particles smaller than 10 μm can be encapsulated into the leaf cuticle, and waxy leaves can increase this effect (Popek et al., 2013).

VERTICAL GREENING SYSTEMS AND GASEOUS POLLUTANTS

VGS can also mitigate the vertical dispersal of gaseous air pollutants in urban street canyons that may be directly affecting plants' health, as well as the quality of water and soil, and the ecosystem services that they support (European Commission, 2013). Ground-level O_3 and NO_x damage plants by reducing their growth rates. Also SO_2 contribute to the acidification of soil, lakes and rivers, causing biodiversity loss. In addition, NO_x emissions disrupt terrestrial and aquatic ecosystems by introducing excessive amounts of nutrient nitrogen (European Environment Agency, 2015). This leads to eutrophication, which is an oversupply of nutrients that can lead to changes in species diversity and to invasions of new species.

Several air pollutants are also climate forcers, which have a potential impact on climate and global warming in the short term (i.e., decades). Tropospheric O_3 is an example of an air pollutant that is a short-lived climate forcer that contributes directly to global warming. Plants can mitigate these effects both via absorption of gases and through their natural ability in decreasing building temperature with a beneficial cooling effect. Together with cutting emission, VGS can be used to reduce health, ecosystem impacts, and the extent of global climate warming.

Also the CO_2 greenhouse gas has an annual variation caused by anthropogenic emissions in urbanized sites that are about ten times that observed among vegetated ecosystems, as recently shown in dense urban and suburban areas (Ward et al., 2015). The choice of evergreen species for VGS, that are active all year long especially during winter when CO_2 and, in general, air pollutants' emission from road transport reach the maximum, must be taken into account when planning nature-based solutions like VGS (Ottelé et al., 2010; Pandey et al., 2016; Perini et al., 2017).

INDIRECT EFFECTS

Providing several ecosystem services, VGS allow reducing the negative effects of anthropic activities on human health with direct and indirect effects. Direct effects were discussed above. In addition VGS can, e.g., indirectly reduce the gas emissions connected to air conditioning, thanks to plants' cooling capacity (Coma et al., 2017); depending on climate, plants, substrate, etc. at district and city scale, (also UHI mitigation) can play an important role (Akbari, 2005; Rosenzweig et al., 2006).

High performances of VGS must take into account plant species resistance to pollutants, high biomass development, leaf morphology and distribution, but also avoiding deleterious effects that can occur when species can be highly allergenic or VOCs emitting. It should be considered that people living in urban areas suffer airborne pollen allergies 20% more frequently than people living countryside (D'Amato et al., 2005; Emberlin, 2008), and the wrong choice of plant species can decrease citizens' life quality instead of improving it.

CONCLUSION

VGS can be integrated in urban areas to improve air quality, acting on gaseous pollutants and PM_x concentration. VGS can reduce adverse health impacts from long term exposure to air pollutants at the same time improving the local urban ecology. Selecting the most suitable plant species is very important, since some species can act as a great sink of air pollutants. Attention will be paid to high performances, low allergenicity, low cost of maintenance, and climatic suitability. The selection of specific plants highly influences the performances of VGS in terms of air quality improvement, affecting the PM resuspension in the air.

REFERENCES

Akbari, H., 2005. Energy Saving Potentials and Air Quality Benefits of Urban Heat Island Mitigation.

Baik, J.-J., Kwak, K.-H., Park, S.-B., Ryu, Y.-H., 2012. Effects of building roof greening on air quality in street canyons. Atmos. Environ. 61, 48−55. Available from: https://doi.org/10.1016/j.atmosenv.2012.06.076.

Cohen, A.J., Ross Anderson, H., Ostro, B., Pandey, K.D., Krzyzanowski, M., Künzli, N., et al., 2005. The global burden of disease due to outdoor air pollution. J. Toxicol. Environ. Health A 68, 1301−1307. Available from: https://doi.org/10.1080/15287390590936166.

Coma, J., Pérez, G., de Gracia, A., Burés, S., Urrestarazu, M., Cabeza, L.F., 2017. Vertical greenery systems for energy savings in buildings: a comparative study between green walls and green facades. Build. Environ. 111, 228−237. Available from: https://doi.org/10.1016/j.buildenv.2016.11.014.

D'Amato, G., Liccardi, G., D'Amato, M., Holgate, S., 2005. Environmental risk factors and allergic bronchial asthma. Clin. Exp. Allergy 35, 1113−1124. Available from: https://doi.org/10.1111/j.1365-2222.2005.02328.x.

Emberlin, J., 2008. Grass, Tree, and Weed Pollen, in: Immunology, A.B.K.M., FRCP, DSc, FRSE, FMedSci Emeritusessor of Allergy and Clinical, MDessor, A.P.K., MD, J.B., Head, P.G.H.Ds., FRCPath, FAA (Eds.), Allergy and Allergic Diseases. Wiley-Blackwell, pp. 942−962.

Escobedo, F.J., Kroeger, T., Wagner, J.E., 2011. Urban forests and pollution mitigation: Analyzing ecosystem services and disservices. Environ. Pollut., Selected papers from the conference Urban Environmental Pollution: Overcoming Obstacles to Sustainability and Quality of Life (UEP2010), 20-23 June 2010, Boston, USA 159, 2078−2087. doi:10.1016/j.envpol.2011.01.010.

European Commission (Ed.), 2013. Building a green infrastructure for Europe. Publ. Office of the European Union, Luxembourg.

European Commission, 2016a. Urban Environment [WWW Document]. <http://ec.europa.eu/environment/urban/index_en.htm> (accessed 10.3.16.).

European Commission, 2016b. Nature-Based Solutions [WWW Document]. <https://ec.europa.eu/research/environment/index.cfm?pg = nbs> (accessed 10.3.16.).

European Environment Agency, 2015. Air quality inEurope - 2015 report.

Janhäll, S., 2015. Review on urban vegetation and particle air pollution − deposition and dispersion. Atmos. Environ. 105, 130−137. Available from: https://doi.org/10.1016/j.atmosenv.2015.01.052.

Lin, M.-Y., Hagler, G., Baldauf, R., Isakov, V., Lin, H.-Y., Khlystov, A., 2016. The effects of vegetation barriers on near-road ultrafine particle number and carbon monoxide concentrations. Sci. Total Environ. 553, 372−379. Available from: https://doi.org/10.1016/j.scitotenv.2016.02.035.

Manes, F., Marando, F., Capotorti, G., Blasi, C., Salvatori, E., Fusaro, L., et al., 2016. Regulating Ecosystem services of forests in ten italian metropolitan cities: air quality improvement by PM10 and O3 removal. Ecol. Indic. 67, 425−440. Available from: https://doi.org/10.1016/j.ecolind.2016.03.009.

Merbitz, H., Fritz, S., Schneider, C., 2012. Mobile measurements and regression modeling of the spatial particulate matter variability in an urban area. Sci. Total Environ. 438, 389−403. Available from: https://doi.org/10.1016/j.scitotenv.2012.08.049.

Nowak, D.J., Hirabayashi, S., Bodine, A., Greenfield, E., 2014. Tree and forest effects on air quality and human health in the United States. Environ. Pollut. 193, 119−129. Available from: https://doi.org/10.1016/j.envpol.2014.05.028.

Ottelé, M., van Bohemen, H.D., Fraaij, A.L.A., 2010. Quantifying the deposition of particulate matter on climber vegetation on living walls. Ecol. Eng. 36, 154−162. Available from: https://doi.org/10.1016/j.ecoleng.2009.02.007.

Pandey, A.K., Pandey, M., Tripathi, B.D., 2016. Assessment of Air Pollution Tolerance Index of some plants to develop vertical gardens near street canyons of a polluted tropical city. Ecotoxicol. Environ. Saf. 134, 358−364. Available from: https://doi.org/10.1016/j.ecoenv.2015.08.028.

Pérez, G., Coma, J., Martorell, I., Cabeza, L.F., 2014. Vertical Greenery Systems (VGS) for energy saving in buildings: a review. Renew. Sustain. Energy Rev. 39, 139−165. Available from: https://doi.org/10.1016/j.rser.2014.07.055.

Pérez-Urrestarazu, L., Fernández-Cañero, R., Franco-Salas, A., Egea, G., 2015. Vertical greening systems and sustainable cities. J. Urban Technol. 22, 65−85. Available from: https://doi.org/10.1080/10630732.2015.1073900.

Perini, K., Ottelé, M., Haas, E.M., Raiteri, R., 2011. Greening the building envelope, facade greening and living wall systems. Open J. Ecol. 1, 1−8. Available from: https://doi.org/10.4236/oje.2011.11001.

Perini, K., Ottelé, M., Giulini, S., Magliocco, A., Roccotiello, E., 2017. Quantification of fine dust deposition on different plant species in a vertical greening system. Ecol. Eng. 100, 268−276. Available from: https://doi.org/10.1016/j.ecoleng.2016.12.032.

Popek, R., Gawrońska, H., Wrochna, M., Gawroński, S.W., Sæbø, A., 2013. Particulate matter on foliage of 13 woody species: deposition on surfaces and phytostabilisation in waxes − a 3-year study. Int. J. Phytoremed. 15, 245−256. Available from: https://doi.org/10.1080/15226514.2012.694498.

Powe, N.A., Willis, K.G., 2004. Mortality and morbidity benefits of air pollution (SO2 and PM10) absorption attributable to woodland in Britain. J. Environ. Manage. 70, 119−128. Available from: https://doi.org/10.1016/j.jenvman.2003.11.003.

Pugh, T.A.M., MacKenzie, A.R., Whyatt, J.D., Hewitt, C.N., 2012. Effectiveness of green infrastructure for improvement of air quality in urban street canyons. Environ. Sci. Technol. 46, 7692−7699. Available from: https://doi.org/10.1021/es300826w.

Roccotiello, E., Perini, K., Cannatà, L., Mariotti, M., 2016. Air Pollution Mitigation via Urban Green Interactions with Particulate Matter. Presented at the III international Plant Science Conference, In press, Roma.

Rosenzweig, C., Solecki, W.D., Slosberg, R.B., 2006. Mitigating New York City's heat island with urban forestry, living roofs, and light surfaces. New York City regional heat island initiative.

Tallis, M., Taylor, G., Sinnett, D., Freer-Smith, P., 2011. Estimating the removal of atmospheric particulate pollution by the urban tree canopy of London, under current and future environments. Landsc. Urban Plan. 103, 129−138. Available from: https://doi.org/10.1016/j.landurbplan.2011.07.003.

Terzaghi, E., Wild, E., Zacchello, G., Cerabolini, B.E.L., Jones, K.C., Di Guardo, A., 2013. Forest filter effect: role of leaves in capturing/releasing air particulate matter and its associated PAHs. Atmos. Environ. 74, 378−384. Available from: https://doi.org/10.1016/j.atmosenv.2013.04.013.

Thönessen, M., Hellack, B., Bördgen, I., 2008. Staubfilterung durch Gehölzblätter und Gehölzbeständ.

von Döhren, P., Haase, D., 2015. Ecosystem disservices research: a review of the state of the art with a focus on cities. Ecol. Indic. 52, 490−497. Available from: https://doi.org/10.1016/j.ecolind.2014.12.027.

Vos, P.E.J., Maiheu, B., Vankerkom, J., Janssen, S., 2013. Improving local air quality in cities: to tree or not to tree? Environ. Pollut. 183, 113−122. Available from: https://doi.org/10.1016/j.envpol.2012.10.021.

Wania, A., Bruse, M., Blond, N., Weber, C., 2012. Analysing the influence of different street vegetation on traffic-induced particle dispersion using microscale simulations. J. Environ. Manage. 94, 91−101. Available from: https://doi.org/10.1016/j.jenvman.2011.06.036.

Ward, H.C., Kotthaus, S., Grimmond, C.S.B., Bjorkegren, A., Wilkinson, M., Morrison, W.T.J., et al., 2015. Effects of urban density on carbon dioxide exchanges: observations of dense urban, suburban and woodland areas of southern England. Environ. Pollut. 198, 186−200. Available from: https://doi.org/10.1016/j.envpol.2014.12.031.

WHO, 2005. Air Quality Guidelines Global Update 2005.

WHO, 2013. Health risks of air pollution in Europe − HRAPIE project.

Wolter, S., Diebel, J., Schroeder, F.-G., 2009. Development of hydroponic systems for urban façade greenery. In: Proceedings of international symposium on soilless culture and hydroponics. Acta Hortic. 393−402. doi:10.17660/ActaHortic.2009.843.53.

Chapter 3.5

Green Roofs for Pollutants' Reduction

Bradley Rowe

Chapter Outline

INTRODUCTION

Plants clean the air in several ways. They directly intercept particulate matter with their leaves, take up gases through their stomata, and break down organic compounds in their plant tissues and in the soil. They also mitigate air pollutants by lowering temperatures by providing shade through transpiration, thus decreasing energy consumption and pollutants such as ozone in the atmosphere. Other chapters in this book discuss how plants perform these duties in urban sites at ground level. Plants on green roofs behave much the same except that the growing conditions are more severe, which limits plant selection and their effectiveness compared to plants at ground level.

AIR POLLUTANTS AND PARTICULATE MATTER

Because of weight restrictions that limit growing substrate depth, many green roofs are planted with drought tolerant succulents such as *Sedum* sp. Plant species vary in their ability to mitigate pollutants. Clark et al. (2005) reported that tobacco (*Nicotiana tabacum*) had 30 times the uptake capacity for NO_2 than the sedum-like succulent (*Kalanchoe blossfeldiana*), which suggests that sedum may not be the optimal choice if the primary purpose of the roof is reducing air pollution. Likewise, Speak et al. (2012) found that grasses *Agrostis stolonifera* and *Festuca rubra* were more effective at capturing

Nature Based Strategies for Urban and Building Sustainability.
DOI: https://doi.org/10.1016/B978-0-12-812150-4.00013-6

particulate matter smaller than 10 μm (PM$_{10}$) than *Plantago lanceolata* or *Sedum album*. They attributed this result to differences in morphology of the above surface biomass. Even so, they calculated that if all the roofs within a 325 ha area in Manchester, United Kingdom, were covered with *Sedum* they would remove 2.3% of the annual PM$_{10}$ for this area (Figs. 1–2).

FIGURE 1 A shallow extensive green roof consisting of *Sedum* sp. on the Plant and Soil Sciences Building at MSU (Michigan State University). *Photo credit: Brad Rowe.*

FIGURE 2 Green roofs with deeper substrate depths like this one on the Molecular Plant Sciences Building at MSU (Michigan State University) allow for plants with greater biomass and will likely provide greater air pollution benefits than shallower roofs. *Photo credit: Brad Rowe.*

Several studies have shown that green roofs lower ambient CO_2 concentrations above and near the roof (Agra et al., 2017; Li et al., 2010; Moghbel and Erfanian Salim, 2017). Tan and Sia (2005) also found that sulfur dioxide and nitrous acid were reduced 37% and 21%, respectively, directly above a green roof in Singapore. Green roofs can also improve air quality at street level. Using a computational fluid dynamics model for a building with a street canyon ratio of one, (Baik et al., 2012) showed that pollutants were removed at street level due to the cooling of air above the building. The cooler air above a green roof flowed into the street canyon below, increasing street canyon air flow that dispersed pollutants near the road. Thus, air quality was improved for pedestrians. The magnitude of air quality improvement depended on the difference between ambient air temperature and the air temperatures above the roof. The greater the difference, the greater the air flow and reduction in air pollutants.

In addition, Currie and Bass (2008) and Yang et al. (2008) both studied the effects of green roofs on air pollution using the urban forest effects dry deposition model in Toronto and Chicago, respectively. The model quantified levels and hourly reduction rates of NO_2, SO_2, CO_2, $PM_{10,}$ and ozone. Results showed that in Chicago, pollutants were removed at a rate of 85 kg ha^{-1} year^{-1} with ozone accounting for 52% of the total followed by NO_2 (27%), PM_{10} (14%), and SO_2 (7%). In Detroit, Clark et al. (2005) estimated that if 20% of all industrial and commercial roofs were covered with *Sedum*, then over 800,000 kg year^{-1} of NO_2 would be removed.

Green roofs also indirectly reduce air pollution. Since they reduce heating and cooling requirements for individual buildings and help mitigate the urban heat island, they reduce energy consumption and emissions from power plants (Getter et al., 2011). In Los Angeles, emissions from coal burning power plants could be reduced by 350 tons of NO_x day^{-1} by reducing the need for air conditioning (Rosenfeld et al., 1998). This value equates to 10% reduction in the precursors to smog.

CARBON SEQUESTRATION AND STORAGE

CO_2 concentrations have increased more than 40% since the dawn of the industrial revolution from approximately 280 ppmv during the 18th century to over 400 ppmv in 2015 (USEPA, 2017b). Coinciding with the increase in CO_2, the average temperature of the Earth has risen by 0.83°C over the past century, and is projected to increase another 0.3−4.8°C within the next hundred years (USEPA, 2017a). Unless we drastically reduce the amount of carbon we put into the atmosphere, climate change will likely become even more pronounced in the future and further disrupt our environment and social order.

Green roofs can play a role in mitigating climate change in two ways. First, plants use atmospheric CO_2 during the process of photosynthesis.

This carbon is incorporated into carbohydrates used for energy and plant structures within the living plant and will eventually become part of the organic matter in the soil as they die. Second, the need to burn fossil fuels to produce energy is reduced because green roofs decrease energy requirements by insulating individual buildings and by mitigating the urban heat island (Li and Babcock, 2014; Rowe, 2011).

Terrestrial carbon sequestration occurs when CO_2 is removed from the atmosphere during photosynthesis and is stored as plant biomass. In addition, carbon is sequestered and stored in the soil as organic compounds as plants or plant parts die and become plant litter that will eventually decompose (Rowe, 2011, 2017). Although plants will continue to sequester carbon as long as they are alive, net carbon sequestration is what's important. When an ecosystem reaches the point when carbon sequestration equals decomposition then there is no further net sequestration. Up until this point, the system can be considered a carbon sink. How long it takes to reach this equilibrium stage is dependent on plant species' composition and diversity, ecosystem age, plant morphology, plant density, climate, and management practices. For example, woody plants such as trees will add greater biomass than an herbaceous perennial and will take a longer time to do so.

To date, most carbon sequestration research has focused on natural or agricultural ecosystems. Less is known about urban landscapes (Marble et al., 2011), although trees provide a significant contribution to the reduction of air pollutants such as CO_2 (Nowak, 2006). However, in many urban areas there is limited space to plant trees or other plant material due to the prevalence of roads, parking lots, and rooftops. For example, 94% of the mid-Manhattan west section of New York is covered with impervious surfaces (Rosenzweig et al., 2006) and 40%−50% of impervious surfaces in urban areas often consist of rooftops (Dunnett and Kingsbury, 2004). These typically wasted spaces provide an opportunity to sequester carbon, but how much can they really sequester?

Two studies to quantify carbon storage on green roofs have been conducted at Michigan State University (Getter et al., 2009; Whittinghill et al., 2014). The first study was conducted on one roof in 20 replicated plots of four different species of *Sedum* grown in monocultures along with a control plot of substrate only (Getter et al., 2009). Here the substrate depth equaled 6.0 cm across all plots and stored carbon was quantified for above- and below-ground biomass, as well as carbon present in the soil substrate. At the end of the second year, above-ground plant and root biomass stored an average of 168 g C m^{-2} and 107 g C m^{-2}, respectively, with differences among species. Carbon content in the substrate averaged 913 g C m^{-2}. In total, this entire green roof stored 1188 g C m^{-2}, however, after subtracting the 810 g C m^{-2} that existed in the original substrate, net carbon sequestration totaled 378 g C m^{-2}. Based on these numbers, if all the roofs in Detroit metropolitan area were covered with a similar green roof, these roofs would store 55,252 tons of carbon, equal to the CO_2 emissions from approximately 10,000 mid-sized sport utility vehicles or trucks (Getter et al., 2009).

It stands to reason that plants with greater biomass would sequester and store more carbon than the *Sedum* in the shallow roof studied above in the Getter et al. (2009) studies. To verify this assumption, Whittinghill et al. (2014) evaluated nine in-ground and four green roof landscape systems with increasing levels of complexity ranging from *Sedum* to woody shrubs. As expected, the three shrub landscape systems contained the greatest amount of carbon followed by the herbaceous perennial and grasses. *Sedum* sequestered the least amount of carbon. This makes sense as wood contains more carbon (4.7%−16.7% more) than other plant structures (Fang et al., 2007). In addition, in most cases the in-ground landscapes sequestered more carbon than the same plants in the green roof system. With a limited volume of soil substrate to exploit, the shallow depth on the green roofs restricted biomass.

IMPROVING CARBON SEQUESTRATION POTENTIAL

The amount of carbon stored can be improved immensely by altering plant selection, substrate depth, substrate composition, and management practices such as supplemental irrigation, fertilization, and the use of power equipment (Rowe, 2017). This holds true whether a landscape is on a roof or at ground level.

The influence of plant species, a function of plant biomass, can be seen when we compare the results of the Getter et al. (2009) and the Whittinghill et al. (2014) studies. The genus *Sedum* and other succulents are often used on green roofs due to their ability to withstand shallow substrate depths and superior drought tolerance (Cushman, 2001; Rowe et al., 2012). Many exhibit crassulacean acid metabolism (CAM), a form of plant metabolism that allows them to conserve water by opening their stomata during the night to take up CO_2 and closing them during the day to reduce transpiration. However, the same mechanisms that provide the drought tolerance can limit growth and reduce their likelihood of sequestering large amounts of carbon. When operating under CAM mode, rates for daily carbon assimilation are half to one third of non-CAM species (Hopkins and Hüner, 2004).

Tied to plant species is substrate depth and composition. Increasing substrate depth would not only provide a larger volume for carbon storage, it also enables a wider plant palette that could include larger perennials and even trees. All of the green roof landscape systems studied in the Whittinghill et al. (2014) study exhibited greater carbon sequestration than what was sequestered in the Getter et al. (2009) study due in part to the greater substrate depth (10.5 cm vs 6.0 cm). Regarding composition, different substrate components may alter plant growth which in turn would influence sequestration (Rowe et al., 2006). Also, using natural materials such as volcanic pumice or recycled materials such as crushed brick, crumb rubber, and construction waste that are locally available can reduce the carbon cost of roof construction (Eksi and Rowe, 2016; Matlock and Rowe, 2016).

Lastly, management practices play a large part in net carbon sequestration and the permanence of the carbon that is sequestered. Supplemental irrigation and fertilization influence plant growth as soil moisture and nutrients are often a limitation in many plant ecosystems. For example, plants in the Whittinghill et al. (2014) study discussed above were irrigated, whereas, those in the Getter et al. (2009) study were not. Different plant species vary in their water use efficiency, nutrient needs, growth and biomass allocation, and decomposition rates (Naeem et al., 1996).

CONCLUSION

Since plants naturally clean the air and sequester carbon, implementing green roofs on a wide scale provides opportunities to utilize these typically unused spaces to address environmental issues such as air pollution. Larger plant material with greater biomass is usually more effective, but limits on the structural weight capacity of many buildings often restricts the depth of the growing substrate and in turn the plant species that can be grown. Shallow green roofs can augment the urban forest, but cannot replace it.

REFERENCES

Agra, H., Klein, T., Vasl, A., Kadas, G., Blaustein, L., 2017. Measuring the effect of plant-community composition on carbon fixation on green roofs. Urban For. Urban Gree. 24, 1−4. Available from: https://doi.org/10.1016/j.ufug.2017.03.003.

Baik, J., Kwak, K., Park, S., Ryu, Y., 2012. Effects of building roof greening on air quality in street canyons. Atmos. Environ. 61, 48−55. Available from: https://doi.org/10.1016/j.atmosenv.2012.06.076.

Clark, C., Talbot, B., Bulkley, J., Adriaens, P., 2005. Optimization of green roofs for air pollution mitigation. Proc. of 3rd North American Green Roof Conference: Greening Rooftops for Sustainable Communities, Washington, DC. 4-6 May 2005.

Currie, B.A., Bass, B., 2008. Estimates of air pollution mitigation with green plants and green roofs using the UFORE model. Urban Ecosyst. 11, 409−422.

Cushman, J.C., 2001. Crassulacean acid metabolism: a plastic photosynthetic adaptation to arid environments. Plant Physiol. 127, 1439−1448.

Dunnett, N., Kingsbury, N., 2004. Planting Green Roofs and Living Walls. Timber Press, Inc., Portland, OR.

Eksi, M., Rowe, D.B., 2016. Green roof substrates: effect of recycled crushed porcelain and foamed glass on plant growth and water retention. Urban For. Urban Gree. 20, 81−88. Available from: https://doi.org/10.1016/j.ufug.2016.08.008.

Fang, S., Xue, J., Tang, L., 2007. Biomass production and carbon sequestration potential in poplar plantations with different management patterns. J. Environ. Manage. 85, 672−679.

Getter, K.L., Rowe, D.B., Robertson, G.P., Cregg, B.M., Andresen, J.A., 2009. Carbon sequestration potential of extensive green roofs. Environ. Sci. Technol. 43 (19), 7564−7570.

Getter, K.L., Rowe, D.B., Andresen, J.A., Wichman, I.S., 2011. Seasonal heat flux properties of an extensive green roof in a Midwestern U.S. climate. Energy Build. 43, 3548−3557. Available from: https://doi.org/10.1016/j.enbuild.2011.09.018.

Hopkins, W.G., Hüner, N.P.A., 2004. Introduction to plant physiology, third ed John Wiley & Sons, New York.

Li, J.F., Wai, O.W.H., Li, Y.S., Zhan, J.M., Ho, Y.A., Li, J., et al., 2010. Effect of green roof on ambient CO2 concentration. Build. Environ. 45 (12), 2644−2651. Available from: https://doi.org/10.1016/j.buildenv.2010.05.025.

Li, Y., Babcock Jr., R.W., 2014. Green roofs against pollution and climate change. A review. Agron. Sustain. Dev. 34, 695−705. Available from: https://doi.org/10.1007/s13593-014-0230-9.

Marble, S.C., Prior, S.A., Runion, G.B., Torbert, H.A., Gilliam, C.H., Fain, G.B., 2011. The importance of determining carbon sequestration and greenhouse gas mitigation potential in ornamental horticulture. HortScience 46 (2), 240−244.

Matlock, J.M., Rowe, D.B., 2016. The suitability of crushed porcelain and foamed glass as alternatives to heat-expanded shale in green roof substrates: an assessment of plant growth, substrate moisture, and thermal regulation. Ecol. Eng. 94, 244−254. Available from: https://doi.org/10.1016/j.ecoleng.2016.05.044.

Moghbel, M., Erfanian Salim, R., 2017. Environmental benefits of green roofs on microclimate of Tehran with specific focus on air temperature, humidity and CO2 content. Urban Clim. 20, 46−58. Available from: https://doi.org/10.1016/j.uclim.2017.02.012.

Naeem, S., Håkansson, K., Lawton, J.H., Crawley, M.J., Thompson, L.J., 1996. Biodiversity and plant productivity in a model assemblage of plant species. Oikos 76 (2), 259−264.

Nowak, D.J., 2006. Air pollution removal by urban trees and shrubs in the United States. Urban For. Urban Gree. 4, 115−123.

Rosenfeld, A.H., Akbari, H., Romm, J.J., Pomerantz, M., 1998. Cool communities: strategies for heat island mitigation and smog reduction. Energy Build. 28, 51−62.

Rosenzweig, C., Solecki, W., Parshall, L., Gaffin, S., Lynn, B., Goldberg, R., et al., 2006. Mitigating New York City's heat island with urban forestry, living roofs, and light surfaces. In: Proceedings of Sixth Symposium on the Urban Environment, Jan 30−Feb 2, Atlanta, GA. <http://amsconfex.com/ams/pdfpapers/103341.pdf>.

Rowe, D.B., Monterusso, M.A., Rugh, C.L., 2006. Assessment of heat-expanded slate and fertility requirements in green roof substrates. HortTechnology 16 (3), 471−477.

Rowe, D.B., 2011. Green roofs as a means of pollution abatement. Environ. Pollut. 159 (8-9), 2100−2110. Available from: https://doi.org/10.1016/j.envpol.2010.10.029.

Rowe, D.B., Getter, K.L., Durhman, A.K., 2012. Effect of green roof media depth on Crassulacean plant succession over seven years. Landscape Urban Plan. 104 (3-4), 310−319. Available from: https://doi.org/10.1016/j.landurbplan.2011.11.010.

Rowe, D.B., 2017. Carbon sequestration and storage. In: Charlesworth, Sue, Booth, Colin (Eds.), Sustainable Surface Water Management. John Wiley & Sons, Ltd., Hoboken, NJ, pp. 193−204.

Speak, F., Rothwell, J.J., Lindley, S.J., Smith, C.L., 2012. Urban particulate pollution reduction by four species of green roof vegetation in a UK city. Atmos. Environ. 61, 283−293. Available from: https://doi.org/10.1016/j.atmosenv.2012.07.043.

Tan, P., Sia, A., 2005. A pilot green roof research project in Singapore. Proc. of 3rd North American Green Roof Conference: Greening Rooftops for Sustainable Communities, Washington, DC. 4-6 May 2005.

USEPA, 2017a. Climate change: basic information. <http://www.epa.gov/climatechange/basics> (accessed 7.03.17.).

USEPA, 2017b. Climate change science: causes of climate change. <https://www.epa.gov/climate-change-science/causes-climate-change> (accessed 7.03.17.).

Whittinghill, L.J., Rowe, D.B., Cregg, B.M., Schutzki, R., 2014. Quantifying carbon sequestration of various green roof and ornamental landscape systems. Landscape Urban Plan. 123, 41–48. Available from: https://doi.org/10.1016/j.landurbplan.2013.11.015.

Yang, J., Yu, Q., Gong, P., 2008. Quantifying air pollution removal by green roofs in Chicago. Atmos. Environ. 42, 7266–7273. Available from: https://doi.org/10.1016/j.atmosenv.2008.07.003.

Chapter 3.6

Green Streets for Pollutants Reduction

Stefano Lazzari, Katia Perini and Enrica Roccotiello

Chapter Outline

INTRODUCTION

Air pollution in major cities is a serious problem that is faced everyday by both public administrations and citizens. Indeed, the presence in the atmosphere of a mix of contaminants (particles, liquids, and gases) can rise up to such a quality and duration that could be injurious not only to human welfare but even to health, as well as animal and plant life. Moreover, air pollution can also lead to environmental problems, such as global warming, acid rain, and deterioration of the ozone layer. Without any intention to be exhaustive, the following air contaminants can be listed: carbon monoxide (mainly due to incomplete combustion of carbon-based fuels by cars), carbon dioxide (mainly due to building heating systems and cars), chlorofluorocarbons (due to air-conditioning systems and refrigeration systems), nitrogen dioxide (from cars and power plants), sulfur oxides (mainly from thermal power plants), particulate matter (again, mainly from cars and power plants), lead (present in petrol, diesel, lead batteries, paints, hair dye products), and metals in general. Each contaminant is characterized by a different level of risk for human health. The European Directive 2008/50/EC and its transposition in Italian Law D.Lgs. 155/2010 dictate the air quality limits that are admitted. For instance, the amount of fuel that cars use on the road, and hence also the CO_2 emissions, in 2014 was around 40% higher than the official

Nature Based Strategies for Urban and Building Sustainability.
DOI: https://doi.org/10.1016/B978-0-12-812150-4.00014-8

measurements (EC, 2009). This counteracts the effect of more stringent regulations for car vehicle emission of 130 gCO_2 km^{-1} in 2015 (EEA, 2016).

Different approaches are adopted to reduce air pollution in an urban environment and several best practices can be suggested. It is worth noticing that focusing on greening street canyons instead of, e.g., urban parks can potentially be more effective, due to the difficulty in finding empty spaces for large green areas in dense cities.

This chapter provides an overview of the main parameters to consider, the most promising strategies, and the most suitable plant species. Computational fluid dynamics (CFD) numerical modeling is fundamental in order to identify effective strategies based on greening urban canyons. Indeed, the numerical modeling approach allows for investigating the effects of greening solutions by easily evaluating different configurations and by avoiding the complex environmental factors that affect field measurements. For this reason, the main issues that have to be taken into account in implementing CFD simulations are listed and commented. Some first results obtained by means of a simple CFD model representing an urban canyon with hedges and trees will be shown and commented.

EFFECTS OF GREENING SOLUTIONS AND MAIN INFLUENCING FACTORS

Several scientific papers deal with the evaluation of air pollutants' reduction that can be achieved by means of a proper disposal of urban greening solutions. An interesting review on this topic is presented in Janhäll (2015), in which it is remarked that design and choice of urban plant species is crucial for air quality improvement. For instance, leaves close to a pollution source can increase air quality, thanks to plant's porosity that allows pollutants finding enough deposition and/or penetration surfaces, and affects also pollutants absorption or dispersion. Moreover, it is shown that also choosing tall or short plants determines their effectiveness, which might also result in an increase of local air pollution levels.

In urban ecosystems, several parameters influence the performances of plant species. The dramatic complexity of a real urban canyon holds, not just from a geometrical standpoint but in the wide range of dynamic working conditions that can arise during the day and over the year (number and speed of vehicles, speed and direction of wind, local temperature changes, and their effect on pollutant transport by convection, behavior of greening solutions, etc.). This also results in difficulties arising when the effects of different plants and configuration have to be forecasted. For this reason, some researches face the problem from an experimental standpoint (Gromke, 2011; Tallis et al., 2011; Pugh et al., 2012) whereas some others adopt a numerical modeling approach (Vos et al., 2013; Li et al., 2006).

Especially in the last few years, there is an increasing interest in the study of air dynamics—typically with CFD simulations—within urban street canyons (Amorim et al., 2013). In Vos et al. (2013), numerical simulations by ENVI-met lead to the "green paradox," namely the opposite effect that has the positioning of trees relative to the pollutant source; for enhancing the local air quality they should be planted far away from the pollutant source, but when the city averaged air quality as the goal, it is recommended to plant trees as close as possible to the pollutant sources. In the picture outlined, authors often do not agree on the benefits that can be obtained through the use of greening solutions in terms of urban air quality improvement and, sometimes, even question that hedges and trees can reduce air pollution.

Another important point is represented by plants arrangement in the greening solution (trees, hedges, green roofs, green walls, vertical greening systems, etc.), plant species, leaf micromorphology, canopy density, leaf area index (LAI) and leaf area density (LAD) that can deeply vary the plant performance with respect to air pollutants' mitigation (Ali-Toudert and Mayer, 2007; Mayer et al., 2009; Perini et al., 2017). Some authors have paid attention also to the shape of the plants (Vos et al., 2013) and characteristics, such as LAI, porosity, etc., (Köhler, 1993) showing that both features play a key role and have to be taken into account in order to properly design the greening solutions.

In fact, plants species can strongly vary the ability to subtract an air pollutant from the atmosphere even if arranged in the same way (e.g., in the same nature-based solutions). The species phenology and physiology can change the plant species performances considering that air pollutants (e.g., CO_2) can cause the closing of stomata, the structure responsible for plant transpiration (Xu et al., 2016). In addition PMX can specifically bind to leaf cuticular waxes resulting in a "barrier effect" on leaf surfaces with a consequent decrease of photosynthetic efficiency (according to a wide research campaign that is currently in progress by Roccotiello). Several authors have considered some of these factors by introducing the air pollution tolerance index and the anticipated pollution index as main parameters to consider alteration of air pollutants with respect to leaf pH of the extract, relative water content, total chlorophyll content, stomatal conductance, and ascorbic acid level (Rai, 2016). On the other hand, further studies are necessary to clarify the plant response to air pollutants and to provide reliable and realistic input parameters for numerical simulations.

In addition, for future greening solutions several plant factors can be taken into account: plant photosynthetic performance and its adaptability and resilience to climate change (i.e., low water requirements), high biomass with a good LAI and LAD and adequate leaf micromorphology for air pollutants' capture (Perini et al., 2017), limited production of volatile organic compounds, low allergenicity, high resistance to plant pests, low cost of maintenance, and ability to act as a strong atmospheric CO_2 sink.

APPROACHING THE NUMERICAL MODELING OF GREENERY PERFORMANCE IN URBAN STREETS

The effects of green streets on air quality, i.e., the amount of NO_x, CO_x, PM_x, etc., i.e., present in the air, can be forecasted by means of models. Generally speaking, the simpler the model is kept, both in geometry description and in physical assumptions, the faster the computational time is to get some insights into the problem's phenomenology. The price to pay, most of the time, consists in a rough reliability of the results, which is likely to lead to further more accurate models. In this sense, the CFD approach to a problem, regardless of the specific subject, can be seen as going down a stairway: from the top, you have an overview of the problem, but cannot distinguish details; as you put an effort in taking further steps, you can see much more details, but at a certain point you are so close that you start losing the view of the full frame. Thus, the best CFD model is always the one that gives the best compromise between accuracy and computational effort.

Coming to CFD simulations of air quality in urban canyons, depending on the specific aim of the study modeler has to decide whether some physical effects are worthy to be considered or not (for instance, the variation of air density and viscosity with temperature and pressure), and whether the mass transport of contaminants by convection has a predominant role on mass transport by diffusion or not. Moreover, the modeler has to decide what geometrical elements really need to be modeled because they noteworthy affect the air flow, and what other can be conveniently neglected. For instance, the presence of aligned and/or continuous buildings, the main permanent obstructions, but also the overall size of the flow domain, the number and shape of air inlet and outlet sections, the shape and disposition of greening solutions, etc., have to be described in each specific case as a compromise between model accuracy and computation effort.

An important issue that has to be faced in building a CFD model for the description of air quality in urban streets is related to the proper description of air flow characteristics. Indeed, due to the expected values of wind velocities within streets, the flow is likely to be turbulent and, thus, requires an accurate choice of the turbulence model to be adopted (Li et al., 2006).

Another crucial decision for the modeler is how to describe the greening solutions. One option is to model plants (green facades, vertical greening systems, green wall, green roofs, hedges, trees, etc.) as porous media, i.e., as materials containing voids through which the air is forced to pass. Indeed, the ensemble of leaves in a plant represents an obstacle to the free flow of air, which is quite similar to the one represented by a porous medium, provided that its main describing parameters, i.e., the porosity and the permeability, are given suitable values according to the specific plant species.

Moreover, according to the specific plant species, and to whether the spatial distribution of leaves is fine or coarse, the produced absorption/blocking mechanism of air pollutants' changes. Thus, the modeler has to associate to each CFD volume that describes the plants a proper mathematical equation, which reliably accounts for pollutant's removal from the air flow by the considered plant species. In other words, a sort of "reduction reaction equation" has to be applied to these volumes and its mathematical expression is crucial for the model.

MODELING THE PERFORMANCES OF GREEN STREETS FOR AIR QUALITY IMPROVEMENT

In order to provide an example of some results that can be obtained by means of a numerical model, together with an insight on the most influencing factors, a simple numerical 3D model that has been adopted to study the effect of greening solutions on air pollution in urban canyons is presented. The model—implemented through the commercial CFD software COMSOL Multiphysics 5.2a (©COMSOL, Inc.)—simulates the effects of pollutant distribution (in particular, CO_2) in relation to evergreen plants arranged in hedges or in a row of trees, as shown in Fig. 1.

Case 1 represents the presence of a continuous hedge composed by dense shrubs, while case 2 represents the presence of trees with five plants uniformly distributed along y-direction. In order to evaluate the effect of the considered greening solution, a clear urban canyon, i.e., with the pollutant source but without any plant, is adopted as reference case (Case 3).

Air flow is assumed in the y-direction, coming in with an undisturbed velocity v_{in} that can vary in the range between 0.5 and 2 m s^{-1}. In the model, vehicles moving along the road are not considered except for their exhaust emissions, which are indeed the source of pollution in the domain. These emissions are located in the dotted volume $V_{exhaust}$ shown in Fig. 1.

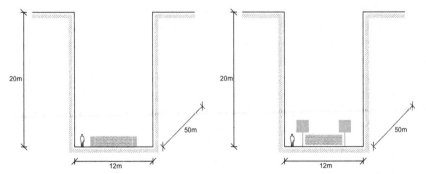

FIGURE 1 A qualitative sketch of the urban canyon cross-section: presence of the hedge on the left (Case 1); presence of the row of trees on the right (Case 2). The useful absorbing part of the greening solutions is painted in *green* (light gray in print versions), while the pollutant source, represented by vehicles exhaust, is in *red dots* (dark gray dots in print versions).

In order to best focus on the effect of greening solutions, it is possible to suppose that air entering the canyon is pollutant free. Moreover, pollutants (in this case CO_2) can be assumed as uniformly generated inside the dotted volume. Other parameters assumed for the simulation, which determine the pollutant emission rate, include the average car speed in the road, the average CO_2 car emission (g km^{-1}), the molar mass of CO_2, and the number of cars in the street.

Plants are modeled as porous materials; with this assumption, the porosity and the permeability are important quantities since they affect the greenery performances. Thus, their values have to be chosen accurately according to the specific plant species that is considered (in the following, the values 0.5 [−] and 1E-10 m^2 have been assumed, by making reference to evergreen shrubs arranged in hedge and composed by, e.g., × *Cupressocyparis leylandii* or *Prunus laurocerasus*). Further studies are required to select new plants species basing on their porosity, permeability, macro- and micromorphological characteristics, and physiological performances to obtain an efficient subtraction of CO_2 from the atmosphere, as in the case of tree carbon storage in US urban areas highlighted by Nowak et al. (2013).

Some results obtained by numerical simulations are reported in Table 1. The table shows that the hedge can have a positive effect in reducing the overall pollution within the urban canyon and in particular in the most critical zone, namely the volume where people walk. The effect is greater as the wind speed is slower, since diffusion becomes more important than convection. On the other hand, the presence of trees does not seem to have a positive effect in the investigated cases. This can be the result of some concurrent phenomena since, for instance, the canopy of the greening solutions is higher than the volume occupied by people, and the trunks enhance

TABLE 1 Results for Wind Speed $v_{in} = 0.5$, 2 m s^{-1}

Pollutant Concentration (mol m^{-3})	$v_{in} = 0.5$ m s^{-1}			$v_{in} = 2$ m s^{-1}		
	Case 3	Case 1	Case 2	Case 3	Case 1	Case 2
Surface at 0.3 m from the ground	2.29E-03	8.96E-04	2.10E-03	2.53E-04	8.43E-05	2.28E-04
Surface at 1.5 m from the ground	8.98E-05	2.02E-08	9.16E-05	8.99E-06	2.35E-10	9.11E-06
People volume up to 1.8 m	1.55E-03	5.04E-05	1.62E-03	1.81E-04	4.66E-06	1.79E-04

the mixing of polluted air. Other quantities play an important role that can lead to different results (width of the street, temperature gradients, etc.) and ask for an accurate tuning of the model through available experimental data.

CONCLUSION

The effect of greening solutions on enhancing the air quality in urban canyons is a very complex problem.

CFD modeling allows simulating the effects of plants on pollutants' concentrations under different conditions, if local investigations are the main purpose. Moreover, the results depend strongly on the specific geometry of the considered canyon and on the assumptions made in the numerical model. On the other hand, especially when the goal is to study the greening solutions effect on bigger scale, such as several streets or districts, the use of a specifically devoted code—as for instance ENVI-met or i-Tree Eco model (Nowak et al., 2013)—is advisable.

Plants may act as either an emission source or as a sink of CO_2 depending on the species and characteristics of the photosynthetic biomass and on the overall physiological performances of the plant species. Among the several aspects to be taken into account when planning greening solutions, some of them are fundamental, as for instance the choice of evergreen species that are active all year long and especially during winter, when CO_2 emission by road transport peaks. Despite no method currently existing that can directly evaluate the CO_2 uptake by greening solutions. Planning nature-based solutions in urban areas with the specific design and choice of the best plant species to be used must be sustainable and balanced to minimize ecosystem disservices and management costs. For this reason further studies are required.

REFERENCES

Ali-Toudert, F., Mayer, H., 2007. Effects of asymmetry, galleries, overhanging façades and vegetation on thermal comfort in urban street canyons. Solar Energy 81, 742–754.

Amorim, J.H., Rodrigues, V., Tavares, R., Valente, J., Borrego, C., 2013. CFD modelling of the aerodynamic effect of trees on urban air. Sci. Total Enviro. 461-462, 541–551.

European Environment Agency, 2016. Explaining Road Transport Emissions — A Non-Technical Guide. European Environmental Agency.

Gromke, C., 2011. A vegetation modeling concept for Building and Environmental Aerodynamics wind tunnel tests and its application in pollutant dispersion studies. Environ. Pollut. 159, 2094–2099.

Janhäll, S., 2015. Review on urban vegetation and particle air pollution – deposition and dispersion. Atmos. Environ. 105, 130–137.

Köhler, M., Fassaden- und Dachbergrunung, 1993. Ulmer Fachbuch Landschafts- und Grunplanung. Stuttgart.

Li, X.-X., Liu, C.-H., Leung, D.Y.C., Lam, K.M., 2006. Recent progress in CFD modelling of wind field and pollutant transport in street canyons. Atmos. Environ. 40, 5640–5658.

Mayer, H., Kuppe, S., Holst, J., 2009. Human thermal comfort below the canopy of street trees on a typical Central European summer day. Meteorology 211−219.

Nowak, D.J., Greenfield, E.J., Hoehn, R.E., Lapoint, E., 2013. Carbon storage and sequestration by trees in urban and community areas of the United States. Environ. Pollut. 178, 229−236.

Perini, K., Ottelé, M., Giulini, S., Magliocco, A., Roccotiello, E., 2017. Quantification of fine dust deposition on different plant species in a vertical greening system. Ecol. Eng. 100, 268−276.

Pugh, T.A.M., Mackenzie, A.R., Whyatt, J.D., Hewitt, C.N., 2012. Effectiveness of green infrastructure for improvement of air quality in urban street canyons. Environ. Sci. Technol. 46, 7692−7699.

Rai, P.K., 2016. Impacts of particulate matter pollution on plants: implications for environmental biomonitoring. Ecotoxicol. Environ. Safety 129, 120−136.

Regulation (EC) No 443/2009 of the European Parliament and of the Council of 23 April 2009 setting emission performance standards for new passenger cars as part of the Community's integrated approach to reduce CO2 emissions from light-duty vehicles (Text with EEA relevance).

Tallis, M., Taylor, G., Sinnett, D., Freer-Smith, P., 2011. Estimating the removal of atmospheric particulate pollution by the urban tree canopy of London, under current and future environments. Landscape Urban Plan. 103, 129−138.

Vos, P.E.J., Maiheu, B., Vankerkom, J., Janssen, S., 2013. Improving local air quality in cities: to tree or not to tree? Environ. Pollut. 183, 113−122.

Xu, Z., Jiang, Y., Jia, B., Zhou, G., 2016. Elevated-CO_2 response of stomata and its dependence on environmental factors. Front. Plant Sci. 7, 657.

FURTHER READING

Decreto Legislativo 13 agosto 2010, n.155 Attuazione della direttiva 2008/50/CE relativa alla qualità dell'aria ambiente e per un'aria più pulita in Europa, Gazzetta Ufficiale n. 216 del 15 settembre 2010 - Suppl. Ordinario n. 217.

Directive 2008/50/EC of the European Parliament and of the Council of 21 May 2008 on ambient air quality and cleaner air for Europe.

Chapter 3.7

Vertical Greening Systems for Acoustic Insulation and Noise Reduction

Gabriel Pérez, Julià Coma and Luisa F. Cabeza

Chapter Outline

INTRODUCTION

Urban green infrastructure involves, among others, all the different components that allow renature cities by means of adding vegetation to buildings (green roofs and all vertical greening systems (VGS)), making cities more attractive and enhancing human well-being (European Commission-Environment, 2012).

Among the different ecosystem services provided by urban green infrastructure to the urban ecosystem and to the citizens, such as support to biodiversity, rain water recovery, energy savings in buildings, etc., those concerning human health improvement stand out. Thus, the contribution of these nature-based solutions to the heat island effect reduction, to the pollution and particulate capture, and to the noise mitigation, are nowadays crucial issues for the built environment.

According to the World Health Organization (2009) findings, noise is the second largest environmental cause of health problems, just after the impact of air quality (particulate matter). A number of adverse health impacts, both direct and indirect, have been linked to exposure to persistent or high levels

Nature Based Strategies for Urban and Building Sustainability.
DOI: https://doi.org/10.1016/B978-0-12-812150-4.00015-X

157

of noise. Night-time effects can differ significantly from daytime impacts—the WHO reports an onset of adverse health effects in humans exposed to noise levels at night above 40 dB (WHO, 2009).

By 2020 it is estimated that approximately 80% of Europeans will be living in urban areas, with road transport being responsible for a significant fraction of environmental pollution, including noise (European Environment Agency, 2014a). Environmental impacts associated with road traffic are projected to affect larger areas and larger numbers of people, with the consequent need for such impacts to be managed in order to mitigate negative environmental impacts in Europe's urban areas (European Environment Agency, 2014b).

According to experts, the improvement actions should consider a twofold objective against noise, mainly acting on the source of noise, which is usually the most effective action, and secondly promoting protection actions from noise (William et al., 1998).

In this context, the possibility to use urban green infrastructure for noise reduction purposes arises, especially VGS for buildings because of their higher exposure to street canyons.

However, while the effect of belts of trees is clearly demonstrated for road traffic noise control, with reductions ranging from 5 to 10 dB (Van Renterghem et al., 2012) the use of vertical vegetation in buildings to reduce the noise at city scale, even at building scale, remains yet undeveloped. Despite this, nowadays with the recent development and consolidation of contemporary VGS, a new possibility to explore has appeared in the field of acoustic insulation and noise reduction for the built environment.

In this chapter, the current state of the art, the main influencing aspects for this purpose, that are the differences between the current construction systems, the influence of plant species, and the operating methods, will be analyzed. Moreover, a short review throughout the main research findings on the topic will be presented to finish with some suggestions for further research in the conclusions.

DIFFERENCES BETWEEN CONSTRUCTION SYSTEMS

VGS integrate a set of solutions that allow vegetation to grow on building's facades. The different followed design strategies have led in different commercial and research systems. Thus, these differences imply different contributions to the ecosystem services provided, consequently also for the acoustic performance.

The biggest difference takes place between the green facades and green walls or living walls, as the former tend to be made only by the layer of vegetation, often using climber plants. In green facades, creepers use traditionally the building facade material as support (direct systems), or sometimes through lightweight structures (indirect systems), in a more contemporary conception.

TABLE 1 VGS Layers and Composition

Vertical Greening Systems		Air Gap	Physical Support for the Plants	Plant Layer
Green facades	Direct	No	–	Climber plants
	Indirect	Yes (space for the support structure)	Light structures (steel wires, mesh, etc.)	Climber plants
Green walls	Plastic modules	Yes (space for the support structure)	Modules/panels (stuffed with substrate)	Climber and hanging shrubs
	Geotextile felts	No	Geotextile felts	Climber and hanging shrubs

Green walls are mainly composed of a set of layers that allow the survival of several sorts of plants, often herbaceous and shrubs. The most common layers in a green wall will be from the inside to the outside, an air gap layer, which often coincide with the space where the anchorage system to the building facade is located. Second, the physical support for plantscan be made by means of geotextile felts or by plastic modules often filled with special gardening substrate. Finally, the outermost layer is the plant layer (Table 1).

As it could be observed in Table 1, the presence of the air gap layer, the use of substrates, the differences between the plants species used, can strongly influence the final acoustic performance of these systems.

INFLUENCE OF PLANT SPECIES

Since there are differences between the typologies of plants used in the different VGS, basically climbing plants for green facades and shrubs for green walls, the acoustic performance could be influenced during the plant growing phases as well as the yearly seasons. Thus, the possibility to use deciduous or perennial plants will influence the contribution for acoustic insulation or noise reduction purposes.

In any case, a good development of the foliage layer will be a desired objective for sound control because the mass of vegetation is a key point for this purpose. Therefore, the selection of species with a good biomass development will be a good option. Usually the most interesting species for this regard are the autochthonous ones, well-adapted to the local climate conditions. Although in some previous studies relating to VGS autochthonous were used, no research was found specifically addressing the acoustic improvement of the VGS performance.

The most commonly used species for green facades research purposes have been Ivy (*Hereda* sp.) as perennial type, and Boston Ivy (*Parthenocissus tricuspidata*) and Wisteria (*Wisteria sinensis*) as deciduous ones. For green walls the number of species is larger than for green facades, and usually are perennial herbaceous and shrubs (Pérez et al., 2014).

Generally, the possibility to create plant catalogs by climate in order to facilitate the decision making during the design phase will be an issue to address for the future.

OPERATING METHODS

From previous studies relating to the use of vegetation, green belts and trees near roads, to reduce sound levels it can be deduced that there are basically three main effects (Van Renterghem et al., 2012). The first effect is diffraction provided by plants, i.e., the sound can be reflected and scattered (diffracted) by plant elements, such as trunks, branches, twigs, and leaves. The second effect is absorption of sound by plants. The mechanical vibrations of plant elements due to the sound waves imply dissipation by converting sound energy to heat. In addition, plants contribute to sound attenuation by thermoviscous boundary layer effects at vegetation surfaces. The third mechanism relates to the destructive interference of sound waves. In this regard, the presence of soil can lead to destructive interference between the direct contribution from the source to the receiver and a ground-reflected contribution, resulting in sound levels reduction. The presence of vegetation leads to an acoustically very porous soil, mainly due to the presence of a litter layer and plant rooting layer. This results in a more pronounced ground effect and produces a shift towards lower frequencies compared to sound propagation over grassland. This acoustical ground effect is more efficient in limiting the typical engine noise frequencies (approximately 100 Hz) of road traffic. In these previous studies it was concluded that a 2 m-high shrub zone with a length of 15 m, implies an average road traffic noise insertion loss of 4.7 dBA for a light vehicle at 70 km h^{-1} at typical ear heights in reference to sound propagation over grassland.

To compare with other strategies, effective artificial noise barriers can reduce noise levels by 5−10 dB, cutting the loudness of traffic noise by as much as one half. For example, a barrier which achieves a 10 dB reduction can reduce the sound level of a typical tractor trailer pass-by to that of an automobile (U.S. Department of Transportation, 2011). According to the current regulations "high noise levels" are defined as noise levels above 55 dB L_{den} (day-evening-night level) and 50 dB L_{night}. Road traffic is the most dominant source of environmental noise with an estimated 125 million people affected by noise levels greater than 55 dB L_{den}, including more than 37 million exposed to noise levels above 65 dB L_{den} (European Environment Agency, 2014b).

55 dB Lden is the EU threshold for excess exposure, indicating a weighted average during the day, evening, and night.

On the other hand, according to Horoshenkov et al. (2013), the acoustic absorption coefficient of plants is controlled predominantly by the leaf area density and the angle leaf orientation. In addition, light-density soils exhibit very high values of acoustic absorption whereas the absorption coefficient of high-density clay base soil is low.

Considering these effects and the measured contributions, it highlights that the most influential factors are the sort of plant species (plant morphology), the green screen shape and dimensions as well as the location in reference to the sound source. Moreover, the presence of soils, substrates, and other materials can have consequences on the acoustic performance. Thus considering the contribution of VGS for acoustic insulation and noise reduction, all these factors must be taken into account.

MAIN RESEARCH FINDINGS

In spite of the fact that the potential contribution of urban green infrastructure for acoustic insulation and noise reduction is generally assumed in nonscientific literature, only few research studies have been previously conducted for this aim. In addition, most of them are laboratory studies that use small samples or even simulations, and only two are in situ experiments.

Wong et al. (2010) evaluated the noise reduction potential of different VGS by means of the in situ measurement. From the results it stands out as a bigger contribution from those systems with a substrate layer due to the absorption effect, especially in the middle frequencies (reductions around 5−10 dB). In addition a contribution was observed, though smaller, at a high frequency spectrum due to the scattering effect of greenery (reductions from 2 to 3.9 dB).

Van Renterghem et al. (2013) carried out a numerical study on the potential of building a greening envelope to achieve quietness in the urban environment. Conclusions highlight the complexity of the substrates used in green walls from the point of view of the acoustic performance due to their high porosity and low density. In addition, the presence of water inside the substrate leads to similar effects of a rigid material, and consequently the absorption properties could become strongly influenced.

According to Horoshenkov et al. (2013), from studies at lab scale, the absorption coefficient of plants is dependent of the foliage density and the angle leaf orientation, whereas light-density soils exhibit very high values of acoustic absorption in reference to high-density clay base soils.

Recently, Azkorra et al. (2015) conducted a laboratory evaluation of a green wall system. The main results were a weighted sound reduction index of 15 dB and a weighted sound absorption coefficient of 0.40. The green wall showed similar or better acoustic absorption coefficient than other common building materials, and its effects on low frequencies were of particular interest because its observed properties were better than those of some current sound absorbent materials at low frequencies. Taking into consideration that the voice frequency was around 60 dB, this corresponds to the frequency at which this modular green facade is more efficient at absorbing sound, so it could be used very effectively in public places for instance restaurants, hotels, and halfway up the street to the passage of people.

Subsequently, Perez et al. (2016) in a complete in situ experiment acoustically assessed two different VGS, a green facade and a green wall. From the results, it can be observed that a thin layer of vegetation (20−30 cm) was able to provide an increase in the sound insulation of 1 dB for traffic noise (in both cases, green wall and green facade), and an insulation increase between 2 dB (green wall) and 3 dB (green facade) for a pink noise. In addition, the acoustic insulation contribution from vegetation (scattering effect) for both greenery systems in high frequencies, as well as from substrate (absorption effect) in the middle frequencies in green walls, were verified (Fig. 1). In this study, the necessity to homogenize the way of studying the acoustic behavior of VGS by following the UNE-EN ISO 140-5 standard, which regulates the acoustic insulation capacity measurement for building and construction elements, was highlighted.

Finally, in the case of the studied green wall, the differences between the good results obtained in previous laboratory studies (Azkorra et al., 2015) and the obtained in situ measurements (Perez et al., 2016), suggest that it is necessary to consider other factors in addition to the vegetation layer to improve the acoustic insulation capacity of VGS. To achieve these improvements, issues such as the mass (thickness and composition of the substrate and vegetation layers), the impenetrability (sealing joints between modules), and the structural insulation (support structure) must be addressed.

Regarding the mass, this measure can be achieved in the case of green walls by improving the composition of substrates used for this purpose. Usually the substrate composition in green walls responds to plant survival necessities (i.e., the provision of water, nutrients, and physical support) as well as weight constraints, but not to supply other ecosystem services such as thermal or acoustic insulation. Another aspect to consider is the possibility of gaining mass in the vegetation layer, either by increasing the thickness or by using plant species with higher foliage density.

In the case of impenetrability, it is known that small fissures can cause big effects on global acoustic insulation. This issue can unlikely be improved in a double-skin facade system which is fully permeable and in where the whole function of acoustic insulation is provided by the vegetation layer. On the contrary, in the case of the green wall, the complete sealing of the joints between modules and in the facade edges would lead to an improvement on sound insulation in terms of impenetrability.

Green wall

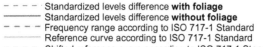

- – – – – Standardized levels difference **with foliage**
- ————— Standardized levels difference **without foliage**
- – – – – Frequency range according to ISO 717-1 Standard
- ————— Reference curve according to ISO 717-1 Standard
- – – – – Shifted reference curve according to ISO 717-1 Standard with foliage
- – – – – Shifted reference curve according to ISO 717-1 Standard without foliage

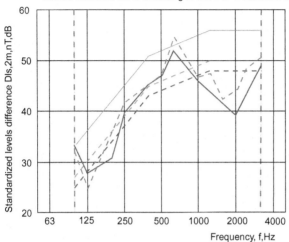

Green facade

- – – – – Standardized levels difference **with foliage**
- ————— Standardized levels difference **without foliage**
- – – – – Frequency range according to ISO 717-1 Standard
- ————— Reference curve according to ISO 717-1 Standard
- – – – – Shifted reference curve according to ISO 717-1 Standard with foliage
- – – – – Shifted reference curve according to ISO 717-1 Standard without foliage

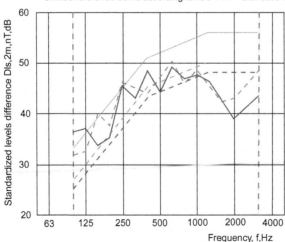

FIGURE 1 Standardized difference of levels D2m, nT. Green wall versus green facade (Perez et al., 2016).

Finally, regarding the structural insulation, it is necessary to consider that a certain physical separation between building elements must be guaranteed in order to prevent the sound transmission. This can be the main aspect to improve the two analyzed VGS because in both cases, green wall and double-skin green facade, lightweight structures anchored directly to the building facade wall were used resulting probably in the existence of acoustic bridges.

CONCLUSION

Among other ecosystem services, VGS can contribute to the acoustic insulation, a noise sound reduction on the built environment. However, the current systems are not designed for this purpose and therefore the real contribution is limited. Despite this, the research conducted until now has revealed the main operational methods when VGS is used for this purpose. Thereby, the acoustic insulation contribution from vegetation (scattering) in high frequencies, both in green facades and in green walls, as well as from substrate (absorption) in the middle frequencies in green walls, were verified. From the results, it can be observed that a thin layer of vegetation (20−30 cm) was able to provide an increase in the sound insulation of 1 dB for traffic noise (in both cases, green wall and green facade), and an insulation increase between 2 dB (green wall) to 3 dB (green facade) for a pink noise.

Future research must deal with studies regarding to the types of plants, the thickness of the vegetation layer, the thickness and composition of the substrate layer, the type of support structure and materials to be used, as well as to take measures to prevent the transmission of sound (structural impenetrability and insulation). In addition, future experiments should be made following international standards of measurement in order to compare experiments and results relating to the different VGS.

ACKNOWLEDGMENTS

The work partially funded by the Spanish Government (ENE2015-64117-C5-1-R (MINECO/FEDER) and ULLE10-4E-1305). The authors would like to thank the Catalan Government for the quality accreditation given to their research group (2014 SGR 123) and to the city hall of Puigverd de Lleida. This project has received funding from the European Commission Seventh Framework Program (FP/2007-2013) under grant agreement Nọ PIRSES-GA-2013-610692 (INNOSTORAGE) and from European Union's Horizon 2020 research and innovation program under grant agreement Nọ 657466 (INPATH-TES). Julià Coma would like to thank the Departament d'Universitats, Recerca i Societat de la Informació de la Generalitat de Catalunya for his research fellowship.

REFERENCES

Azkorra, Z., et al., 2015. Evaluation of green walls as passive acoustic insulation system for buildings. Appl. Acoust. 89, 46−56.

European Commission-Environment, 2012. The Multifunctionality of Green Infrastructure. Science for Environment Policy. In-depth Reports.

European Environment Agency, 2014a. Analysing and managing urban growth. <http://www.eea.europa.eu/articles/analysing-and-managing-urban-growth>.

European Environment Agency, 2014b. Noise in Europe 2014. ISBN 978-92-9213-505-8.

Horoshenkov, K.V., et al., 2013. Acoustic properties of low growing plants. J. Acoust. Soc. Am. 133 (5), 2554−2565.

Pérez, G., et al., 2014. Vertical Greenery Systems (VGS) for energy saving in buildings: a review. Renew. Sustain. Energy Rev. 39, 139−165.

Pérez, G., et al., 2016. Acoustic insulation capacity of Vertical Greenery Systems for buildings. Appl. Acoust. 110, 218−226.

U.S. Departament of Transportation. Federal Highway Administration, 2011. <https://www.fhwa.dot.gov/environment/noise/noise_barriers/design_construction/keepdown.cfm>.

Van Renterghem, T., et al., 2012. Road traffic noise shielding by vegetation belts of limited depth. J. Sound Vibrat. 331, 2404−2425.

Van Renterghem, T., et al., 2013. The potential of building envelope greening to achieve quietness. Build. Environ. 61, 34−44.

William, J., et al., 1998. Architectural Acoustics. Principles and Practice. John Wiley & Sons, Inc, Hoboken NJ, ISBN 0-471-30682-7.

Wong, N.H., et al., 2010. Acoustics evaluation of vertical greenery Systems for building walls. Build. Environ. 45, 411−420.

World Health Organisation, 2009. Europe Night noise guidelines for Europe.

Chapter 3.8

Green Roofs for Acoustic Insulation and Noise Reduction

Timothy Van Renterghem

Chapter Outline

INTRODUCTION

Although city densification is often considered a need in the viewpoint of the growing world population and ecological challenges (European Environment Agency, 2010), it puts at the same time strong pressure on the urban environmental quality. The noise issue is one of the major challenges to be tackled in the urban environment, and is unlikely to become less prominent in the years to come. In a city, there is a combination of a large number of sound sources and a large number of potentially affected persons, either indoors or in the public space. This negatively affects human health and quality of life following the World Health Organization (Fritschi et al., 2011).

In the built environment, building skins are typically acoustically rigid or close to rigid (like glass, concrete, tiles, bricks, and pavements) (Cox and D'Antonio, 2004). As a consequence, there is often a strong increase in the sound pressure level due to the multiple reflections in between opposing building facades and on the street surface. This street amplification not only affects the most-exposed facade (the building side facing the source), but also increases sound pressure levels at shielded facades directly

Nature Based Strategies for Urban and Building Sustainability.
DOI: https://doi.org/10.1016/B978-0-12-812150-4.00016-1

connected to the source canyon. In addition, (rigid) roof reflections might lead to pressure doubling ($+6$ dB) during diffraction. A closed row of houses could be a rather efficient noise barrier. However, the combination of a highly trafficked urban street (canyon), street reverberation, and roofs acting as perfect acoustic mirrors, exposure levels are often too high to fully benefit from the quiet side effect (Öhrström et al., 2006).

For dwellings along major flight paths in the vicinity of airports, sound energy entering through the roof system might be high. Although relevant contributions during a flight-over might come from other sound paths than those strictly interacting with the roof, especially in dense cities there can be a substantial benefit from further increasing acoustical roof insulation. Other common noise sources that might benefit from transmission reduction through the roof are air-conditioning units and rain impact sounds (that can be prominent on, e.g., metal roofs (Dubout, 1969)).

In this chapter, it will be shown that green roofs have good sound absorption. In contrast to common (porous) absorbing materials, green roof materials are specifically designed to be placed outdoors during a long period. At the same time, they are able to reduce sound transmission through the roof system on top of the acoustic insulation provided by the basic roof.

GREEN ROOF ABSORPTION CHARACTERISTICS

The green roof's constituting layers influencing the sound absorption characteristics are the porous growing substrate, air voids, water storage/retention fabrics, and the plant layer. Together, they form a multilayered system with a complex acoustic behavior. Furthermore, their properties significantly change over time due to water dynamics, plant community development, and compaction.

The growing substrates in extensive green roofs are mostly granular materials with open pores, the latter being essential for sound penetration and interaction with the particles, leading to sound absorption. Connelly and Hodgson (2015) measured separately the absorption characteristics of the constituting parts of growing substrates. Sand was shown to have the lowest absorption coefficient in such mixtures, while pumice and especially organic matter make substrates reasonably good sound absorbers. Increasing the organic fraction leads to a rise in the overall sound absorption (Connelly and Hodgson, 2015).

In Fig. 1, an assembly of measured absorption characteristics (at normal incidence) of green roof systems is presented. As with common porous materials, sound absorption generally increases with frequency. The wide variety in behavior suggests that substrates can be engineered to optimize sound absorption.

The influence of the plant layer on absorption is less obvious. Although it was recently shown that specific plants are able to absorb sound reasonably

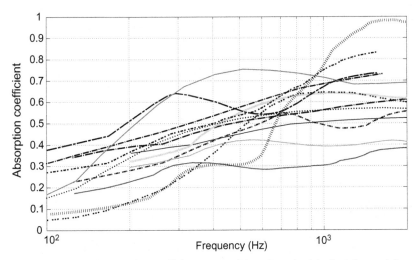

FIGURE 1 Measured absorption coefficient at normal impedance (mainly dry substrates) for a wide variety of green roofs as reported in literature (Connelly and Hodgson, 2015; Liu and Hornikx, 2015; Nilsson et al., 2015; Van Renterghem et al., 2013; Yang et al., 2012).

well (Horoshenkov et al., 2013), the acoustic system formed by plants on top of a porous material is more complex. Measurements (Connelly and Hodgson, 2015; Ding et al., 2013; Nilsson et al., 2015) typically show that in the low-frequency range, there is a slight enhancement of the absorption of the plant-substrate system, while at higher frequencies plants lead to a decreased absorption relative to uncovered substrates. A large pack of leaves on a porous substrate, however, was measured to increase the absorption coefficient at all sound frequencies (Attal et al., 2016).

Adding water to any porous material deteriorates its absorbing properties. Various complex effects might appear like a reduction in the effective layer depth, substrate particles swelling, clogging of pores and exchange of water in between the different layers constituting the green roof system. When fully saturated, the substrate surface approaches a perfectly reflecting plane. Impedance tube measurements at Ghent University during (unforced) evaporation (indoors) of an initially fully saturated green roof substrate clearly show (see Fig. 2) the increase in absorption coefficient with decreasing soil moisture content. The green roof system considered consisted of a thin root barrier membrane, a 25-mm thick polypropylene water evacuation fabric (however, no water could be evacuated due to its positioning in the fully enclosed impedance tube), a thin filter membrane, a 25-mm thick water retention mat, and finally a 60-mm substrate. Especially in the higher frequency range, a large variation, covering the full range between no absorption at all and full absorption, is observed.

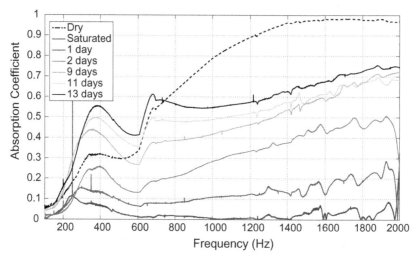

FIGURE 2 Absorption characteristic of a green roof system during unforced evaporation (impedance tube measurement, normal incidence, in lab).

SOUND DIFFRACTION OVER GREEN ROOF

Building Geometries of Interest

In order to benefit from green roof absorption, there has to be a dominant sound path interacting with the roof. The two building typologies shown in Fig. 3 are examples where a green roof could potentially reduce the acoustical facade load, either by a single diffraction or by two subsequent diffractions at the (horizontal) roof edges (double diffraction case). In isolated building setups (see Fig. 4), however, diffractions around the vertical building edges could become the dominant sound contributions toward the shielded side. Clearly, a green roof will hardly help to reduce sound pressure levels there. Another application, studied by Yang et al. (2012), deals with green roof systems positioned on low-profiled structures in urban streets, e.g., on top of underground parking lots.

Diffraction Over Absorbing Surfaces

Diffraction theory (Hadden and Pierce, 1981) stipulates that the absorption characteristics of the faces constituting the diffracting object have an important effect on the sound pressure levels in the shielded zone. The largest effects are expected in a double diffraction case, in which sound waves propagate at grazing angle over the green roof. Following simplified ray-theory, this leads to canceling of the waves shearing over the roof and those reflected from the soft material (Embleton, 1996). In practice, this canceling

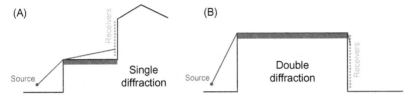

FIGURE 3 The single (A) and double (B) diffraction geometries in cross-section.

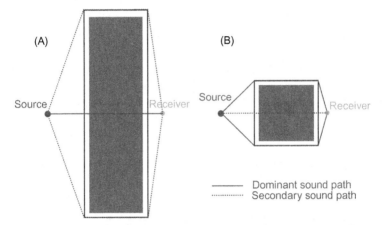

FIGURE 4 Schematic of the dominant and secondary sound paths near an isolated building in top view. In (B), no significant sound reduction is expected by the green roof.

will be incomplete as the green roof's material is not perfectly absorbing, since sound waves arrive at the roof's edge after diffraction, and due to the presence of ground waves that are prominent mainly in the lower frequency range (Embleton, 1996).

Numerical Analysis and Parameter Study

Full-wave numerical techniques, capturing all those wave phenomena, are useful to study the basic parameters with relation to the potential noise reduction obtained by a green roof in an idealized urban geometry. A comparison with common rigid roofs can easily be made.

The diffraction over a building, the green roof absorption characteristics, a road traffic noise spectrum, and also the A-weighting (used to account for the sensitivity of the human ear in environmental noise studies) are all strongly frequency-dependent. The overall noise reduction in dBA provided by a green roof in, e.g., a road traffic noise case, is therefore not easily estimated and such specific studies are therefore considered here.

Double Diffraction Cases

Effect of Sound Frequency

Simulations with the finite-difference time-domain method (Van Renterghem and Botteldooren, 2008, 2009), including a detailed and validated modeling of the growing substrate (Van Renterghem et al., 2013), showed positive effects mainly at the octave bands with center frequencies above 500 Hz. At low sound frequencies, the noise reducing performance of a green roof is limited since the porous substrate (just like any porous absorber) still has a large impedance; as a result, the rigid roof response is obtained.

Effect of Roof Cover Fraction

Van Renterghem and Botteldooren (2008) predicted a linear relationship between the fraction of the roof covered by the green roof substrate, and the decrease in sound pressure level at an indirectly exposed receiver canyon. With increasing sound frequency, this slope increases; for a given increase in green roof cover, the sound reduction at high frequencies will be stronger.

Effect of Substrate Layer Thickness

Detailed numerical analysis of the sound field on top of the roof (Van Renterghem and Botteldooren, 2008) showed a distinct effect of substrate layer thickness in extensive green roofs. The sound waves diffracting over the green roof, and those reflecting from within the substrate or from the roof membrane below it, interfere. When a sufficiently pronounced destructive interference appears just above the green roof, a maximum sound level reduction at the shielded building facade is found as the sound fields at these places are strongly linked.

With increasing sound frequency, the optimal layer thickness shifts towards lower values, while this destructive interference becomes more pronounced (Van Renterghem and Botteldooren, 2008). Numerical simulations show that a green roof modeled by a porous rigid-frame model (with a flow resistivity of $10 \, kPa \, sm^{-2}$ and a porosity of 0.40) gives an optimum noise reduction for the 1 kHz octave band for a layer thickness of 10 cm ($-10 \, dB$ relative to a rigid roof). For the 500 Hz octave band, the optimum layer depth shifts to 20 cm but becomes somewhat less pronounced ($-6 \, dB$ relative to a rigid roof). At lower frequencies, this minimum lies outside the range of commonly used substrate depths. In the case of denser substrates, such minima were not found in the simulations reported; sound waves are unable to sufficiently penetrate the substrate to see similar layer depth effects as discussed before (Van Renterghem and Botteldooren, 2008).

Road Traffic Noise Reduction of Flat Green Roofs

Although high frequencies can be strongly reduced by a green roof, the overall road traffic noise reduction, keeping in mind the low-pass filtering due to double-edge diffraction, will be smaller. Predicted flat green roof reductions, relative to rigid roofs and averaged over all facades forming a closed courtyard (Van Renterghem et al., 2013), are typically near 3 dBA (averaged over vehicle speeds between 30 and 70 km h^{-1} and assuming a fleet composition consisting of 95% light and 5% heavy vehicles). The green roof insertion loss is rather uniform over the shielded facades.

Vehicle speed has a strong impact on the relative importance of the rolling and engine noise contribution to the overall noise produced by road vehicles. Given the (relatively) more intense low frequencies at limited vehicle speeds, a somewhat lower green roof noise reduction is obtained than at higher vehicle speeds (Van Renterghem and Botteldooren, 2009). Note, however, that the absolute levels are lower at limited vehicle speeds. The green roof noise reduction for heavy vehicles is also somewhat smaller than for light vehicles, given the relative stronger low-frequency content.

Effect of Roof Shape

Road traffic noise shielding in an urban setting is largely influenced by the specific roof shape, even under the assumption of a similar building volume (Van Renterghem and Botteldooren, 2010). Simulations (Van Renterghem and Botteldooren, 2009; Van Renterghem et al., 2013) showed that the positive effect of placing a green roof on a nonflat roof is stronger than on a flat roof; a flat roof is already close to the optimum shielding. On a 10 m wide symmetric ridge roof, the green roof insertion loss is predicted to exceed 7 dBA (Van Renterghem et al., 2013), averaged over the shielded facade for urban road traffic. This can be explained by the additional interaction length between sound waves and the green roof. A ridge roof equipped with a green roof was predicted to outperform the shielding of a flat rigid roof (Van Renterghem and Botteldooren, 2009; Van Renterghem et al., 2013).

Single Diffraction Case

Similar to double diffraction cases, the sound reduction of a green roof will increase with sound frequency and interaction length (Van Renterghem and Botteldooren, 2009). For heavy vehicles and lower vehicle speeds, the insertion of a green roof (relative to a rigid roof) will be smaller too (Van Renterghem and Botteldooren, 2009). The influence of substrate depth, however, might be different.

In case of a building extension with a facade directly overlooking the green roof, noise reduction will depend on the position of the source, the building extension height, and the position on the facade. At directly exposed parts of the facade (in a high-frequency approach, a straight line can then be drawn between the source and the facade), no reduction relative to an identical rigid building extension is expected. At lower and thus shielded parts, significant road traffic noise reduction is possible by the green roof (exceeding 5 dBA at 70 km h^{-1}, see Van Renterghem and Botteldooren, 2009). Given a single edge diffraction, higher frequencies are relatively more important, and these are reduced to a larger extent by the green roof. This explains the typically larger road traffic noise reduction than for the double diffraction case. The overall exposure along a facade for a single diffraction case will be the energetically-weighted average over a relevant part of it, e.g., near a window.

Case Studies

Case Studies: Dry Substrates

A set of in-situ measurements of sound propagating over flat, extensive green roofs is described in detail in Van Renterghem and Botteldooren (2011) and summarized in Table 1. Measurements in front of the building's facade were each time performed just before and just after the placement of the green roof (mainly dry state) with an identical source-receiver configuration, allowing a direct estimate of its acoustical effect. There was a wide variety of layer built-up, substrate thickness, propagation path length interacting with the green roof, vegetation type and cover fraction. The measurement results, summarized to a single-number green roof road traffic noise insertion loss in Table 1, show consistency with the findings from the numerical parameter studies described before. For less shielded receivers, a change in interference pattern is often observed, leading to positive or negative effects in specific frequency ranges, relative to a nonvegetated (rigid) roof top. For the double diffraction cases the green roof improvement showed to be less frequency-dependent. These measurements show the range of noise reductions that can be expected for the current practice of green roofs, not specifically designed to tackle environmental noise.

Another flat green roof in situ measurement in the Netherlands (Van Maercke et al., 2013) and a street canyon scale model at scale one tenth (Jeon et al., 2013) confirmed the noise reduction of about 2−4 dBA in enclosed courtyards.

Case Studies: Wet Substrates

An in-situ diffraction experiment (Van Renterghem and Botteldooren, 2014) near the edge of a building (single diffraction case) equipped with an extensive green roof, subjected to natural precipitation, showed that sound propagation was especially sensitive to the substrate's volumetric water content (VWC) in the frequency range between 250 and 1250 Hz. The difference in the green roof's noise attenuation between a relatively dry (near 10% VWC)

TABLE 1 Overview of the In-Situ Measurements

	Case Number	Substrate Depth (mm)	Propagation Path Length Interacting With Green Roof (m)	Vegetation Cover	Green Roof Road Traffic Noise Insertion Loss (30–70 km h^{-1}, 5% Heavy Traffic)	
					Low Microphone Position (dBA)	High Microphone Position (dBA)
Single diffraction cases	1	20–30	8	> 75% (*Sedum* + mosses)	4.1	1
	2	50–60	2.5	< 5% (*Sedum* shoots)	2.3	−2.4
	3	180	4.5	50% (grasses)	5.5	2.1
Double diffraction cases	4	30–40	25	> 90% (*Sedum*)	3.1	2.2
	5	80–100	25	< 5% (*Sedum* shoots)	3.4	5.1

Source: From Van Renterghem, T., Botteldooren, D., 2011. In-situ measurements of sound propagating over extensive green roofs. Building and Environment 46, 729–738, with a calculation of the global green roof road traffic noise insertion loss.

and fully saturated state ranged up to 10 dB. At low frequencies, the absorption coefficient is limited anyhow and the presence of water in the substrate had no impact. At high sound frequencies, the somewhat raised (rigid) roof edge at that building allowed sound to propagate in a more or less straight line from this roof edge toward the microphone. Consequently, there is less interaction with the green roof and the VWC has a limited effect only. When combing the experimental attenuations with a typical road traffic noise spectrum, the impact of the substrate's water content on the A-weighted total road traffic sound pressure level abatement was predicted to be at maximum 1.5 dBA.

Large Scale Effects of Green Roofs

The aforementioned sections focus on the noise reduction by a green roof positioned on a building directly connecting a source canyon and a receiver canyon/courtyard or facade. A relevant question is whether green roofs have large scale effects as well. Especially very low frequencies have the ability to propagate easily through the atmosphere, not hindered by atmospheric absorption processes and hardly shielded by even large obstacles. After propagation over several kilometers, only the low-frequency part of the spectrum remains. Although the sensitivity of the human ear for such low frequencies is rather limited, this effect can be easily heard in busy and dense urban environments, especially at locations shielded from nearby sound sources. This effect is often referred to as the "city hum." Although green roofs absorb low frequencies to a limited extent only, extended interaction path lengths with a multitude of green roofs could potentially limit this city hum. Research on large scale effects of green roofs on urban noise is however lacking.

SOUND TRANSMISSION REDUCTION (ROOF INSULATION)

Due to their relative large surface mass density, their low stiffness, and pronounced damping properties, a green roof could be an interesting sound insulation material from a theoretical point of view.

Reported scientific research assessing the additional transmission loss by a green roof, on top of a standard roof system, is rather scarce. Measurements on small boxes under controlled conditions by Kang et al. (2009) showed the significant increase in sound insulation when a poorly performing roof membrane was equipped with a green roof; the measured additional transmission loss showed to be strongly frequency dependent (frequencies were considered between 50 Hz and 5 kHz) and ranged from 10 to 40 dB. An anecdotic in-situ measurement at full scale (Yzewyn, 2008) showed a near 6 dB decrease in transmission at the 250 Hz-octave band by a green roof (thickness 40–60 mm), relative to part of the roof covered with 40 mm thick concrete tiles.

Two elaborate studies have been reported in peer-reviewed scientific literature: Connelly and Hodgson (2013) studied sound insulation in a specifically designed test-facility for green roofs. Galburn and Scerri (2017) experimented in a vertical (standard) sound transmission suite in an acoustic lab.

Connelly and Hodgson (2013) reported an increased transmission loss relative to nonvegetated reference roofs up to 10 and 20 dB in the low and mid frequency range. Galbrun and Scerri (2017) measured lightweight green roofs on top of an 18-mm thick plywood panel (9.3 kg m^{-2}). The measured total sound reduction index was in between 20 dB (at 63 Hz) and 30 dB (at 1 kHz), leading to a maximum weighted sound reduction index R_w equal to 35 dB. Connelly and Hodgson pointed at the increased acoustic insulation of the roof system at low frequencies, which is otherwise difficult to achieve posthoc. Transmission loss was found to increase with substrate depth in a nonlinear way (Connelly and Hodgson, 2013); with increasing substrate depths between 50 and 150 mm, the additional transmission loss was larger at high frequencies. In both experiments, moisture content did not have a significant influence on the transmission loss, as was also found earlier by Kang et al. (2009). The presence of a plant community influences both mass loading and substrate porosity to some extent; as a result, transmission loss could (slightly) increase or decrease based on species choice (Connelly and Hodgson, 2013). Galbrun and Scerri (2017), similar to Kang et al. (2009), reported no net effect of placing a vegetation layer on top of the substrate.

Relating to classical plate theory, a roof system equipped with a green roof gives lower than expected additional insulation in the mass-controlled frequency range (Connelly and Hodgson, 2013). Due to the strong internal damping and the near absence of coincidence effects, the additional sound transmission loss by a green roof strongly increases at high frequencies. By allowing a 50 mm cavity filled with mineral wool below the green roof (Galbrun and Scerri, 2017), a mass-spring-mass system is constructed, largely enhancing the overall acoustic insulation (an increase in R_w of 13 dB, relative to the absence of such a cavity, was measured).

CONCLUSION

Green roofs are able to reduce sound exposure near or inside a building by mitigating diffracting sound waves over (parts of) roofs and by reducing sound transmission through the roof system. Road traffic noise shielding at a quiet side of a dwelling has been studied in detail, by means of numerical simulations and by in-situ measurements, showing consistent and useful noise reductions.

Research on sound transmission through green roofs is rather scarce and is predominantly based on nonstandardized measurements. The possibility to enhance low-frequency transmission loss is especially worth mentioning. In addition, the presence of an acoustically damped cavity below the green roof should be considered in future applications.

The contribution of the different layers constituting the green roof, for both the acoustic absorption and transmission loss, is not fully understood. Increasing such knowledge could further improve their acoustic performance.

REFERENCES

Attal, E., Cote, N., Haw, G., Pot, G., Vasseur, C., Shimizu, T., et al., 2016. Experimental characterization of foliage and substrate samples by the three microphone two load method. Proceedings of Internoise 2016, Hamburg, Germany.

Connelly, M., Hodgson, M., 2013. Experimental investigation of the sound transmission of vegetated roofs. Appl. Acous. 74, 1136−1143.

Connelly, M., Hodgson, M., 2015. Experimental investigation of the sound absorption characteristics of vegetated roofs. Build. Environ. 92, 335−346.

Cox, T., D'Antonio, P., 2004. Acoustic Absorbers and Diffusers: Theory, Design and Application. Taylor and Francis, London and New York.

Ding, L., Van Renterghem, T., Botteldooren, D., Horoshenkov, K., Khan, A., 2013. Sound absorption of porous substrates covered by foliage: experimental results and numerical predictions. J. Acous. Soc. Am. 134, 4599−4609.

Dubout, P., 1969. The sound of rain on a steel roof. J. Sound Vibrat. 10, 144−150.

Embleton, T., 1996. Tutorial on sound propagation outdoors. J. Acous. Soc. Am. 100, 31−48.

European Environment Agency, 2010. The European environment, state and outlook 2010, urban environment.

Fritschi, L., Brown, L., Kim, R., Schwela, D., Kephalopoulos, S., 2011. Burden of disease from environmental noise: a quantification of healthy life years lost in Europe. WHO Regional Office for Europe.

Galbrun, L., Scerri, L., 2017. Sound insulation of lightweight extensive green roofs. Build. Environ. 116, 130−139.

Hadden, J., Pierce, A., 1981. Sound diffraction around screens and wedges for arbitrary point source locations. J. Acous. Soc. Am. 69, 1266−1276.

Horoshenkov, K., Khan, A., Benkreira, H., 2013. Acoustic properties of low growing plants. J. Acous. Soc. Am. 133, 2554−2565.

Jeon, J.-Y., Jang, H.-S., Kim, Y.-H., 2013. Examine and quantify the acoustic effectiveness of vegetation in urban spaces through 1:10 scale model, technical report, the HOSANNA project.

Kang, J., Huang, H., Sorrill, J., 2009. Experimental study of the sound insulation of semiextensive green roofs. Proceedings of Internoise 2009, Ottawa, Ontario, Canada.

Liu, C., Hornikx, M., 2015. Determination of the impedance of vegetated roofs with a doublelayer Miki model. Proceedings of Euronoise 2015, Maastricht, The Netherlands.

Nilsson, M., Bengtsson, J., Klæboe, R., 2015. Environmental Methods for Transport Noise Reduction. CRC Press, Taylor & Francis Group, Boca Raton, FL, US.

Öhrström, E., Skånberg, A., Svensson, H., Gidlöf-Gunnarsson, A., 2006. Effects of road traffic noise and the benefit of access to quietness. J. Sound Vibrat. 295, 40−59.

Van Maercke, D., Defrance, J., Maillard, J., Anselme, C., Mandon, A., Altreuther, B., et al., 2013. Holistic acoustic design and perceptual evaluation: case studies and data collection, technical report, the HOSANNA project.

Van Renterghem, T., Botteldooren, D., 2008. Numerical evaluation of sound propagating over green roofs. J. Sound Vibrat. 317, 781–799.

Van Renterghem, T., Botteldooren, D., 2009. Reducing the acoustical façade load from road traffic with green roofs. Build. Environ. 44, 1081–1087.

Van Renterghem, T., Botteldooren, D., 2010. The importance of roof shape for road traffic noise shielding in the urban environment. J. Sound Vibrat. 329, 1422–1434.

Van Renterghem, T., Botteldooren, D., 2011. In-situ measurements of sound propagating over extensive green roofs. Build. Environ. 46, 729–738.

Van Renterghem, T., Botteldooren, D., 2014. Influence of rainfall on the noise shielding by a green roof. Build. Environ. 82, 1–8.

Van Renterghem, T., Hornikx, M., Forssen, J., Botteldooren, D., 2013. The potential of building envelope greening to achieve quietness. Build. Environ. 61, 34–44.

Yang, H.-S., Kang, J., Choi, M.-S., 2012. Acoustic effects of green roof systems on a low-profiled structure at street level. Build. Environ. 50, 44–55.

Yzewyn, T., 2008. Reductie van geluids blootsteling door groendaken (in Dutch, "Reduction of noise exposure by green roofs"). Master thesis. Ghent University, Belgium.

Chapter 3.9

Green Streets
for Noise Reduction

Ana M. Lacasta, Angelina Peñaranda and Inma R. Cantalapiedra

Chapter Outline

INTRODUCTION

Noise pollution in urban environments is a frequent cause of discomfort, health, and psychological problems (IGCB, 2010). Potential effects of noise include speech interference, ear discomfort, sleep disturbance alterations in concentration capacity, decrease of productivity, and problems in children's learning (Kang, 2006).

Although there are many sources of noise related with people activities and machinery, road traffic is the most important urban noise. The outdoor noise level that arrives to a receptor depends on the kind and speed of the vehicle, the distance between the source and the receiver, likewise the obstacles between them and the characteristics of the environment that can affect the sound propagation, besides their subjective effects. Many of the measures usually used to control noise in highways or industrial environments, as high noise barriers, cannot be used in dense urban emplacements, because of space limitations, safety, or visual impacts.

Using vegetation in the reduction of urban noise is a concept that has gained much attention around the world. Studies on tree belts (Fang and Ling, 2003; Islam et al., 2012) showed important noise attenuations, being their density, height, length, and width the most effective factors. Width of vegetation belts is also a significant noise reduction factor, because of the increment in the sound absorption and dissipation with larger acoustic pathway.

Nature Based Strategies for Urban and Building Sustainability.
DOI: https://doi.org/10.1016/B978-0-12-812150-4.00017-3

In urban streets, when the integration of wide tree belts is not a feasible solution, the disposal of hedges and dense shrubs can be a very useful option in reducing noise (Fang and Ling, 2003; Kalansuriya et al., 2009; Van Renterghem et al., 2015). Another possibility is to combine vegetation with solid walls (greenery barriers) (Dunnett and Kingsbury, 2004; Wong et al., 2010; Daltrop and Hodgson, 2012; Hodgson et al., 2013). Both plants and air between the vegetation increase the acoustic attenuation of the wall. Such greenery barriers can be placed in the streets near to the source and the receiver, increasing their effectiveness. Vegetation can reduce the traffic noise level, especially in narrow streets with hard facades. While multiple reflections in buildings lead to an amplification of the noise, the acoustic absorption by plants, placed along the street or covering facades and roofs, avoid such amplification (Van Renterghem et al., 2015).

In addition to the physical noise reduction, vegetation by itself affects noise perception positively. Many studies indicate that an urban sound scene is not perceived in isolation, but other sensory modalities such as vision interact with auditory information and modify the sound perception (Viollon et al., 2002). In this sense, the vision of vegetation elements reduces the noise annoyance (Van Renterghem and Botteldooren, 2016). Other studies indicate that the perception of natural soundscapes, such as bird sounds, decreases the perceived level of traffic noise (De Coensel et al., 2011; Hong and Jeon, 2013; Preis et al., 2015).

This paper summarizes some relevant outcomes found in literature about the physical reduction of noise in green streets, as well as those related with the improvement in the sound perception in presence of vegetation. An example is also presented, corresponding to a greenery barrier in a main artery of the city of Barcelona.

NOISE SHIELDING BY HEDGES AND GREENERY BARRIERS

Hedges and dense shrubs can be considered as low-height noise barriers. They can be effective provided that their height is greater than the position of the receiver, since the source of traffic noise is typically located close to the street surface (Fang and Ling, 2003; Van Renterghem et al., 2015). In order to limit diffraction produced in low barriers, they should be placed close to the traffic lanes, and receivers should be close to the barrier. Kalansuriya et al. (2009) analyzed the effect of roadside vegetation of different characteristics on the reduction of road traffic noise levels. They obtained an average noise reduction of 4 dB(A), necessary vegetation being at least 1.5 m thick in order to achieve a reduction of 5 dB(A). The spectral noise reduction depends on the plant species. Fan (2010) analyzed the noise-reducing spectrum of six hedges, concluding that noise attenuation is improved by combining plants with reciprocal noise-reducing spectra, or with spectra similar to the source noise.

Many investigations have been done about the effectiveness of low-height noise barriers by means of scale modeling and numerical approaches (Martin and Hothersall, 2002; Van Renterghem et al., 2015; Kang, 2006). Fig. 1 shows, as an illustrative example, the screening effect of barriers of height $H = 0$, 1.5, 3, and 5 m, placed at 10 m from the center of the traffic line. In this example, the noise levels perceived by a receiver located near the barrier (1.5 m to the right) and at the height of 1.5 m from the ground are of 64.2, 62.3, 58.4, and 47.6 dB(A) respectively. These results indicate that significant noise attenuation can be obtained with low barriers (1.5 m) although, as expected, effectiveness is clearly improved for medium height (3 m).

A medium-height solid barrier can be integrated in a green street by covering with vegetation, transforming it into a greenery barrier. A number of

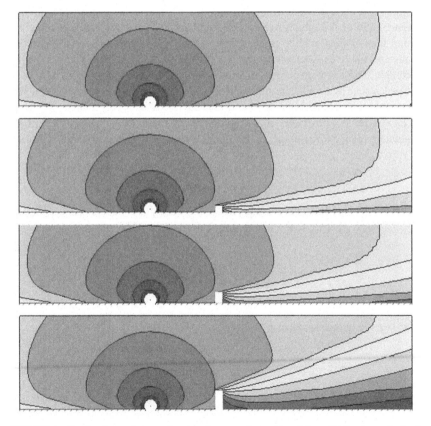

FIGURE 1   Predicted sound maps showing the screening achieved by solid barriers, obtained with the environmental noise prediction software CadnaA. From top to bottom, they correspond to barrier height of 0, 1, 1.5, and 3 m. The white circle indicates the traffic line, which is 10 m at the left of the barrier. Color from *red* (dark gray in print versions) to *green* (light gray in print versions) shows a decrease in noise level.

studies shows that the addition of vegetation in a solid barrier can increase the noise reduction (Dunnett and Kingsbury, 2004; Wong et al., 2010; Daltrop and Hodgson, 2012; Hodgson et al., 2013). The plants and the wall have opposite spectral behavior regarding noise transmission, blocking the sound in the lower and higher frequency ranges, respectively (Dunnett and Kingsbury, 2004). In addition, the attenuation produced by the air between the vegetation and the wall can increase the acoustic insulation provided by the wall. Wong et al. (2010) analyzed eight vertical greenery systems, evaluating the insertion loss, defined as the difference between the sound pressure level at the same point behind the wall, before and after the addition of the greenery system. Their results showed stronger attenuation at low to moderate frequencies, whereas weaker attenuation was observed in the high-frequency spectrum.

A successful example of integration of a medium-height green barrier in urban context is shown in Fig. 2A. The photograph corresponds to Diagonal Avenue, one of the main arteries of Barcelona, and presents many of the elements that characterize a green street. It has a central walkway for pedestrians and bicycles, separated from the traffic lanes by two lines of trees. On each side, there is a tram lane covered with grass, three motorway lanes, and the pedestrian sidewalk. On the right of the image, a greenery barrier can be observed. It is 2.6 m high and separates the road from a public park. The green barrier is composed of a concrete wall 25 cm thick with an external vegetated layer (toward the road) of Bougainvillea species, about 100 cm

FIGURE 2 (A) Stretch of Diagonal Avenue, where a greenery barrier 500 m long (on the right of the image) was used for acoustic measurements; (B) transverse detail of the concrete wall with the external vegetated layer; (C) internal face.

thick, and an ivy-draped internal face (toward the park) about 50 cm thick (see Fig. 2B and C). The vegetated barrier achieves noise to be noticeably lower in the inner path than in the external sidewalk. Measurements have been performed in order to quantify this noise reduction.

Fig. 3 shows a 2 minutes evolution of the noise level, at the external (road) and internal (park) sides of the greenery barrier. The global noise

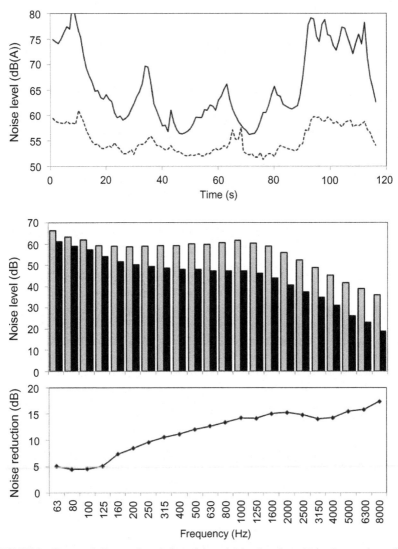

FIGURE 3 Top panel: Temporal evolution of the noise level registered, simultaneously, at the two sides of the greenery barrier. Middle panel: spectral distribution of the noise levels measured at the external side (gray bars) and internal side (black bars). Bottom panel: spectral distribution of the noise reduction achieved by the barrier.

levels, averaged over all the evaluation time, at both sides of the barrier, were of the 68.6 and 55.3 dB(A). The frequency noise distributions for each 1/3-octave band is also presented in Fig. 3, as well as the noise reduction, calculated as the difference of the two noise levels is showed in Fig. 3 (Bottom panel). A noise reduction is observed for all the frequencies, being higher at medium and at high frequencies it is about 5 dB for frequencies lower that 125 Hz, and they increase continuously until about 15 dB at 2000 Hz. At higher frequencies it is maintained in the range 14−18 dB. These results indicate that medium green barriers can be an effective solution in urban areas.

SOUND ABSORPTION BY VEGETATION

Many studies have been done to determine the absorption of sound by vegetation. Horoshenkov et al. (2013) studied the effect of leaf morphology and area on the acoustic absorption coefficient in an impedance tube to quantify the ability of absorption of a series of low growing plants with a relatively high leaf area density, and with and without a soil substratum. The larger leaf area density and dominant angle (defined as the angle of the leaf orientation from the vertical direction), the higher values of the acoustic absorption coefficient. The type of soil in which a plant is growing also affects the ability of the plant to absorb sound. Soils with high permeability and low density are more effective.

The pioneer work of Aylor (1972), analyzed the effects of adding different types of vegetation and ground to determine the noise reduction, concluding that foliage is effective especially at the higher frequencies, where scattering is enhanced. Ground attenuates sound for lower frequencies, in which scattering is not effective being more efficient in limiting typical engine noise frequencies (near 100 Hz) of road traffic.

In another interesting experiment, branches of pine trees were brought into a small reverberant chamber to determine their sound absorption mechanism, and the acoustics attenuation is found to be due to the thermoviscous absorption in the surrounding air's boundary layers (Burns, 1979). Other authors studied the vegetation absorption effect on the insertion loss of building walls in an urban setting and determined the sound absorption coefficient of the vertical greenery system in the reverberation chamber. AzKorra et al. (2015) found in a module-based green wall a coefficient of approximately constant in all the spectra analyzed with a weighted value of 0.4. Lacasta et al. (2016) performed in situ measures on the same type of module-based green wall. The absorption coefficient had an average value of 0.7 and frequency dependent values higher than those previously obtained in the laboratory. In any case the values are higher than other buildings' materials and furnishings.

Low-height vegetal noise barriers (0.64 m width and 0.92 m height) placed in the middle and in each side of the street canyon can reduce the noise due to absorption (Van Renterghem et al., 2012). They considered different arrangements of such screens, for receivers on walkways and distributed over the building facades, obtaining a reduction of the road traffic noise. Therefore, absorption was shown to be essential to obtain positive effects, even when both sources and receivers are located in the same reverberant space.

IMPROVEMENT OF URBAN SOUNDSCAPE PERCEPTION BY VEGETATION

The evaluation of noise effects is a complex problem due to the disciplines that take part into it: acoustic, physiology, psychology, sociology, statistics, etc. It is well known that pure tones and/or impulsive noise increase discomfort. High levels of noise in a pleasant surrounding can be appreciated with less annoyance that a lower level in an unpleasant environment due to subjective evaluation of noise. The perception of the sound includes therefore, not only the energy propagation of the physical waves, but also their interaction with other environmental factors such as light, color, temperature, humidity, natural sounds, visual environment, between others. The visual accessibility of the source by the receiver can affect the acoustic sensation.

Perception of particular sources of noise pollution, such as traffic, is modified by what people expect to hear and by the visual surrounding where it is perceived. In this sense one of the aspects that affects the noise perception is the vegetation. It makes a considerable difference in people's evaluation of an urban context by substantially improving, not only visual quality, but in many settings, expected acoustic quality as well. The availability of nearby trees, green landscapes, gardens, places to walk, etc., has influence in the well-being of citizens decreasing their stress, including the annoyance caused by an uncomfortable noise.

The evaluation of interrelation between vegetation and sound perception is the subject of extensive research. Some of them are based in experiments on the annoyance perceived by people whose homes are closer or not to green spaces. Other have been performed in laboratory experiments, on selected samples of people, under control conditions of level and type of sound and with or without images of different landscapes.

Gidlöf-Gunnarsson and Ohrstr (2007), as participants in the research program "Soundscape Support to Health," explored the relationship between green-area availability and noise response. They found that residents with access to green areas do not perceive noise as a neighborhood problem. An opposed answer was presented in residents with worse access, showing two times greater disturbance. In the same way, Van Renterghem and Botteldooren (2016) studied, through surveys, the effect of the outdoor vegetation on perception of traffic noise annoyance in dwellings. Although all facades were exposed to the same noise level, the percentage of people who expressed discomfort perception was twice lower in those who could see the vegetation through the windows.

It was reported that landscape plants are thought to be effective in noise reduction (Yang et al., 2011). Their studies from a questionnaire survey show that 90% of their subjects believed that landscape plants contribute to noise reduction and that 55% overrated the plants' actual ability to attenuate noise. Eighty percent of the participants indicated that plant hedges were the most effective noise barriers, better than those of concrete and plastic. Additional analysis of results obtained with electroencephalograms showed significant difference in human physiological responses to vegetation and traffic views. It was observed that visual stimuli partially influence the psychological perception of acoustic perceptions. In particular, landscapes with vegetation, water, and other natural elements have a positive influence on physiological health and psychological well-being regardless of what kind of urban sounds accompany the visual observations (Yang et al., 2011).

This interrelation was corroborated by other studies that analyzed the influence of visual vegetation, natural sounds and both, in the traffic noise evaluation. Acoustical stimulus only, visual stimulus only, and both simultaneously were applied to surveys. The most unfavorable situation appeared when traffic noise was only presented, decreasing the unpleasant perception if natural sounds were included as soundscape elements. Bird sounds were more useful for enhancing soundscape quality (Hong and Jeon, 2013; Viollon et al., 2002; Preis et al., 2015; De Coensel et al., 2011). Audiovisual experiments with road traffic noise were perceived less stressful and less unpleasant when the visual setting was greener or less urban.

CONCLUSION

It has been shown that green streets contribute to the reduction of noise annoyance, especially in traffic noise. The sound produced in streets suffers reflections when reaching the surfaces of the buildings in its propagation, and returning to the street. Other urban elements such as ground, cars, street furniture, can behave in the same way by increasing the level noise. Trees, grass, ground, hedges, green barriers interposed in the path of sound propagation, absorb part of it by diminishing the sound coming to a receiver or to the facades. The facades covered with plants or green roofs have the same effect. In this work the results obtained from measures in a main arteria of Barcelona city showing the shielding efficacy of greenery medium-height barriers are presented. The insertion loss obtained was between 5 and 15 dB (A) depending on the frequency. Green streets therefore act as noise-reducing devices in addition to their visual attractive as well as the improvement in the sound perception in presence of vegetation.

REFERENCES

Aylor, D., 1972. Noise reduction by vegetation and ground. J. Acous. Soc. Am. 51, 197−205.

Azkorra, Z., Pérez, G., Coma, J., Cabeza, L.F., Bures, S., Álvaro, J.E., et al., 2015. Evaluation of green walls as a passive acoustic insulation system for buildings. Appl. Acous. 89, 46−56.

Burns, S.H., 1979. The absorption of sound by pine trees. J. Acous. Soc. Am. 65, 658−61.

Daltrop, S., Hodgson, M., 2012. Scale-model investigation of the effects of surface absorption and nearby foliage on noise-barier performance. Can. Acous. 40 (4), 41−48.

De Coensel, B., Vanwetswinkel, S., Botteldooren, D., 2011. Effects of natural sounds on the perception of road traffic noise. J. Acous. Soc. Am. 129 (4), EL148−EL153.

Dunnett, N., Kingsbury, N., 2004. Planting Green Roofs and Living Walls, vol. 254. Timber Press, Portland, OR.

Fan, Y., 2010. The investigation of noise attenuation by plants and the corresponding noise-reducing spectrum. J. Environ. Health 72 (8), 8.

Fang, C.F., Ling, D.L., 2003. Investigation of the noise reduction provided by tree belts. Landscape Urban Plan. 63 (1), 187−195.

Gidlöf-Gunnarsson, A., Ohrstr, E., 2007. Noise and well-being in urban residential environments: the potentialrole of perceived availability to nearby green areas. Landscape Urban Plan. 83, 115−126.

Hodgson, M., Daltrop, S., Peterson, R., Benedict, P., 2013, June. Compliance and vegetated-barrier acoustical testing in a purpose-built sound-transmission suite. In Proceedings of Meetings on Acoustics (Vol. 19, No. 1, p. 040083). Acoustical Society of America.

Hong, J.Y., Jeon, J.Y., 2013. Designing sound and visual components for enhancement of urban soundscapes. J. Acous. Soc. Am. 134 (3), 2026−2036.

Horoshenkov, K.V., Khan, A., Benkreira, H., 2013. Acoustic properties of low growing plants. J. Acous. Soc. Am. 133 (5), 2554−2565.

Interdepartmental Group on Costs and Benefits Noise Subject Group (IGCB(N)) Second report, 2010. Valuing the human health impacts of environmental noise exposure.

Islam, M.N., Rahman, K.S., Bahar, M.M., Habib, M.A., Ando, K., Hattori, N., 2012. Pollution attenuation by roadside greenbelt in and around urban areas. Urban For. Urban Gree. 11 (4), 460−464.

Kalansuriya, C.M., Pannila, A.S., Sonnadara, D.U.J., 2009. Effect of roadside vegetation on the reduction of traffic noise levels. In Proceedings of the Technical Sessions (Vol. 25, No. 2009, pp. 1−6).

Kang, J., 2006. Urban Sound Environment. CRC Press, Boca Raton, FL.

Lacasta, A.M., Penaranda, A., Cantalapiedra, I.R., Auguet, C., Bures, S., Urrestarazu, M., 2016. Acoustic evaluation of modular greenery noise barriers. Urban For. Urban Gree. 20, 172−179.

Martin, S.J., Hothersall, D.C., 2002. Numerical modelling of median road traffic noise barriers. J. Sound Vibrat. 251 (4), 671−681.

Preis, A., Kociński, J., Hafke-Dys, H., Wrzosek, M., 2015. Audio-visual interactions in environment assessment. Sci. Total Environ. 523, 191−200.

Van Renterghem, T., Botteldooren, D., 2016. View on outdoor vegetation reduces noise annoyance for dwellers near busy roads. Landscape Urban Plan. 148, 203−215.

Van Renterghem, T., Botteldooren, D., Hornikx, M., Jean, P., Defrance, J., Smyrnova, Y., et al., 2012. Road traffic noise reduction by vegetated low noise barriers in urban streets. Euronoise, Praga.

Van Renterghem, T., Forssén, J., Attenborough, K., Jean, P., Defrance, J., Hornikx, M., et al., 2015. Using natural means to reduce surface transport noise during propagation outdoors. Appl. Acous. 92, 86–101.

Viollon, S., Lavandier, C., Drake, C., 2002. Influence of visual setting on sound ratings in an urban environment. Appl. Acous. 63 (5), 493–511.

Wong, N.H., Tan, A.Y.K., Tan, P.Y., Chiang, K., Wong, N.C., 2010. Acoustics evaluation of vertical greenery systems for building walls. Build. Environ. 45 (2), 411–420.

Yang, F., Bao, Z., Zhu, Z., 2011. An assessment of psychological noise reduction by landscape plants. Int. J. Environ. Res. Public Health 8, 1032–1048.

Chapter 3.10

Vertical Greening Systems to Improve Water Management

Gabriel Pérez and Julià Coma

Chapter Outline

INTRODUCTION

Vertical greening systems (VGS), covering all types of vegetation integration on buildings' facades, are systems highly dependent on water supply for plant development. The different systems (see Chapter 2.1: Vertical Greening Systems: Classifications, Plant Species, and Substrates) also have different water requirements, ranging from the most extensive (green facades) to the intensive ones (green walls or living wall systems).

Within the context of sustainable construction, the responsibility in the use of water resources requires VGS designs under an efficient and conservative use of water approach. This means not only the reduction of water demand but the reuse of wastewater such as rain and gray waters. Related to the urban concerns, such as the rainwater management in cities, the increment of vertical green surfaces, as well as green roofs, is sustainable strategies that alleviate this problem, specially in densely urban areas.

In spite of that, the overall water balance established in a VGS is basically configured by inputs from irrigation and outputs from plants; also the influence of evapotranspiration (from plants and the substrates), which basically depends on the exposed area, the plant species, the substrate properties and the climatology, must be considered as water losses. Moreover,

Nature Based Strategies for Urban and Building Sustainability.
DOI: https://doi.org/10.1016/B978-0-12-812150-4.00018-5

evapotranspiration process implies valuable effects from the thermal energy balance by the provision of extra cooling effect to the building.

In addition, recently interesting studies about new applications of water management in VGS have emerged, being two of the most promising ones the use of VGS as biofilters for wastewater treatment as well as a complementary element for aquaponics production in small spaces.

In this chapter the main issues relating to water management in VGS are reviewed, such as the necessity to efficiently use that resource, the possibility of reusing rain water and gray water which also implies the possibility to runoff control in urban areas, the cooling effect through evapotranspiration, and the use of VGS as biofilters or in aquaponics systems.

WATER USE EFFICIENCY

The main differences between irrigation systems for VGS are closely related to the different developed design strategies over the time. Thus, it can be found from simple irrigation lines in the lower parts of a green facade providing water to climbing plants, to the high-tech irrigation systems that require control strategies, to provide water and fertilizers to the most contemporary living walls. In Chapter 2.2, Vertical Greening Systems: Irrigation and Maintenance, these issues are addressed from an in-depth technical standpoint.

During this process, one of the biggest overcame challenges has been the water distribution on a vertical surface. Nowadays a high level of technology in VGS watering systems has been achieved. However, as Riley (2017) stated, two main barriers still remain to be overcome: the irrigation efficiency and water distribution homogeneity through the wall surface, since often areas with lack or excess of irrigation can be observed in the on-going projects. Considering the high exposure of the buildings facades, any small problem in the water supply can become a great visual problem due to the influence over the plants, a fact that can compromise the confidence of owners and technicians in the future usage of these systems.

Therefore, technological improvements and more research will be necessary to minimize the risk of bankruptcy of VGS due to imbalances in the water supply systems. The main research areas to improve this aspect are:

- To establish irrigation sectors by zoning the wall at different heights and/ or depending on the distribution of plants, and to place drip lines so that water can flow vertically and horizontally.
- To control the operational variables, such as the duration and frequency of the irrigation, as well as the water flow and working pressure (depending on number and types of emitters).
- To develop growing medium and substrates with improved water retention and distribution capabilities, when applied both vertically and

horizontally, while they avoid saturation, keeping simultaneously oxygenated the roots zone.

- To reuse wastewater, by means of recovering the excess of water in VGS, capturing rain water (e.g., in nearby roofs) and reusing gray water, from buildings and industry, once they are previously treated.

RAIN AND GRAY WATER REUSE

As it is highlighted above, the efficient use of reusing wastewater may be a key point for the future establishment of VGS, especially for big projects in which a large quantity of water is required to maintain lots of square meters of greening surfaces.

It is clear that while the capture of rain water on roofs (horizontal) is an easily achieved function, it is not the case in building facades (vertical). Therefore, here the combination of these two typologies of systems stands out as the most promising strategy to solve the lack of potential of VGS to capture rain water.

In addition, the possibility to recover gray water from buildings could be a source of water easily usable for urban green infrastructure, thus for VGS. Gray water is all wastewater generated in households or office buildings without fecal contamination, i.e., sinks, showers, baths, clothes washing machines, or dish washers. Gray water, which does not contains large quantities of pathogens, can be easily treated and reused onsite for toilets, landscape, and other nonpotable uses (Li et al., 2009).

In any case, the establishment of a recovery system whether it's rain or gray water, will involve the installation of filters and tanks to prior treatment of these waters and also to store it for later use. Although this fact will determine the necessary initial investment, it could be the guarantee of the future survival of VGS, especially in big projects.

Moreover in those climates or places where runoff water resulting from heavy rainfall events can be a problem, some types of VGS properly connected to the horizontal collecting surfaces (e.g., green roofs) can help control this punctual excess of water.

Currently, some on-going VGS projects in which the recovery of water, both from rain and gray water, have been taken into account can be found.

Los Cabos International Convention Center (ICC), in Los Cabos, Mexico, is a commercial building built in 2012, which hosts a 2700 m^2 living wall. In this project, a wastewater treatment system was also installed to recycle treated water back to the building for toilet flushing and irrigating the living wall (Biomicrobics and ICA Construcción Urbana, 2012).

The city of London enjoys interesting features such as the vertical garden on London's Tooley Street which uses no power and is sustained only by rain water. What makes this project unique is the ability to harvest enough rain water in storage tanks concealed behind the planting, to sustain the

FIGURE 1 Left: London's Tooley Street vertical garden. Right: Rubens Palace hotel green wall.

living wall for up to 6 weeks. This innovation allows the plants to absorb the water in a controlled way via capillary action, negating the need for a pressurized irrigation system that would require power and a water supply (Treebox, 2015) (Fig. 1, Left). Another iconic project in London is the 350 m^2 green wall at Rubens Palace hotel. In this project, built in 2013, plants are irrigated by harvested rain water that is caught in dedicated storage tanks on the roof (Treebox, 2013) (Fig. 1, Right).

Finally, another example is the Santalaia building, in Bogota, Colombia, a multifamily residential building built in 2015 that includes a 3000 m^2 living wall, in which the irrigation system not only has been designed using humidity sensors to optimize water consumption but it includes a water treatment plant in order to achieve zero water waste. In this project, the gray water from the apartments' showers is used for VGS irrigation (Groncol, 2017).

COOLING EFFECT; EVAPOTRANSPIRATION

Both the plant transpiration process and evaporation from wet substrate play an important role in the whole water balance in VGS.

This evapotranspiration capacity depends on multiple factors such as the amount of available water, the types of plants and substrates as well as the orientation and sun exposure, among others (Perini et al., 2017). Also, the climate conditions are highly influencing, since in dry environments or windy conditions evapotranspiration will increase.

For evapotranspiration occurrence a quantity of energy is necessary. This physical process named "evaporative cooling" implies a quantity of 2450 Joule of each gram of evaporated water (latent heat of vaporization of water at 20°C). This energy is captured from the atmosphere, resulting in a cooling of the nearby surroundings.

This effect is desirable from the viewpoint of passive cooling energy savings in buildings as well as a means to improve environmental conditions at city scale, but it can be counterproductive from the point of view of water management, since it implies a higher spending of this resource. Therefore, systems that optimize the achievement of various ecosystem services related to the efficient use of water, i.e., energy savings and heat island effect reduction, at the same time that the suitable provision of water for plant development, must be designed.

Despite the importance of this cooling effect, there is a lack of in-depth studies of its contribution by means VGS in the built environment. Some authors have addressed the issue of the contribution of VGS for energy savings in buildings, but hardly ever the contribution of the different involved effects has been provided, i.e., the shadow effect, the cooling effect, the insulation effect and the wind barrier effect (Pérez, 2014).

Recently, Davis et al. (2015) have addressed the study about the potential of vertical gardens as evaporative coolers by means of the adaptation of FAO-56 Penman Monteith Equation from the theoretical point of view. The equation was modified such that it could be used to predict the theoretical evaporation rate from air flowing in the space between the substrate and the surface onto which a vertical garden is attached. Even though the results were not the expected ones, with future improvements of these experiments, possibly the use of VGS as evaporative cooling systems can become a reality in the future.

In a more experimental approach, Perini et al. (2017) concluded that the evaporative performance of a VGS is strongly influenced by the evaporation from the panels, in addition to the water transpiration from plants. Regarding to this, the foliage density will have great influence in the final performance, especially in summer periods. According to the data recorded, *Hedera helix* showed higher cooling capacities than the other tested plants.

WASTEWATER TREATMENT; BIOFILTERS

A step beyond the reuse of rain and gray waters in VGS, previously discussed, is the use of green walls as biofilters for the treatment of wastewater, either gray or other types.

Bussy (2009) proposed one household scale way of treating gray water for dry areas in the countryside, in Sweden. The big innovative part of this treatment system lies in a green wall (height: 1.55 m, thickness: 0.20 m, width: 2.40 m), filled with small gravels, in which some different plants grew up (tomato, salad, thyme, campanula, carnation, and strawberry). The cleaned water from this treatment showed a reduction of 95% on phosphorus, 75% on organic matter and around 50% on nitrogen, so that treated water could be reused for human consumption.

Svete (2012) examined the potential to combine the existing wastewater treatment technology of intermittent media filter with green walls. Three separate wall sections were constructed to monitor the treatments effects of filter material choice and presence of vegetation. Under a daily dosing rate of nearly $1000\,L\,m^{-2}$ the system achieved greater average reduction rates of over 95%, 80%, 90%, 30%, and 69% for 5-day biochemical oxygen demand

(BOD$_5$), chemical oxygen demand (COD), total suspended solids (TSS), total nitrogen, and total phosphorus, respectively, as well as approximately two log unit reduction (approx. 98%) of bacteria indicator *Escherichia coli*.

Wolcott (2015) investigated the use of green walls in treating brewery wastewater. The green wall consisted of panels filled with lightweight recycled glass substrate and *Golden pothos* plant type. Preliminary testing suggests that these green walls can reduce some high strength organic components from wastewater. Turbidity and BOD can be reduced up to 50% in 6 hours or less, but no firm conclusions could be found in reference to the removal of nitrogen and phosphorous.

Masi (2016), within NaWaTech project, conducted research to study the potentiality of vegetated gardens for gray water, from an office building, treatment, and recycling.

The pilot green wall comprises two parallel units and the feeding of the treatment unit happens through an hourly flush of 10 L of gray water. The discharge was directly allowed to flow into the garden next to the walls. Each individual treatment unit consists of a 12×6 matrix of pots (6 pots in a column and 12 pots in a row). Each pot has a top surface of 0.01 m^2. The pots have been planted of the following genus *Abelia, Wedelia Portulaca, Alternenthera, Duranta,* and *Hemigraphis*.

The experimental analysis has been divided in two phases (Table 1). Initially, the pots have been filled with an inert planting material—LECA (light expanded clay aggregates)—to ensure that the only nutrient source for the plants is the gray water. In a second phase the LECA has been mixed with two different other mediums (sand and coconut fibers) in order to slow the water flow, increase residence time, and favor a greater biofilm development.

The effluent quality was in all samples already suitable for reuse by land irrigation according to the Indian regional and national regulations; in the last samples collected in Phase 2 the effluent reached an appropriate quality for reuse by flushing toilets (considering in the treatment scheme a further disinfection by an UV lamp).

Recently Harsha et al. (2017) have conducted a specific research on the use of living walls for gray water treatment. This study presents the development of a low energy and low maintenance gray water treatment technology: a living wall system, employing ornamental plants (including vines) grown in a sand filter on a side of a building to treat shower, bath, and washing basin wastewaters. The results indicate that the use of ornamental species (e.g., *Canna lilies, Lonicera japonica*, ornamental grape vine) can contribute to pollutant removal. Vegetation selection was found to be particularly important for nutrient removal. While a wider range of tested plant species was effective for nitrogen removal (>80%), phosphorus removal was more variable (13%−99%) over the study period, with only a few tested plants being effective—*Carex appressa* and *C. lilies* were the best performers.

TABLE 1 Influent and Effluent From the Two Experimental Phases

	Phase 1		Phase 2		Phase 3	
	LECA		LECA-Coconut		LECA-Sand	
No Samples	14		10		6	
	Influent mg L^{-1}	Effluent mg L^{-1}	Influent mg L^{-1}	Effluent mg L^{-1}	Influent mg L^{-1}	Effluent mg L^{-1}
COD	84.9 ± 2.5	69.4 ± 1.7	58.3 ± 16.5	48.8 ± 35.1	63.9 ± 21.3	44.7 ± 22.8
BOD$_5$	39.7 ± 3.2	29.9 ± 2.5	21.1 ± 8.2	17.4 ± 14.0	23.0 ± 10.5	14.4 ± 10.3
NH$_4^+$_N	–	–	3.6 ± 1.3	1.9 ± 2.3	3.4 ± 1.2	1.0 ± 10.3
TKN	–	–	5.7 ± 1.0	7.3 ± 4.2	4.9 ± 1.5	5.0 ± 4.7
SS	–	–	62.5 ± 24.1	47.0 ± 22.8	61.2 ± 28.1	52.6 ± 22.5

Source: Adapted from Masi, F., 2016. Green walls for greywater treatment and recycling in dense urban areas: a case-study in Pune. J. Water Sanit. Hyg. Dev. 6, 342–347.

It was also found that phosphorus removal can be compromised over the longer term as a result of leaching. Excellent suspended solids and organics removal efficiencies can be generally achieved in these systems (>80% for TSS and >90% for BOD) with plants having a relatively small impact. Columns had an acceptable infiltration capacity after 1 year of operation.

Under case study approach, a very interesting example of application is the "Espai Tabacalera" in the city of Tarragona (Catalonia, Spain), in which a 2500 m² green wall was built in 2011, directly linked to a 6000 m² nearby park. The project integrates these two features to manage not only the rain but also the gray water. The project implies the purification in situ of wastewater so that part of the vegetal wall acts as a filter in the tertiary treatment of water purification and the rest of the wall takes advantage of water components that are generated from the purification (e.g., nitrates) as nutrients for the green wall and the rest of the garden (Fig. 2).

In order to obtain the best performance of the drip irrigation system, a remote management to modify the irrigation schedule, to control the water consumption, and to detect failures has been implemented.

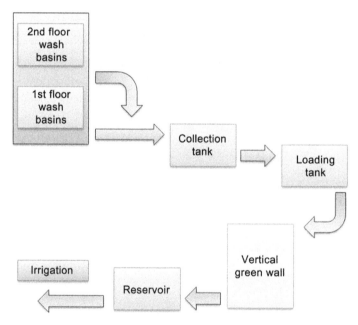

FIGURE 2 Sketch of the gray water treatment unit. *Adapted from Masi, F., 2016. Green walls for greywater treatment and recycling in dense urban areas: a case-study in Pune. J. Water Sanit. Hyg. Dev. 6, 342–347.*

AQUAPONICS

Aquaponics is an integrated system that links hydroponic production of plants with recirculating aquaculture production of fish into a sustainable agriculture system that uses natural biological cycles to supply nitrogen and minimizes the use of nonrenewable resources, thus providing economic benefits that can increase over time. The advantages of linking crop production and the culture of fish are shared start-up, operating, and infrastructure costs; recirculating tank waste nutrients and water removal by plants, thus reducing water usage and waste discharge to the environment; and increasing profit potential by simultaneously producing two cash crops.

Currently it is possible to find simple commercial aquaponic systems that allow growing plants vertically in small spaces by means of the adaptation of this technology under a VGS approach.

Recently some authors have smartly suggested a combination of aquaponics and green walls, not only thinking exclusively about agricultural production but for aesthetic purposes, both for decorative aquariums as for the plants. Thus, Fernández-cañero et al. (2015) have conducted the first experiments of a vertical aquaponics system for ornamental purposes (Fig. 3).

CONCLUSION

VGS can positively contribute to water management both at building and urban scale. Between the most promising ecosystem services provided by VGS relating to water management, the cooling effect, due to the evapotranspiration from plants and substrates, and the rain water runoff control stand out in an urban sustainable context.

Moreover, VGS are usually highly dependent of a water supply whereas few opportunities to capture this resource are possible in the vertical plane. This drawback is currently one of the most important barriers for the final deployment of these systems in the built environment. Facing this challenge, besides the irrigation systems efficiency, some strategies such as the

FIGURE 3 "Espai Tabacalera" project. Tarragona (Catalonia, Spain).

connection of VGS with the nearby roofs in which there is a chance for rain water capture and storage, or the possibility to treat or reuse gray water from buildings and industry, stand out as the most promising ones.

Finally, new opportunities to increase the contribution of green walls to the provided ecosystem services can be found in their use as biofilters. Regarding this idea, promising results on treating wastewaters as well as complementary features for aquaponics can be found in recent researches.

Generally speaking, great efforts in research must be done in the future in reference to the benefits and weaknesses relating to water management in VGS.

ACKNOWLEDGMENTS

This study has received funding from European Union's Horizon 2020 research and innovation program under grant agreement No 657466 (INPATH-TES), from the European Commission Seventh Framework Program (FP/2007-2013) under grant agreement No PIRSES-GA-2013-610692 (INNOSTORAGE). The work is partially funded by the Spanish Government (ENE2015-64117-C5-1-R (MINECO/FEDER)). GREA is certified agent TECNIO in the category of technology developers from the government of Catalonia. The authors would like to thank the Catalan Government for the quality accreditation given to their research group (2014 SGR 123).

REFERENCES

Biomicrobics and ICA Construcción Urbana, 2012. Available at: <http://www.biomicrobics.com/products/fast-wastewater-treatment-systems/microfast/> and <https://www.ica.com.mx/es_ES/web/ica/international-convention-center-los-cabos> (accessed April 2017).

Bussy, E., 2009. An Innovative Way to Treat Wastewater. Strasbourg & Uppsala: Engees (Ecole national de Genie de l'Eau et de l'Environment de Strassbourg) and Uppsala University.

Davis, M.M., et al., 2015. The potential for vertical gardens as evaporative coolers: An adaptation of the Penman Monteith Equation. Build. Environ. 92, 135−141.

Fernández-cañero, R., et al., 2015. International Conference on Living Walls and Ecosystems Services. University of Greenwich 6−8 July 2015.

Groncol, 2017. Available at: <http://groncol.com/> (Accessed April 2017).

Harsha, S.F., et al., 2017. Designing living walls for greywater treatment. Water Res. 110, 218−232.

Li, F., et al., 2009. Review of the technological approaches for grey water treatment and reuses. Sci. Total Environ. 407 (11), 3439−3449.

Masi, F., 2016. Green walls for greywater treatment and recycling in dense urban areas: a case-study in Pune. J. Water Sanit. Hyg. Dev. 6, 342−347.

Pérez, G., 2014. Vertical Greenery Systems (VGS) for energy saving in buildings: a review. Renew. Sustain. Energy Rev. 39, 139−165.

Perini, K., et al., 2017. Vertical greening systems evaporation measurements: does plant species influence cooling performances? Int. J. Vent. 16 (2), 152−160.

Riley, B., 2017. The state of the art of living walls: lessons learned. Build. Environ. 114, 219−232.

Svete, L.E., 2012. Vegetated Greywater Treatment Walls: Design Modifications for Intermittent Media Filters. Master Thesis. As: Department of Mathematical Sciences and Technology (IMT) Norwegian University of Life Sciences (UMB).

Treebox, 2013. Available at: <http://www.treebox.co.uk/news/rubens-at-the-palace-hotel-unveils-one-of-londons-largest-and-most-colourful-living-walls.html> (Accessed April 2017).

Treebox, 2015. Available at: <http://www.hortweek.com/world-first-vertical-rain-garden-installed-london-bridge/landscape/article/1365392> (Accessed April 2017).

Wolcott, S., 2015. Performance of green walls in treating brewery wastewater. 2015. International conference. Living walls and ecosystem services. University of Greenwich, London. 6−8 July 2015.

FURTHER READING

Tyson, R., 2011. Opportunities and challenges to sustainability in aquaponic systems. HorTechnology 21 (1), 6−13.

Vivers Ter, 2011. Available at: <https://regaber.com> and <http://www.v-ter.com> (Accessed April 2017).

Chapter 3.11

Green Roofs to Improve Water Management

Anna Palla and Ilaria Gnecco

Chapter Outline

INTRODUCTION

Green roofs are determinant for the hydrologic restoration in urban areas and an interesting alternative to more conventional practices; they provide a way for roofs to be converted into pervious areas and used beneficially rather than contributing to storm water management problems (Palla et al., 2009). The overall aim of this chapter is to analyze and discuss the interactions between green roofs and the storm water management (quantity and quality issues) in the urban environment. In the first section the hydrologic response of the green roof is compared to the one of the reference impervious rooftop in order to assess their performance in the quantitative management of storm water. In the second section the impact of these systems on the quality of storm water is examined in order to assess whether they may act as a sink or a source of pollutants. In the conclusion section, the need for future research on the effectiveness of green roofs in the storm water management are discussed together with suggestions for their integration in the water mitigation and climate adaptation plans in urban areas.

HYDROLOGIC PERFORMANCE

The hydrologic response of green roofs tends to mimic the natural one by restoring on-site infiltration, storage, and evapotranspiration. Green roofs operate by means of retention and detention processes throughout the

Nature Based Strategies for Urban and Building Sustainability.
DOI: https://doi.org/10.1016/B978-0-12-812150-4.00019-7

stratigraphy components. The retention process involves the permanent storage of storm water in the substrate and drainage layers and the subsequently lost for evapotranspiration due to the requirements of the vegetation. The detention process deals with the temporal storage of storm water and its slow release; the detention volume determines the attenuation and delay of storm water runoff peaks at the inflow into the drainage network. Fig. 1 illustrates the impact of green roofs in storm water management through the retention and detention processes.

Various metrics have been introduced in the technical and scientific literature to describe the hydrologic performance of green roofs by means of retention and detention indexes.

As for the retention indexes, the storm water volume reduction (i.e., overall retention) is generally assessed as a relative percentage difference between the rainfall and outflow volumes. The overall retention may be evaluated at a different temporal scale ranging from the annual to the single event temporal scales. The annual analysis is one of the most investigated hydrologic performances and surely the most consistent. The annual storm water volume reduction may range from 40% to 80% of the total rainfall volume with the actual magnitude of retention being a function of the green roof structure (the layout of layers and the related depths), the climatic conditions, and the amount of precipitation (Palla et al., 2010). In general, the thickness of the substrate, as well as its characteristics (such as void ratio, hydraulic conductivity, etc.), prove to be the most important factors affecting the volume retention (Czemiel Berndtsson, 2010). The vegetation, in terms of plant coverage has an effect in aiding overall retention through rainfall interception and mainly through evapotranspiration: greater areal plant coverage and water need, higher transpiring plants (Poë et al., 2015).

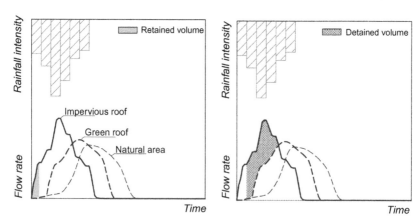

FIGURE 1 Hydrologic response of a green roof with respect to the reference impervious roof and the natural conditions. Graphical representations of the retained and detained volumes are also provided.

Focusing on storm water management, the green roof performance is analyzed at the event scale. In Fig. 2, a review of the overall retention at the event scale is reported for full-scale green roof systems. Data reported in Fig. 2 show that the retention varies between 12% and 74.6%; this range of variation is greater than what was reported for the annual analysis being mainly affected by the climatic variables at the event scale (Cipolla et al., 2016). The authors point out that the definition of the "event" has a notable influence on the number of considered events and consequently on the evaluated average retention performance at the event scale.

As for the detention indexes, the peak reduction and the outflow delay are generally assessed at the event scale as illustrated in Fig. 3. In order to calculate the peak flow reduction a model of the impervious roof has to be implemented so that the reference rooftop behavior is made available for comparison purposes.

The peak flow reduction is then calculated as the percentage difference between the outflow peak of the green roof and the reference impervious roof (see Fig. 3). Peak attenuation is mainly due to the storage capacity of the substrate (short of the field capacity) and the drainage layers. The peak reduction ranges between 60% and 80% with greater percent reductions associated with deeper green roofs (e.g., Bliss et al., 2009). The outflow delay is determined as the difference in time between the hydrograph and hyetograph centroids (see Fig. 3). The delay is due to the time needed for the substrate to start draining (wetting time) and to the extra-time for the complete drainage of the system, indeed the green roof outflow continues for an extended period of time after rainfall has stopped (see also Fig. 1). The observed green roof delay values, in the order of magnitude of the hour, are relevant in view of the usual concentration times of urban catchments (e.g., Palla et al., 2012).

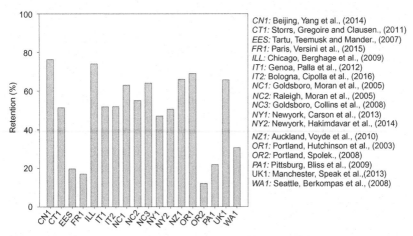

FIGURE 2 Average retention values observed at the rainfall event scale for full-scale experimental sites (after Cipolla et al., 2016).

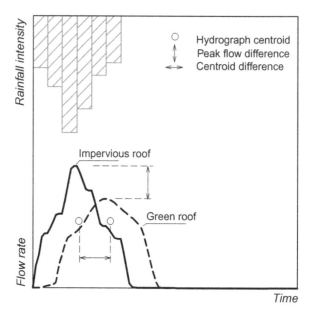

FIGURE 3 Hydrologic response of a green roof with respect to the reference impervious roof and graphical representations of the peak reduction and the outflow delay.

In the event-scale analysis, the soil water content of the substrates immediately before a rain event significantly influences both water retention and detention. Moreover, the regeneration of a green roof's retention and detention capacity is mainly due via evapotranspiration since soil water content depends upon soil—water characteristics and plant interactions (e.g., Berretta et al., 2014; Fassman and Simcock, 2011).

THE IMPACT ON WATER QUALITY

The impact of green roof on water quality is a relevant research topic, indeed it is widely recognized that the adsorption capacity and the leaching tendency of the substrate components as well as the phytoremediation ability of the plants need to be deeply investigated and accounted, in order to optimize the environmental benefits of green roofs.

Green roofs act as a storage device for both quantity and quality aspects. Focusing on the latter, pollutants from atmospheric deposition (mainly affected by urban catchment features) are accumulated on the surface during the dry period and in the substrate layer during low-intensity rainfall events that are fully retained within the green roof stratigraphy. Due to the evapotranspiration process, some of the dissolved constituents precipitate and bind to the substrate particles, subsequently precipitated components can be washed off by medium-to-high-intensity rainfall events and transported by the subsurface outflow to the drainage system and eventually to the receiving water body.

Due to the complexity and range of the hydrologic and chemical processes occurring within the green roof substrates during a rainfall event, it clearly emerges that the quality of leachate flow from green roofs depends on several factors including environmental conditions (e.g., precipitation regime), urban catchment features (e.g., land use characteristics), and the specific green roof characteristics (Palla et al., 2010). The green roof design plays a relevant role in affecting the pollutant leaching/adsorption mainly due to the following aspects:

- the building technique including the type of vegetation, depth, and composition of each layer of the stratigraphy (i.e., the soil layer and the drainage layer),
- the maintenance operations including irrigation, fertilization, and grass trimming.

Recently, experimental studies documented in the literature investigated the quality of green roof subsurface outflows in order to evaluate the green roof performance in reducing the pollutant load discharged into receiving water bodies during a rainfall event. Particular attention has been posed firstly on the leaching of nutrients and especially of nitrogen (e.g., Teemusk and Mander, 2007; Berndtsson et al., 2009) from the green roof stratigraphy since the increase in nutrient discharges may cause eutrophication into the receiving surface waterbodies (i.e., depletion of shallow water oxygen and potentially reducing the aquatic ecosystem). Later, the leaching of heavy metals have been investigated (e.g., Gregoire and Clausen, 2011; Alsup et al., 2011); metals have become a major concern since they tend to bioaccumulate in the water bodies thus potentially causing acute or chronic toxic effects on the aquatic ecosystem.

Particular attention needs to be posed when comparing water quality data collected from different experimental studies since the site-specific features (e.g., small plot) as well as the monitoring program characteristics (e.g., use of artificial rainfall) can significantly affect the quality data (e.g., Gnecco et al., 2013).

Focusing on nutrients, the concentration of nitrogen is linked to the properties of green roof elements as well as the maintenance practices. Bacterial activity in the substrate, fertilization operations, and wet/dry atmospheric deposition are source of nitrogen. Nitrogen forms in the green roof leachate are mainly nitrate-nitrogen (NO_3-N) and ammonium-nitrogen (NH_4-N); NO_3-N is negatively charged thus easily subjected to leaching while NH_4-N tends to be particle-bound to soil particles being positively charged. Table 1 provides an example of experimental data reported in the literature, in particular the concentration range of the different nitrogen form (nitrate-nitrogen, ammonium-nitrogen, nitrite-nitrogen ,and total nitrogen) observed in rainfall and green roof outflow. Experimental data points out the wide range of variation of nitrogen leachate and results are even contrasting: Berndtsson et al. (2009) and Gregoire and Clausen (2011) demonstrated that green roof applications contribute to decrease the total nitrogen concentration in storm water discharges, while other studies such as Aitkenhead-Peterson et al. (2011) observed a substantial release of nitrogen form.

The release of metals has been investigated to address the influence of green roof on storm water quality with respect to traditional rooftops, since it is well documented in the literature that metallic building components used in traditional rooftops (such as zinc claddings or rolled copper sheets) determine relevant concentration values of heavy metals even in the order of 1 mg L^{-1} (e.g., Gnecco et al., 2005). Table 2 reports the range of metals concentrations observed in rainfall and green roof outflow; based on data reported in the literature, zinc, copper, and iron have been selected being three heavy metals typically associated to storm water runoff in urban areas, while calcium and potassium are linked to the dissolution of substrate minerals. Results concerning the leaching of calcium and potassium are generally consistent thus confirming that green roofs act as source of these pollutant constituents even if the concentration values strongly depend on the nature of the substrate and the rainfall conditions. On the contrary, experimental data point out that the leaching of metals from green roof substrates is a complex biogeochemical process and no clear evidence on the behavior of metals emerges based on research study results (see Table 2).

In spite of the behavior of green roof as a sink/source of pollutant constituents it is not univocally defined, it can be assessed that the pollutant load associated with the green roof outflow is generally limited mainly when compared with the one observed in storm water runoff from traditional rooftops. Green roofs definitely have more positive than negative effects even on the quality issue of storm water.

CONCLUSION

Green roofs are increasingly used as a sustainable urban drainage system due to their positive impact on the storm water management. The mitigation consists in reducing the total outflow volume, while delaying the initial time of runoff and distributing the outflow over a longer time period. The retention, detention, and infiltration processes promote the occurrence of adsorption and dissolution mechanisms throughout the green roof components generally limiting the total pollutant mass delivered on an event basis.

However, several gaps in the actual developing and design approaches emerge including the need of extended field and experimental data collection over different climatic conditions and spatial scales, the modeling of the

TABLE 1 Concentrations of Nitrogen (Nitrate-Nitrogen, NO_3-N; Ammonium-Nitrogen, NH_4-N; Nitrite-Nitrogen, NO_2-N; Total Nitrogen, TN) Observed in Rainfall (Rain) and Green Roof Outflow (GR) Reported in the Literature

Site Location	NO_3-N (mg L^{-1})		NH_4-N (mg L^{-1})		NO_2-N (mg L^{-1})		TN-N (mg L^{-1})	
	Rain	GR	Rain	GR	Rain	GR	Rain	GR
Tartu (Estonia)	0.18–0.09	0.42–0.8	<0.015–0.22	0.12–0.33	n.a.	n.a.	0.6–1.3	1.2–2.1
Lund (Sweden)	1.03	0.07–0.11	1.08	0.08–0.15	n.a.	n.a.	2.65	0.59–2.31
Texas (TX, USA)	0.07 ± 0.03	6.6 ± 1.2	0.08 ± 0.008	0.10 ± 0.02	n.a.	n.a.	n.a.	n.a.
Storrs (CT, USA)	0.265[a]	0.369[a]	0.101[b]	0.023[b]	0.265[a]	0.369[a]	0.51	0.49
Eight sites (Estonia)	0.09–0.25	0.005–0.85	0.015–0.22	0.01–0.30	n.a.	n.a.	0.4–1.3	0.4–4.9
Taipei city (Taiwan)	1.24	9.01	0.28[b]	0.48[b]	n.a.	n.a.	n.a.	n.a.
Adelaide (Australia)	<1	1–40	1.01–1.10	1–16.5	0.07–0.10	0.02–3.5	n.a.	n.a.

Tartu (Estonia) data are Teemusk and Mander (2007); Lund, (Sweden) from Berndtsson et al. (2009); Texas (TX, USA) from Aitkenhead-Peterson et al. (2011); Storrs (CT, USA) from Gregoire and Clausen (2011); Eight sites (Estonia) from Teemusk and Mander (2011); Taipei city (Taiwan) from Chen (2013); Adelaide (Australia) from Beecham and Razzaghmanesh (2015).

n.a., not available.

[a]Indicates concentration values of NO_3 + NO_2-N.

[b]Indicates concentration values of NH_3-N.

TABLE 2 Concentrations of Metals (Zinc, Zn; Iron, Fe; Copper, Cu; Calcium, Ca; Potassium, K) Observed in Rainfall (Rain) and Green Roof Outflow (GR) Reported in the Literature

Site Location	Zn (mg L^{-1})		Fe (mg L^{-1})		Cu (mg L^{-1})		Ca (mg L^{-1})		K (mg L^{-1})	
	Rain	GR	Rain	GR	Rain	GR	Rain	GR	Rain	GR
Lund (Sweden)	0.015–0.043	0.043–0.279	0.015–0.128	0.053–0.075	0.001–0.009	0.032–0.149	2.66–5.33	5.33–77.33	0.137–0.823	1.921–4.118
Storrs (CT, USA)	0.006–0.059	0.006–0.054	n.a.	n.a.	<0.005	0.006–0.008	n.a.	n.a.	n.a.	n.a.
Edwardsville (IL, USA)	n.a.	0.034–0.137	n.a.	0.005–0.338	n.a.	n.a.	n.a.	n.a.	n.a.	n.a.
Singapore	n.a.	n.a.	n.d.	0.043–0.113	n.d.	0.037–0.056	<1	30.4–34.6	8–9	36.1–39.9
Genoa (Italy)	0.015–0.225	0.008–0.073	n.d.–0.401	0.084–0.200	0.016–0.092	0.013–0.101	0.628–2.68	3.54–5.52	0.236–0.867	2.56–3.88
Adelaide (Australia)	n.a.	n.a.	n.a.	n.a.	n.a.	n.a.	n.d.	4–140	0–0.46	0.52–6.45

Lund (Sweden) data are from Berndtsson et al. (2009); Storrs (CT, USA) from Gregoire and Clausen (2011); Edwardsville (IL, USA) from Alsup et al. (2011); Singapore from Vijayaraghavan et al. (2012); Genoa (Italy) from Gnecco et al. (2013); Adelaide (Australia) from Beecham and Razzaghmanesh (2015). n.d., not detected; n.a., not available.

hydrologic performance at the urban catchment scale, and the development of decision support tools incorporating green roofs in the urban planning process in order to promote their widespread implementation. Indeed green roofs appear to be particularly relevant in the storm water management plan because roof areas represent a significant part of the impervious surfaces in urban areas (between 40% and 50%; Versini et al., 2015). Modeling results at the urban catchment scale confirm the role of green roofs in restoring the critical components of the natural flow regime (e.g., Palla and Gnecco, 2015) even if a minimum implementation area is required in order to obtain noticeable hydrologic performance.

Finally, the observed hydrologic performance and storm water quality benefits suggests that green roofs should be largely adopted as a sustainable practice to deliver more climate-resilient and liveable cities. The ability of the green roofs, as modified natural systems, to absorb and rebound from weather extremes and continue to function (resilience) has to be nowadays recognized both in the technical regulations and policy making processes.

REFERENCES

Aitkenhead-Peterson, J.A., Dvorak, B.D., Voider, A., Stanley, N.C., 2011. Chemistry of growth medium and leachate from green roof systems in south-central Texas. Urban Ecosys. 14 (1), 17−33.

Alsup, S.E., Ebbs, S.D., Battaglia, L.L., Retzlaff, W.A., 2011. Heavy metals in leachate from simulated green roof systems. Ecol. Eng. 37 (11), 1709−1717.

Beecham, S., Razzaghmanesh, M., 2015. Water quality and quantity investigation of green roofs in a dry climate. Water Res. 70, 370−384.

Berndtsson, J.C., Bengtsson, L., Jinno, K., 2009. Runoff water quality from intensive and extensive vegetated roofs. Ecol. Eng. 35, 369−380.

Berretta, C., Poë, S., Stovin, V., 2014. Reprint of moisture content behaviour in extensive green roofs during dry periods: the influence of vegetation and substrate characteristics. J. Hydrol. 516, 37−49.

Bliss, D.J., Neufeld, R.D., Ries, R.J., 2009. Storm water runoff mitigation using a green roof. Environ. Eng. Sci. 26, 407−417.

Chen, C.F., 2013. Performance evaluation and development strategies for green roofs in Taiwan: a review. Ecol. Eng. 52, 51−58.

Cipolla, S.S., Maglionico, M., Stojkov, I., 2016. A long-term hydrological modelling of an extensive green roof by means of SWMM. Ecol. Eng. 95, 876−887.

Czemiel Berndtsson, J., 2010. Green roof performance towards management of runoff water quantity and quality: a review. Ecol. Eng. 36 (4), 351−360.

Fassman, E., Simcock, R., 2011. Moisture measurements as performance criteria for extensive living roof substrates. J. Environ. Eng. 138 (8), 841−851.

Gnecco, I., Berretta, C., Lanza, L.G., La Barbera, P., 2005. Storm water pollution in the urban environment of Genoa, Italy. Atmos. Res. 77 (1−4), 60−73.

Gnecco, I., Palla, A., Lanza, L.G., La Barbera, P., 2013. The role of green roofs as a source/sink of pollutants in storm water outflows. Water Res. Manag. 27 (14), 4715−4730.

Gregoire, B.G., Clausen, J.C., 2011. Effect of a modular extensive green roof on stormwater runoff and water quality. Ecol. Eng. 37, 963–969.

Palla, A., Gnecco, I., 2015. Hydrologic modeling of Low Impact Development systems at the urban catchment scale. J. Hydrol. 528, 361–368.

Palla, A., Gnecco, I., Lanza, L.G., 2009. Unsaturated 2-D modelling of subsurface water flow in the coarse-grained porous matrix of a green roof. J. Hydrol. 379 (1-2), 193–204.

Palla, A., Gnecco, I., Lanza, L.G., 2010. Hydrologic restoration in the urban environment using green roofs. Water 2, 140–154.

Palla, A., Gnecco, I., Lanza, L.G., 2012. Compared performance of a conceptual and a mechanistic hydrologic model of a green roof. Hydrol. Process. 26 (1), 73–84.

Poë, S., Stovin, V., Berretta, C., 2015. Parameters influencing the regeneration of a green roof's retention capacity via evapotranspiration. J. Hydrol. 523, 356–367.

Teemusk, A., Mander, Ü., 2007. Rainwater runoff quantity and quality performance from a greenroof: the effects of short-term events. Ecol. Eng. 30 (3), 271–277.

Teemusk, A., Mander, Ü., 2011. The influence of green roofs on runoff water quality: a case study from Estonia. Water Resour. Manag. 25 (14), 3699–3713.

Versini, P., Ramier, D., Berthier, E., de Gouvello, B., 2015. Assessment of the hydrological impacts of green roof: from building scale to basin scale. J. Hydrol. 524, 562–575.

Vijayaraghavan, K., Joshi, U.M., Balasubramanian, R., 2012. A field study to evaluate runoff quality from green roofs. Water Res. 46 (4), 1337–1345.

FURTHER READING

Bengtsson, L., 2005. Peak flows from thin sedum-moss roof. Nord. Hydrol. 36, 269–280.

Berghage, R., Beattie, D., Jarrett, A., Thuring, E., Razaei, F., O'Connor, T., 2009. GreenRoofs for Stormwater Runoff Control. EPA/600/R-09/ 026.

Berkompas, B., Marx, K., Wachter, H., Beyerlein, D., Spencer, B., 2008. A study of green roof hydrologic performance in the cascadia region. In: Low Impact Development for Urban Ecosystem and Habitat Protection, Proceeding of the 2008 International Low Impact Development Conference, Westin Seattle, WA, USA, November16-19, 2008, 1-10. doi:10.1061/41009(333)8.

Carson, T.B., Marasco, D.E., Culligan, P.J., McGillis, W.R., 2013. Hydrological performance of extensive green roofs in New York City: observations and multi-year modeling of three full-scale systems. Environ. Res. Lett. 8, 1–13.

Collins, K., Hunt, W.F., Hathaway, J.M., 2008. Hydrologic comparison of four types of permeable pavement and standard asphalt in Eastern North Carolina. J. Hydrol. Eng. 13, 1146–1157.

Hakimdavar, R., Culligan, P.J., Finazzi, M., Barontini, S., Ranzi, R., 2014. Scale dynamics of extensive green roofs: quantifying the effect of drainage area and rainfall characteristics on observed and modeled green roof hydrologic performance. Ecol. Eng. 73, 494–508.

Hutchinson, D., Abrams, P., Retzlaff, R., Liptan, T., 2003. Stormwater monitoring two ecoroofs in Portland, Oregon, USA. In: Proceedings of the Greening Rooftops for Sustainable Communities, Chicago, IL, USA, May 2003.

Moran, A.C., Hunt, W., Smith, J., 2005. Green roof hydrologic and water quality performance from two field sites in North Carolina. Manage. Watersheds Hum. Nat. Impacts 1–12. Available from: https://doi.org/10.1061/40763(178)99.

Speak, A.F., Rothwell, J.J., Lindley, S.J., Smith, C.L., 2013. Rainwater runoff retentionon an aged intensive green roof. Sci. Total Environ. J. 461–462, 28–38.

Spolek, G., 2008. Performance monitoring of three ecoroofs in Portland, Oregon. Urban Ecosyst. 11, 349–359.

Voyde, E., Fassman, E., Simcock, R., 2010. Hydrology of an extensive living roof under subtropical climate conditions in Auckland, New Zealand. J. Hydrol. 394, 384–395.

Yang, W., Li, D., Sun, T., Ni, G., 2014. Saturation-excess and infiltration-excess runoff on green roofs. Ecol. Eng. 74, 327–336.

Chapter 3.12

Green Streets to Improve Water Management

Paola Sabbion

Chapter Outline

INTRODUCTION

Urban areas are covered with impervious surfaces, which contribute to increase rainwater runoff. In conditions of natural ground cover (0% impervious surface), runoff is less than 10%, whereas in dense urban areas (75%–100% impervious surfaces) runoff percentage can be up to 55% of rainfall. In these conditions vegetation planted in green streets can efficiently contribute to reduce runoff, increasing the rate and volume of water infiltration into the soil. Trees' canopy intercepts rainfall, and foliage can dispose up to 40% of rainfall through evapotranspiration. Pervious surfaces can slow the water flow to the ground, where plant roots increase the capacity and rate of infiltration, reducing surface flows and pollution (Bartens et al., 2008).

Moreover, untreated stormwater runoff disturbs natural hydrologic processes by causing habitat damage, toxicity, eutrophication, especially during the first hour of intense storm events when water flow has even greater pollutant concentrations. The reduction of the ecological quality of water is an emerging issue worldwide. This main objective passes through the control of urban stormwater runoff, the main source of urban water pollution. For these reasons, in many cities improving water quality and aquatic ecosystem health is a main goal, especially where conventional stormwater management approaches have failed to pursue environmental preservation due to a lack of

Nature Based Strategies for Urban and Building Sustainability.
DOI: https://doi.org/10.1016/B978-0-12-812150-4.00020-3

capacity to address the variations to the flow regime caused by conventional drainage. Green streets can contribute to tackle these issues since natural systems can filter pollutants improving water quality (Benedict and McMahon, 2001).

New approaches emerging worldwide are aimed to develop an integrated urban water management in cities, using green and sustainable technologies. Various strategies include very similar solutions under different names: stormwater Best Management Practices (BMPs) and Low Impact Development (LID) in the United States. and Canada; Blue-green Cities and Sustainable Urban Drainage Systems (SuDS) in the United Kingdom; Water Sensitive Urban Design (WSUD) in Australia; Low Impact Urban Development and Design (LIUDD) in New Zealand. The term, "stormwater control measures" (SCM) is emerging to define the wide variety of global terminology, which is still growing, recently including the concepts of nature-based solutions in Europe and of "sponge city" in China (Madsen et al., 2017).

BENEFITS OF STREETS PLANTING ON STORMWATER MANAGEMENT

Urban trees and hedges, pocket parks, rain gardens, and bioswales are the most important components of the new sustainable approach to stormwater management, since they tend to integrate green and blue infrastructure within urban planning, shifting from traditional control solutions to new approaches focused on local collection and distribution, slower flows, and increased permeability. Green streets are in fact part of green infrastructures, an interconnected network of green spaces able to conserve and provide natural ecosystem functions and related benefits (Benedict and McMahon, 2012). Green streets thus can be defined as a constructed network based on the combination of pervious surfaces, soil, and vegetation to reduce and treat stormwater runoff at its source (Thompson and Sorvig, 2007).

Green streets can incorporate soil, vegetation (trees, shrubs, and herbs), permeable pavements, and engineered systems to design more sustainable surfaces of streets, parking lots, and sidewalks in a context-adaptable process. This strategy is aimed to mimic the hydrologic regime typical of predevelopment conditions, in which runoff volume is reduced at the source through natural processes of infiltration and storage. Green streets, including different features, can significantly act on both the quantity and quality of urban impervious surfaces' runoff before it reaches water bodies (Davis et al., 2012).

Planting trees is a main strategy among a range of natural stormwater control measures. Trees in fact can intercept rainfall, remove a considerable amount of water from the soil via evapotranspiration, and enhance infiltration. Moreover, trees can be useful in dense urban fabric, since they provide important functions in a small footprint. Trees act as reservoirs of water absorbing and purifying rainfall. Canopy interception loss, transpiration, and improved infiltration are constant benefits provided at different times: canopy interception is significant during a storm event, while transpiration performance continues in the period of time between rain events (Berland et al., 2017).

Tree species can determine different canopy interception rates, depending on the amount of water collected and on tree seasonal changes. A closed-canopy forests interception loss can reach 18%−29% of total rainfall for broadleaf and approximately 18%−45% for coniferous forests (Berland et al., 2017). In fact, considering surface water storage capacity of 23 Californian species, it was found that conifers can store more water on leaves and stems than broadleaf (Xiao and McPherson, 2016). Moreover, there are important differences in surface interception by species with different characteristics. For example, *Fagus grandifolia* can intercept an average of approximately 500 L per storm event, while *Liriodendron tulipifera* can intercept 650 L (Van Stan et al., 2015). It is important to consider the characterization of how trees interact with stormwater in different geographic and climatic conditions, according to leaf-on period and storm event season.

CONTROL AND TREATMENT COMPONENTS FEATURED INTO GREEN STREETS

Green streets can be implemented both in new neighborhoods as well as by retrofitting existing streets through a design that incorporates different components in a network of green and blue infrastructures. They can include infiltration systems and vegetative biofilters. Most biofilters include a range of species, varying in size from rushes to large shrubs or small trees, native to the geographical context. The most suitable plants for street features are low-maintenance herbaceous species, which must be able to tolerate pollutants and periodic inundation (EPA, 2004).

Green streets components can exploit biofiltration systems directing stormwater runoff into a pervious and vegetated area for treatment. Green streets can make a stormwater infrastructure, where canopy intercepts and evapotranspires rain while roots take up soil moisture, increasing infiltration and metabolizing pollutants. Excess water that does not infiltrate the soil, exceeding the capacity of one system, can bypass it and flow into an existing street inlet. A large number of studies have explored the efficiency of green streets components in reducing runoff and pollutants (Table 1).

Green streets can include the following systems:

Stormwater tree pits: similar to traditional street tree pits, but engineered to have increased growing space, and to receive and treat runoff. Stormwater tree pits are useful in streetscape retrofits when existing soil is very

TABLE 1 Stormwater Management Measures Suitable in Green Streets and Main Performances (Based on The Low Impact Development Center, n.d.)

Stormwater Management Measure	Reduction in Runoff Volume	Reduction of Peak Flow	Recharge of Groundwater	Reduction of Pollutants
Stormwater tree trench	+	+	+ +	+ +
Infiltration trench	+	+	+	+
Vegetative buffer strip	+	+	+	+ +
Vegetated swale or bioswale	+	+	+	+ +
Rain garden	+ +	+ +	+ +	+ +
Stormwater basin	−	+ +	+ +	+
Stormwater wetland	−	+ +	−	+ +
Private garden/ landscaping	+ +	+ +	+ +	+ +
Permeable Pavement	+ +	+ +	+ +	+

+ +, very effective; +, effective; −, no impact.

compacted and underground space is limited. Connecting multiple tree pits by soil paths or drains can increase soil volume for both trees and stormwater management (Denman et al., 2016).

Stormwater tree trenches: a system of trees connected by an underground infiltration structure put under the sidewalk to manage the incoming runoff. This structure is composed by an excavated trench along the sidewalk, filled with gravel and topped off with soil. Runoff flows to the tree trench and is stored between the gravel, watering trees, and slowly infiltrating through the soil (Bartens et al., 2008).

Infiltration trenches: linear areas consisting of gravel or pervious pavement that absorb runoff in urban areas. Filled with rocks, they are designed to either infiltrate runoff or slow its flow into the sewer system. They should be integrated with sidewalks, curbs, and other street features. Usually buffer strips are combined with vegetated buffer strips to prevent channelization and erosion (New York Department of Design and Construction, 2005).

Vegetative buffer strips: linear areas of land with a vegetative cover. Filter strips are used to treat runoff from pervious surfaces such as roads and small parking lots. Dense vegetative cover facilitates sediment attenuation and pollutant removal. Effective for overland sheet flow, they are not very effective in case of concentrated and high-volume flows (EPA, 2004).

Vegetated swales or bioswales: long, shallow vegetated depressions or open grass-lined channels used to slow the speed of surface runoff and allow stormwater to infiltrate into the ground. They are used for both infiltration and filtration to provide pretreatment before runoff is discharged into treatment system (EPA, 2004).

Rain gardens: vegetated areas collecting runoff from impervious surfaces. A rain garden uses soil and herbs to slow down, removes pollutants from runoff, and allows it to infiltrate into the soil (EPA, 1999). This system can be applied to both new and existing contexts, to reduce the rate and quantity of stormwater entering the sewer system and minimizing localized flooding. It is suitable for highly-impervious areas, such as parking lots, road medians, and street right-of-ways (DNR, 2009) (Fig. 1).

Stormwater basins: vegetated depressions designed to store and infiltrate runoff. They include retention basins, or wet ponds, which collect and store runoff in a permanent pool that removes pollutants through biological uptake; detention basins, which collect and slow the flow before releasing it; and infiltration basins, designed to directly infiltrate runoff into soil (City of Philadelphia, 2017).

Stormwater wetlands: similar to retention basins, wetlands collect runoff and store it in a permanent pool. They are designed to imitate natural wetlands, also providing habitat for wildlife. Stormwater wetlands can bring a great aesthetic appeal, but they need specific conditions to be integrated into dense urban fabric (City of Philadelphia, 2017).

FIGURE 1 A rain garden along a commercial street (drawing by P. Sabbion).

TABLE 2 Average Effectiveness of Main Green Streets Components on Water Control (based on Liu et al., 2017; Ahiablame et al., 2012)

Stormwater ManagementComponent	Reduction of Runoff	Reduces Pollutants
Vegetated swale or bioswale	NA	TSS (30%–98%)TP (24%–99%)TN (14%–61%)Pb (75%)Zn (68%–93%)
Rain garden	48%–97%	TSS (47%–99%)TP (− 3% to 99%) NO_x–N (1%–83%)NH3–N (− 65% to 82%)TKN (26%–80%)TN (32%–99%)Cu (43%–99%)Pb (31%–98%) Zn (62%–99%), Fecal Coliform (71%–97%)Oil/Grease (83%–97%)
Permeable pavement	50%–93%	TSS (58%–94%)TP (10%–78%) NH3–N (75%–85%)TKN (75%–100%)Cu (20%–99%)Pb (74 to 99%)Zn (73%–99%)Fecal Coliform (98%–99%)

Private gardens: they can incorporate urban stormwater management features such as permeable pavement, trees, and native vegetation; providing rain gardens and saving water using detention practices such as rain barrels.

Permeable or porous pavements: although this system is a nonvegetated green infrastructure, it is a very important component as it allows water to infiltrate into the ground. They can include pervious asphalt, pervious concrete, and interlocking pavers. They have an upper porous surface and an underground stone or gravel reservoir for temporary storage. They can also be effective in nitrogen removal (Collins et al., 2010).

Different components may have different results on both runoff reduction and pollutants removal (Table 2). This is due to local design standards, installation quality, and local conditions (soils, climate, and vegetation type) differences (Liu et al., 2017).

GREEN STREETS AND STORMWATER MANAGEMENT IN CITIES

Green streets implementation is growing in cities worldwide as a strategy to counteract soil erosion and improve water quality, increasing soil water storage potential through runoff reduction and transpiration (Bartens et al., 2008).

Instead of traditional runoff management—discharging water from impervious surfaces directly into storm sewer systems as gutters, drains, and pipes into rivers and streams—green streets provide a more sustainable water management at a lower cost. In many cases, in fact, during heavy rainstorms, combined sewers receive high flows that treatment plants are unable to handle. The mix of stormwater and untreated wastewater may discharge directly into the waterways, causing combined sewer overflows (CSOs). Green streets implementation is a consolidated practice in many cities to manage rainwater runoff, to reduce CSOs, and promote urban revitalization.

In the city of Melbourne, Australia, Water Sensitive Urban Design (WSUD) practices are promoted through specific guidelines. In particular, according to the Program "202020 Vison," Australian cities aim to increase tree canopy cover by 20% by 2020. The city of Melbourne is pursuing to increase its tree canopy cover (which was 13% in 2014) to 40% by 2040. Through urban forestry the city will become a water sensitive city with increased liveability and healthier urban ecosystems (Dobbie, 2016).

In Asia many cities are widely using runoff infiltration as a stormwater management practice. In Tokyo, Japan—a city where stormwater management practices are well consolidated—there is an artificial infiltration system that is extremely effective on runoff control. In Bangkok, Thailand, the Bangkok Metropolitan Administration is focused on structural measures in terms of stormwater runoff management, and it is strongly implementing a set of measures involving the population in the construction of infiltration facilities at public and private scale of intervention (Saraswat et al., 2016).

In the United States many cities have a consolidated experience with stormwater management measures. To tackle the water issue, Chicago has recently started a great landscaping and streetscape program, the "Green Alleys" initiative. According to this strategy all city's alleys will be provided with permeable recycled pavement, growing vegetation along roads, in parking lots, and in urban open spaces. Each green alley will be constructed to allow rainfall runoff infiltrating flow into the subsoil (City of Chicago et al., nd).

New York City's initiative for stormwater management is aimed to a cost-effective reduction of CSOs and water quality improvement, while enhancing urban public space quality and values. High Performance Infrastructure Guidelines, delivered by the Department of Design and Construction in 2005, were oriented to the progressive implementation of BMPs into the city priority investments. In 2010, New York City released the NYC Green Infrastructure Plan, that is aimed to achieve better water quality by reducing CSO volume and capturing rainfall from 10% of impervious surfaces through green infrastructure (The City of New York, 2016).

Philadelphia launched the "Green City Clean Waters" initiative in 2011. It is a current program aimed to achieve federal water quality standards by managing runoff from impervious surfaces citywide utilizing green stormwater management practices. To do so, Philadelphia Water Department have worked with the Mayor's Office of Transportation and Utilities, the Streets Department, Philadelphia Parks and Recreation, and other public bodies and partners to develop detailed design guides for green streets implementation (City of Philadelphia, 2017).

Portland, a city characterized by a very high annual rainfall rate, has faced the issue since 1991, through a strategy aimed to improve walkability that resulted in Portland Pedestrian Design Guidelines, a national model for pedestrian design initiatives. Stormwater management is currently faced through the Green Streets Initiative, that is part of a wider long-range planning initiative, "Metro 2040." In 2007, the Portland City Council approved a green street resolution to promote and incorporate the use of green street facilities in public and private development (City of Portland, 2007).

Seattle in recent years has promoted street greening to retrofit residential streets (Fig. 2). In particular, Seattle's 2010 CSO Reduction Plan delivered a series of guidelines and pilot projects to reduce the amount of stormwater that enters the sewer system. Seattle's pilot Street Edge Alternatives Project was completed in 2001. Two years of monitoring have showed that it has reduced the total volume of runoff by 99% (Seattle Public Utilities, 2001).

In Washington, DC, within the planning program "SW Ecodistrict" (a mixed-use neighborhood plan that incorporates green streets with cycle-pedestrian greenways, green roofs, and urban agriculture), stormwater flow is controlled, treated, and recycled in a system of green and blue infrastructures. The city has also promoted the Green Streets Initiative to treat runoff before it reaches the city's main water bodies (National Capital Planning Commission, 2013).

FIGURE 2 Vegetated swale in the high-point neighborhood in Seattle, WA (Clarion Associates More: USEPA Environmental-Protection-Agency).

Finally, in Canada the city of Vancouver is also working to grow its green capital. In particular, the city aims to guarantee an urban greenway network 140 km long. Green streets will be designed to integrate rain gardens and other components to reduce surface flooding and treat runoff. The city is also developing an urban forest strategy, which will provide 150,000 trees by 2020 (Vancouver, 2013).

CONCLUSION

According to new stormwater control measures, green streets should be implemented at various spatial scales and integrated into the restoration of habitat corridors, including river basins and wetlands. To make this possible, it is crucial to reduce, control, and treat stormwater runoff as close to its source as possible. An integrated planning approach to urban stormwater management based on the implementation of green streets as stormwater infrastructure should be addressed to reduce, treat, and control stormwater runoff as close to its source as possible. CSOs reduction, reusing of stormwater promotion, water quality, and urban hydrology improvement are the main goals, resulting in cost-savings and property values increasing (New York City, 2012).

Appropriate strategies should consider the opportunity to downsizing the sewer infrastructure, resulting in significant cost-savings for municipalities and minimizing the use of potable water while enhancing the quality of urban and street landscape. The opportunity to involve residents in the construction of water management features on private properties could be a further step toward a more sustainable runoff management. A long-term efficiency of green streets for stormwater management should be valued according to storm events intensity and occurrence, seasonal changes, and watershed general conditions. A good planning capacity and constant maintenance activity are crucial to good success of the process. Researchers, city planners, and decision makers should work together to plan new green streets and to monitor existing ones to achieve better decisions in cities' watershed plans and projects.

REFERENCES

Ahiablame, L.M., Engel, B.A., Chaubey, I., 2012. Effectiveness of low impact development practices: literature review and suggestions for future research. Water Air Soil Pollut. 223, 4253–4273. Available from: https://doi.org/10.1007/s11270-012-1189-2.

Bartens, J., Day, S.D., Harris, J.R., Dove, J.E., Wynn, T.M., 2008. Can urban tree roots improve infiltration through compacted subsoils for stormwater management? J. Environ. Qual. 37, 2048. Available from: https://doi.org/10.2134/jeq2008.0117.

Benedict, M.A., McMahon, E.T., 2001. Green infrastructure: smart conservation for the 21st century.

Benedict, M.A., McMahon, E.T., 2012. Green Infrastructure: Linking Landscapes and Communities. The Conservation Found. Island Press, Washington, DC.

Berland, A., Shiflett, S.A., Shuster, W.D., Garmestani, A.S., Goddard, H.C., Herrmann, D.L., et al., 2017. The role of trees in urban stormwater management. Landscape Urban Plan. 162, 167–177. Available from: https://doi.org/10.1016/j.landurbplan.2017.02.017.

City of Chicago, Daley, R.M.-M., Byrne, T.G., Attarian, J.L., nd. The Chicago Green Alley Handbook. An Action Guide to Create a Greener, Environmentally Sustainable Chicago.

City of Philadelphia, 2017. Green Stormwater Infrastructure Tools.

City of Portland, 2007. Tour map of some of Portland's Green Streets.

Collins, K.A., Hunt, W.F., Hathaway, J.M., 2010. Side-by-side comparison of nitrogen species removal for four types of permeable pavement and standard asphalt in Eastern North Carolina. J. Hydrol. Eng. 15, 512–521. Available from: https://doi.org/10.1061/(ASCE) HE.1943-5584.0000139.

Davis, A.P., Stagge, J.H., Jamil, E., Kim, H., 2012. Hydraulic performance of grass swales for managing highway runoff. Water Res. Spec. Issue Stormwater urban Areas 46, 6775–6786. Available from: https://doi.org/10.1016/j.watres.2011.10.017.

Denman, E.C., May, P.B., Moore, G.M., 2016. The potential role of urban forests in removing nutrients from stormwater. J. Environ. Qual. 45, 207–214.

DNR, 2009. Iowa Stormwater Management Manual.

Dobbie, M., 2016. Greening Cities with an Urban Forest across Both Public and Private Domains. The Nature of Cities.

EPA, 2004. Stormwater Best Management Practice Design Guide:

EPA, 1999. Storm Water Technology Fact Sheet: Vegetated Swales.

Liu, Y., Engel, B.A., Flanagan, D.C., Gitau, M.W., McMillan, S.K., Chaubey, I., 2017. A review on effectiveness of best management practices in improving hydrology and water quality: needs and opportunities. Sci. Total Environ. 601–602, 580–593. Available from: https://doi.org/10.1016/j.scitotenv.2017.05.212.

Madsen, H.M., Brown, R., Elle, M., Mikkelsen, P.S., 2017. Social construction of stormwater control measures in Melbourne and Copenhagen: a discourse analysis of technological change, embedded meanings and potential mainstreaming. Technol. Forecast. Soc. Change 115, 198–209. Available from: https://doi.org/10.1016/j.techfore.2016.10.003.

National Capital Planning Commission, 2013. SW Ecodistrict.

New York City, 2012. The State of the Harbor.

New York Department of Design and Construction, 2005. High performance infrastructure guidelines: best practices for the public right-of-way: New York City, October 2005. New York City Department of Design + Construction: Design Trust for Public Space, [New York].

Saraswat, C., Kumar, P., Mishra, B.K., 2016. Assessment of stormwater runoff management practices and governance under climate change and urbanization: An analysis of Bangkok, Hanoi and Tokyo. Environ. Sci. Policy 64, 101–117. Available from: https://doi.org/ 10.1016/j.envsci.2016.06.018.

Seattle Public Utilities, 2001. Street Edge Alternatives.

The City of New York, 2016. Green Infrastructure Performance Metrics Report.

The Low Impact Development Center, n.d. Evaluation of best management practices for highway runoff control. Low Impact Development Design Manual for Highway Runoff Control (LID Design Manual).

Thompson, J.W., Sorvig, K., 2007. Sustainable Landscape Construction: A Guide to Green Building Outdoors. Island Press, Washington, DC.

Van Stan, J.T., Levia, D.F., Brett, J.R., 2015. Forest canopy interception loss across temporal scales: implications for urban greening initiatives. Prof. Geogr. 67, 41−51. Available from: https://doi.org/10.1080/00330124.2014.888628.

Vancouver, C., 2013. City greenways: Improving connections across Vancouver [WWW Document]. <http://vancouver.ca. http://vancouver.ca/streets-transportation/city-greenways.aspx> (accessed 29.05.17).

Xiao, Q., McPherson, E.G., 2016. Surface water storage capacity of twenty tree species in Davis, California. J. Environ. Qual. 45, 188−198. Available from: https://doi.org/10.2134/jeq2015.02.0092.

Chapter 3.13

Vertical Greening Systems as Habitat for Biodiversity

Flavie Mayrand, Philippe Clergeau,
Alan Vergnes and Frédéric Madre

Chapter Outline

INTRODUCTION

Cities are inhospitable for many species and contain habitats deeply modified or created by human activities but in some ways similar to more natural ones (Lundholm, 2015). Different ways exist to modify the built environment and make it more compatible with biodiversity. Among them, vertical greening systems (VGS) show a great potential due to the large wall surface that is currently available in cities (Darlington, 1981) and expected to be by 2030 (World Bank, 2009). This chapter reviews biodiversity of each wall system and its specific drivers. Perspectives for design and research are proposed to greatly improve the walls' habitat quality and their biodiversity.

FLORA

By studying spontaneous flora under temperate and tropical climates, publications insist on the high potential of walls for supporting spontaneous biodiversity. Li et al. (2016) inventoried 159 species belonging to 77 families on

Nature Based Strategies for Urban and Building Sustainability.
DOI: https://doi.org/10.1016/B978-0-12-812150-4.00021-5

tropical retaining walls. Under the European latitude, studies followed this similar trend by inventorying between 207 and 300 species of vascular plants and 60 mosses (Duchoslav, 2002; Lisci et al, 2003).

They are mainly native, common, and similar to the ground-level ruderal greenery systems (e.g., Cervelli et al., 2013), forest understories and clearings (Láníková and Lososová, 2009). Invasive species represent a variable part of taxa, from 6.9% to 35%. Jim (2014) showed that four plant types exist according to their occurrence on walls and their rooting habits: key stones species which denote the signature wall cover; companion species with similar rooting traits as key species allowing them to grow successfully on walls but with a lower occurrence; accessory species and incidental species which are fewer and do not show any adaptive traits to grow on vertical habitats and occur only by chance dispersal. Key stone species are very few and belong to few families (e.g., Asteracea, Poacae) (Li et al, 2016; Nedelcheva, 2011). Woody individuals dominate under tropical climate (Jim and Chen, 2010) but remain rare under more temperate conditions (Qiu et al., 2016).

Walls are also habitats for climbing and cascading plants. Recent publication takes an inventory of climbers for green facades (GF) (Dover, 2015; Jim, 2015b). They can grow up the wall by twining around support materials ("indirect climbers") or climb up the wall directly through adhesive pads ("direct climbers"). Ten direct climbers have been shortlisted under tropical conditions. Under temperate climate, Dover (2015) mainly indexed deciduous species belonging to eight genera with *Hedera* as the most well-known. All are light to shade-tolerant making them suitable in cities where exposure is conditioned by architecture. Indirect climbers perform much better than direct climbers overall (Fig. 1). Plants usually used for climbing could also be used for cascading down the wall (Van Bohemen et al., 2008). Very few species have been proposed up to now to do so (e.g., *Parthenocissus quinquefolia*).

Living walls (LWS) allow the integration of a wide variety of plants compared to other systems, by covering the whole wall surface with different life forms and in different states of development. For instance, the vertical garden at the Caixa Forum Museum in Madrid was designed with 250 different species, from mosses to shrubs (Beatley, 2011). Evergreen and perennial species are preferred and chosen according to the desired aesthetic effect in terms of variation in color, texture, foliage forms and density, vitality, and growth. Although recent studies attempt to select wild plants adapted to specific climatic conditions (e.g., Martensson et al., 2014), exotics are nearly systematically selected and spontaneous species eliminated by maintenance. In continuous systems, a large number of plants incurs a high probability of poor performance and more need for replacement (Jim, 2015a).

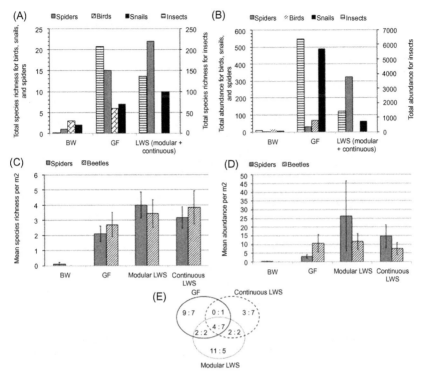

FIGURE 1 Performance grade of 20 climbers in relation to the establishment rate, growth rate, growth density, and flowering. *From Jim, C.Y., 2015b. Assessing growth performance and deficiency of climber species on tropical greenwalls. Landscape Urban Plan. 137, 107–121.*

FAUNA

Research is still scarce but shows that all forms of green walls could be habitats for fauna. Steiner (1994) inventoried 60,000 taxa of invertebrates on stony walls, and over 10% of the British spiders are mural species (Darlington, 1981). Among arthropods, good dispersers such as winged insects and species carried by the wind such as spiders are over represented. Studies on snails and arthropods (Chiquet, 2014; Madre et al., 2015) found that the use of walls by fauna is dependent on the resource provided (e.g., nesting habitat, food, and protecting areas). VGS generally shelter a much more abundant and diverse fauna than bare walls (Fig. 2A and B), and they offer more habitat to preys and predators so as the plant cover is diverse and litter accumulates (e.g., by plants decaying). Thus, modular LWS show the highest occurrence of beetles and spiders (Fig. 2C and D), but GF are also alternative habitats for generalist and common invertebrates, but also for rare and specialist species. GF are also highly visited by specialist flower visitors (e.g., honeybees, bumble bees, bristly flies, and butterflies) for resources

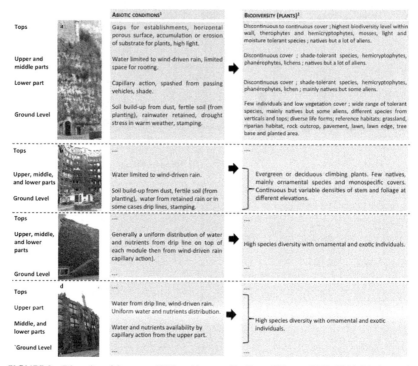

		Abiotic conditions¹	Biodiversity (plants)²
Tops		Gaps for establishments, horizontal porous surface, accumulation or erosion of substrate for plants, high light.	Discontinuous to continuous cover ; highest biodiversity level within wall, therophytes and hemicryptophytes, mosses, light and moisture tolerant species ; natives but a lot of aliens.
Upper and middle parts		Water limited to wind-driven rain, limited space for rooting.	Discontinuous cover ; shade-tolerant species, hemicryptophytes, phanérophytes, lichens ; natives but a lot of aliens.
Lower part		Capillary action, spashed from passing vehicles, shade.	Discontinuous cover ; shade-tolerant species, hemicryptophytes, phanérophytes, lichen ; mainly natives but some aliens.
Ground Level		Soil build-up from dust, fertile soil (from planting), rainwater retained, drought stress in warm weather, stamping.	Few individuals and low vegetation cover ; wide range of tolerant species, mainly natives but some aliens, different species from verticals and tops; diverse life forms; reference habitats: grassland, riparian habitat, rock outcrop, pavement, lawn, lawn edge, tree base and planted area.
Tops		...	...
Upper, middle, and lower parts		Water limited to wind-driven rain.	Evergreen or deciduous climbing plants. Few natives, mainly ornamental species and monospecific covers.
Ground Level		Soil build-up from dust, fertile soil (from planting), water from retained rain or in some cases drip lines, stamping.	Continuous but variable densities of stem and foliage at different elevations.
Tops		...	...
Upper, middle, and lower parts		Generally a uniform distribution of water and nutrients from drip line on top of each module then from wind-driven rain capillary action).	High species diversity with ornamental and exotic individuals.
Ground Level		...	...
Tops		...	...
Upper part		Water from drip line, wind-driven rain. Uniform water and nutrients distribution.	High species diversity with ornamental and exotic individuals.
Middle, and lower parts		Water and nutrients availability by capillary action from the upper part.	
Ground Level		...	...

FIGURE 2 Diversity of fauna on VGS and bare walls (from Chiquet, 2013, 2014; Madre et al., unpublished results). *BW*, bare wall; *GF*, green facade; *LWS*, living wall systems.

(nectar, pollen) (Garbuzov and Ratnieks, 2014). Besides, the abundance of detritivores, herbivores, and predators depends on the climbing species and the wall age (Chiquet, 2014).

Little information is available on the benefits to vertebrates of green walls. Stony walls have long been described as roosting sites and habitats for cliff-nesting species (lizards and birds) (Wheater, 2015). No data are available for LWS from our knowledge but GF supply foods (e.g., invertebrates and berries), perching, and breeding sites for birds (Chiquet et al., 2013). A seasonal effect is observed between evergreens and deciduous with more birds inventoried on evergreen facades in winter. Their function as nesting sites for birds remains nonassessed. Mammals (bats, voles, shrews) are likely to exploit walls but too few studies investigated these taxa (Johnston and Newton, 2004).

LOCAL DRIVERS

Vertical surfaces are some of the most extreme habitats. As steep slope and new building materials prevent substrates from accumulating, limited microhabitats are available for plants and animals (Dover, 2015; Lisci et al., 2003). They are cracks, open joints, ledges with quantities, and types of released nutrients from blocks according to (1) structural factors of walls (age, hardness of construction materials, inclination, aspect, color, heterogeneity of material, occurrence of intersection lines between vertical surfaces or vertical and horizontal surfaces), (2) weathering (exposure, wind, moisture), (3) maintenance intensity (e.g., cleaning, gardening, pruning), (4) pollution and interaction with animals (Jim, 2013; Wheater, 2015). Walls are consequently habitats where species with wide ecological amplitude and stress-tolerance coexist (Duchoslav, 2002; Jim and Chen, 2010; Madre et al., 2015). Overall, species are light to shade-tolerant, indicating either warmly dry or freshly moist soils according to the similarity of VGS to natural habitats (vegetated rainfall or cliff) (Fig. 2E). They also reveal as rather nutrient poor to a moderately rich habitat. As common species form a significant part of the species pool on walls, a weak correlation between the flora composition and habitat attributes is reported (Qiu et al., 2016). Therefore, stochastic events critically influence the floral biodiversity of stony walls.

Wall bases, middle parts, upper parts, and tops are inconsistently impacted by local factors leading to heterogeneous covers on walls (Fig. 3). Tops are more subjected to erosion and accumulation, supplying nutrients to plants (Wheater, 2015). Wall bases accumulate substrate from roads and pavements, but they receive less incident radiation than tops and upper parts.

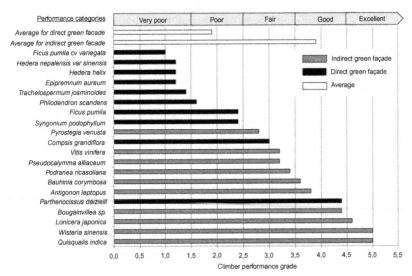

FIGURE 3 Abiotic specifications and plant biodiversity within stony walls (A), GF (B), modular LWS (C), continuous LWS (D) (from [1]Dover, 2015; Wheater, 2015; [2]Duchoslav, 2002; Jim, 2015a,b; Jim and Chen, 2010; Nedelcheva, 2011; Qiu et al., 2016; modified from Dover, 2015, photographs ©Madre).

They are also the wettest part of the walls gaining moisture from the soil through runoff and capillary action. Vertical surfaces also show a distinct response to exposure and maintenance.

Local drivers impact the richness and abundance of plants in different manners (Table 1). Wall size positively influences mature tree occurrence while habitat connectivity mainly drives smaller vascular plants (herbs, shrubs, and tree seedlings), and lichen and mosses are related to the resource supply. The persistence of pioneer species demonstrates that abiotic conditions tend to remain quite stable over-time turning initial and specialist colonizers into long-term residents (Jim, 2014). Ecological conditions in VGS are simpler because they are driven by engineering. In LWS, damp and fresh conditions are produced by a frequent, steady and widely dispersed drop-by-drop irrigation throughout the entire green wall (see Chapter 2.2: Vertical Greening Systems: Irrigation and Maintenance), evapotranspiration, and shade (see Chapter 3.1: Energy Performances and Outdoor Comfort).

THE LANDSCAPE CONFIGURATION

Very few studies investigate how surroundings may influence, or be influenced by, wall biodiversity. Research on stony and masonry walls showed that walls closed to existing plant and animal communities tend to be quickly colonized, affecting richness, abundance as well as life forms in plant assemblages (Jim, 2014; Li et al., 2016). Herbs, shrubs, and tree seedlings seem to be mainly determined by the connectivity of green walls with their surroundings while the lichen-moss strata and mature trees might be more related to resource supply and habitat size respectively (Table 2). In cities, the large pool of plant species disperses mainly by wind and animals (birds, bats, ants) (e.g., Li et al., 2016). The contiguity of urban areas generates continuous supplies of seeds to walls (Jim and Chen, 2010) and leads to the occurrence of mainly accidental species (Duchoslav, 2002). Isolated walls show a slower vegetation development than those forming a part of networks, which suggests that enhanced environmental conditions in surrounding increase plant biodiversity of walls (Shimwell, 2009).

Green walls' microfauna is also significantly influenced by the landscape configuration. In their study, Madre et al. (2015) showed the percentage of green areas around green walls had a significant and positive impact on the richness and abundance of taxa with low dispersal abilities. In a lesser extent, landscape configurations appeared to impact communities of microfauna with strong dispersal capabilities (spiders) by positively increasing their abundance in green walls. Research on birds found that direct GF are associated with 4.5 times more birds on immediate vegetation surrounding green walls than on vegetation surrounding bare walls (Chiquet et al., 2013). Those findings suggest that green walls could play a critical role in the ecological functioning of very fragmented areas by being a refuge for declining

TABLE 1 Drivers for Presence ([1]), Abundance ([2]), and Both ([3]), of Plant Life Forms on Retaining Walls

	Tree	Hacked Tree	Tree Seedling	Shrub	Herb	Lichen and Moss
Water-nutrient availability	—	—	Exposure[3], weathering, integrity[2]	Weathering, integrity[2]	Weathering[3], integrity[2]	Weathering, surface, moisture[2]
Habitat connectivity	Land use at toe[2]	Land use at crest[1]	Land use[2]	Land use at crest[1]	Land use[2]	
Structural characteristics	Joint type, wall ledge, masonry, material, stone size, uniformity, presence of cracks[2]		Maintenance[1], joint type, masonry material, beam[2]	Masonry material, stone shape, exposure, maintenance[3], joint type, beam[2]	Maintenance, exposure, beam[3], joint type, masonry material[2]	Maintenance[2]
Size of the wall patch	Wall width, elevation at toe[2]	Wall height[1] Wall width[2]	Wall width[1] Elevation at toe[2]	Wall width[1] Wall height, elevation at toe[2]	Wall height[2] Wall Elevation at toe[3]	Elevation at toe[2]

From Jim, C.Y., 2013. Drivers for colonization and sustainable management of tree-dominated stonewall ecosystems. Ecol. Eng. 57, 324–335; Jim, C.Y., Chen, W.Y., 2010. Habitat effect on vegetation ecology and occurrence on urban masonry walls. Urban For. Urban Green. 9, 169–178; Qiu, Y., Chen, B., Song, Y., Huang, Z., Wan, L., Huang, C., et al., 2016. Composition, distribution and habitat effects of vascular plants on the vertical surfaces of an ancient city wall. Urban Ecosyst. 19, 939–948.

TABLE 2 Recommendations to Design VGS to Function as a Habitat for Biodiversity (In Response to Disadvantages Reviewed in Jim, 2015a)

Feature	VGF	Disadvantages	Recommendations
Support	Modular LWS	Limited substrate volume for individual plants and limited chance for roots to extend and share substrates.	Enlarge rooting by proposing new containers.
Growing media	Modular LWS	Frequent and difficult soil management operations.	Stimulate the microbiological activity in substrates instead of supplying fertilizers.
			Choose plants with lower nutrient requirements, eventually with symbiotic interaction capacity.
			Leave a part of the decayed plants materials to accumulate organic matter.
Green coverage	GF	Limited choice of climbers for diverse climatic zones.	Additional research for native climbers needed.
	LWS	A large repertoire of exotic species.	Using native species and natural habitats as templates for assemblages.
	GF	Less likely to attract a wide range of wildlife.	Increase the functional diversity by selecting plants with different functional traits (e.g., flowering periods, plants visited by animals for nectar, pollen, or fruits, both deciduous and evergreen species, annual and perennial species).
			Allow the growth of spontaneous plants.
			Allow the coverage changing along the year through ecological processes (seasonal effect).
	LWS	Disturbance of companion wildlife because of frequent maintenance.	Decrease the intensity of maintenance by tolerating few spontaneous species with no impact on the system safety, and the seasonal effect.
	LWS	High probability of poor performance and more need for replacement	Choose native plants adapted to constraint environment.
			Plant diverse assemblages of vegetation according to the wall height and the gradient of ecological conditions (e.g., exposure and wind).
			Seedlings rather than pregrown plants.
Water	LWS	High consumption of irrigation water	Improve flows (e.g., water and nutrients) within the entire wall.
			Choose species with lower water requirements.
			Use rain water for irrigation.

LWS, Living Wall System; GF, Green Facade

species (Shimwell, 2009), and enhancing environmental conditions for wildlife in city center. Because a recent work showed that the walls' height impacts the dispersal of aerial plankton in towns (Vergnes et al., 2017); types and quality of VGS and surroundings need to be investigated with an objective of connectivity.

PERSPECTIVES

Cliff faces should be used as habitat template to render VGS even more compatible with biodiversity (Lundholm, 2015). For them to function as a habitat involves enhancing the habitat quality by incorporating abiotic heterogeneity, introducing a more diversified plant cover, and enhancing the water supply management (Wheater, 2015).

Stony and masonry walls are the only systems incorporating significant heterogeneity in abiotic conditions. Water availability for plants is only due to the porosity of the building materials, but the local drivers participate in high biodiversity levels when maintenance is cautious. Then, stony walls remain interesting for biodiversity in urban planning projects. VGS are more contrasted. GF are not satisfying enough in terms of heterogeneity of abiotic conditions and in general, very few species are used, leading to monospecific covers. Contrary to GF, LWS are largely diversified plant systems, but exotic and ornamental taxa as well as systematic removal of spontaneous species could nonetheless be counterproductive for biodiversity. In addition, LWS provide homogeneous abiotic conditions whereas water retention capability can differ within the wall areas depending on the mat material (Jorgensen et al., 2014). In combination with a high maintenance, they are low hospitable systems for spontaneous plants and animals except for invertebrates. In modular LWS, the habitat function is also likely to be controlled by the container type and the actual substrate quantity available for species.

Opportunities exist to enhance ecological functioning at wall scale. Wheater (2015) suggested changing materials and design to provide breeding habitat, food, resting and roosting surfaces, hibernating and aestivating cover. Without references in the literature, we suggest new designs and practices (Table 2). Supporting engineering, research needs to be done on the value of green walls for vertebrates, according to seasonality, local drivers, and landscape configuration. Biotic functioning of substrates should also be investigated in order to achieve their ecological performance. Even if some exotic plants can be beneficial to pollinators (Salisbury et al., 2015), prioritizing natives should be an optimal management strategy to secure plant survival and conservation of local biodiversity (Deguines et al., 2016). Qualifying the performance of climbing and nonclimbing plants under diverse climates is then required. Finally, understanding the role of VGS in the species flows remains critical. None of the studies examine the role of green walls for dispersal within fragmented areas. They are often said as part of stepping stone corridors for horizontal dispersal because of their linear structures, but no study has effectively determined the mechanisms at play and nor the connectivity with other urban green spaces.

CONCLUSION

VGS have recently been developed in response to many environmental and social challenges. In addition to stony and masonry walls, they are often presented as new stepping stone habitats for urban biodiversity. Wall biodiversity is driven by microclimatic conditions as well as landscape configurations. We stressed the value of each green wall system as a habitat for plants and animals, but to different extents. Major weaknesses of VGS would remain a low diversity of native plants linked to a high maintenance. Technical enhancements and research are needed to optimize the habitat function of green walls and species conservation in cities, and to maximize other ecosystem services.

ACKNOWLEDGMENTS

This study was supported by the French National Research Agency (ANR ECOVILLE).

REFERENCES

Beatley, T., 2011. Biophilic urban design and planning. In: Beatley, T. (Ed.), *Biophilic Cities. Integrating Nature Into Urban Design and Planning*. Island Press, Washington, DC, pp. 83–129.

Cervelli, E.W., Lundholm, J.T., Du, X., 2013. Spontaneous urban vegetation and habitat heterogeneity in Xi'an, China. Landscape Urban Plan. 120, 25–33.

Chiquet, C., 2014. The animal biodiversity of green walls in the urban environment. PhD thesis. Staffordshire University.

Chiquet, C., Dover, J.W., Mitchell, P., 2013. Birds and the urban environment: the value of green walls. Urban Ecosyst. 16, 453–462.

Darlington, A., 1981. Ecology of Walls. Heinemann Educational Books edition, London.

Deguines, N., Julliard, R., de Flores, M., Fontaine, C., 2016. Functional homogenization of flower visitor communities with urbanization. Ecol. Evol. 6, 1967–1976.

Dover, J.W., 2015. Green walls. In: Dover, J.W. (Ed.), Green Infrastructure. Incorporating Plants and Enhancing Biodiversity in Buildings and Urban Environments. Routledge, London.

Duchoslav, M., 2002. Flora and vegetation of stony walls in East Bohemia (Czech Republic). Preslia Praha 74, 1–25.

Garbuzov, M., Ratnieks, F.L.W., 2014. Ivy: an underappreciated key resource to flower-visiting insects in autumn. Insect Conserv. Div. 7, 91–102.

Jim, C.Y., 2013. Drivers for colonization and sustainable management of tree-dominated stone-wall ecosystems. Ecol. Eng. 57, 324–335.

Jim, C.Y., 2014. Ecology and conservation of strangler figs in urban wall habitats. Urban Ecosyst. 17, 405–426.

Jim, C.Y., 2015a. Greenwall classification and critical design-management assessments. Ecol. Eng. 77, 348–362.

Jim, C.Y., 2015b. Assessing growth performance and deficiency of climber species on tropical greenwalls. Landscape Urban Plan. 137, 107–121.

Jim, C.Y., Chen, W.Y., 2010. Habitat effect on vegetation ecology and occurrence on urban masonry walls. Urban For. Urban Gree. 9, 169–178.

Johnston, J., Newton, J., 2004. Building green. A guide to using plants on roofs, walls and pavements. Greater London Authority 2004.

Jorgensen, L., Bodin Dresboll, B., Thorup-Kristensen, K., 2014. Root growth of perennials in vertical growing media for use in green walls. Sci. Horticulturae 166, 31–41.

Li, X., Yin, X., Wang, Y., 2016. Diversity and ecology of vascular plants established on the extant world-longest ancient city wall of Nanjing, China. Urban For. Urban Gree. 18, 41–52.

Lisci, M., Monte, M., Pacini, E., 2003. Lichens and higher plants on stone: a review. Int. Biodet. Biodeg. 51, 1–17.

Lundholm, J.T., 2015. Urban cliffs. In: Douglas, I., Goode, D., Houck, M.C., Wang, R. (Eds.), The Routledge Handbook of Urban Ecology, second ed Routledge, New York, NY.

Láníková, D., Lososová, Z., 2009. Rocks and walls: natural versus secondary habitats. Folia Geobot. 44, 263–280.

Madre, F., Clergeau, P., Machon, N., Vergnes, A., 2015. Building biodiversity: vegetated façades as habitats for spider and beetle assemblages. Global Ecol. Conserv. 3, 222–233.

Martensson, L.M.L.-M., Wuolo, A.A., Fransson, A.M.A.-M., Emilsson, T.T., 2014. Plant performance in living wall systems in the Scandinavian climate. Ecol. Eng. 71, 610–614.

Nedelcheva, A., 2011. Observations on the wall flora of Kyustendil (Bulgaria). Eur. J. BioSc. 5, 80–90.

Qiu, Y., Chen, B., Song, Y., Huang, Z., Wan, L., Huang, C., et al., 2016. Composition, distribution and habitat effects of vascular plants on the vertical surfaces of an ancient city wall. Urban Ecosyst. 19, 939–948.

Salisbury, A., Armitage, J., Bostock, H., Perry, J., Tatchell, M., Thompson, K., 2015. Enhancing gardens as habitats for flower-visiting aerial insects (pollinators): should we plant native or exotic species? J. Appl. Ecol. 52, 1156–1164.

Shimwell, D.W., 2009. Studies in the floristic diversity of Durham walls. Watsonia 27, 323–338.

Steiner, W.A., 1994. The influence of air pollution on moss-dwelling animals: 1. methodology and the composition of flora and fauna. Rev. Suisse Zool. 101, 533–556.

Van Bohemen, H.D., Fraaij, A.L.A., Ottele, M., 2008. Ecological engineering, green roofs and the greening of vertical walls of buildings in urban areas. Ecocity World Summit Proc. 2008, 1–10.

Vergnes, A., Le Saux, E., Clergeau, P., 2017. Preliminary data on low aerial plankton in a large city center, Paris. Urban For. Urban Gree. 22, 36–40.

Wheater, C.P., 2015. Walls and paved surfaces: urban complexes with limited water and nutrients. In: Douglas, I., Goode, D., Houck, M.C., Wang, R. (Eds.), The Routledge Handbook of Urban Ecology, second ed. Routledge, New York, NY.

Word Bank, 2009. The World Bank urban and local governement strategy: concept and issues note. Urban strategy paper concept note. Washington, DC. World Bank.

Chapter 3.14

Green Roofs as Habitats for Biodiversity

Manfred Köhler and Kelly Ksiazek-Mikenas

Chapter Outline

INTRODUCTION

Green roofs are important examples of anthropogenic habitats (Oberndorfer et al., 2007). Their vegetation is similar to that of natural habitats, such as cliffs. These urban anthropogenic versions of rocky cliffs are characterized by drought, wind, and shallow, nutrient- poor substrates, which limit the species spectrum to only a few specialists (Larson et al., 2004). However, design and maintenance can extend the species spectrum. Green roofs offer a chance to develop solutions for buildings that offer habitats for specialized flora and fauna in cities.

Green roofs can act as "stepping stones" or connections for wildlife in urban areas. For example, some bird species can utilize green roofs as secondary feeding and nesting habitats. Both generalist and specialist insect species can be found on green roofs. The abundance of individuals is related to several key factors, such as the amount of green space and nesting structures in the surroundings, age of the roof and microscale niches provided by different materials (Kadas, 2006; MacIvor and Lundholm, 2011). These factors can all be important in supporting biodiversity.

An overview is important to understand how green roof projects are crucial aspects for supporting a diversity of higher trophic levels in urban areas. Now, the use of additional native and diverse plant selections opens up

Nature Based Strategies for Urban and Building Sustainability.
DOI: https://doi.org/10.1016/B978-0-12-812150-4.00022-7

further opportunities to design green roofs with higher ecological values. Long-term observations of plant species growing on green roofs can help us learn about ways the vegetation changes and provides resources to other organisms. The term "biodiversity" has been used since the late 1980s, (Harper and Hawksworth, 1994; Gyllin, 2004), and encompasses the following aspects:

- Species diversity: taxonomic abundance, or species richness;
- Ecosystem diversity: habitat diversity, and interaction with nonliving aspects of the environment;
- Genetic diversity: frequencies and distribution of genes and inherited traits.

All three aspects of biodiversity could be supported in urban habitats if green roofs were designed with these considerations in mind (Ksiazek, 2015).

The preferred plant species for extensive green roofs are succulents of the genera *Sedum* or *Allium*. Species from these two genera are adapted to the environmental conditions on the roofs, including high drought, heat, and wind. Over the past several decades, green roof installation has leaned toward simplicity, defined by the depth of growing media, which is typically about 10−12 cm. Intensive green roofs with 18 cm growing media or more, provide additional opportunities to support biodiversity. Several factors can help to enhance the species richness on green roofs. Typical extensive green roofs support a simple diversity of vegetation, which needs very low maintenance. However, these roofs could be developed as biodiverse roofs if small changes were made. Intensive roofs or roof gardens are garden structures accessible by the building inhabitants and used as an amenity space. The plant spectrum for these roofs with deeper growing media is much wider; it also includes shrubs and small trees. Green roofs in both categories can be developed into biodiverse roofs that provide a myriad of habitats for plants and animals.

Today, knowledge backed by data exists about the natural succession of communities on extensive green roofs. Green roofs can incorporate habitats for biodiversity in cities by increasing the availability of living space for a number of organisms (Francis and Lorimer, 2011). The remainder of this chapter will focus on the spatial resource of green roofs, which currently are only seldom used for biodiversity purposes. Case studies of extensive green roofs as well as an intensive green roof are included. This chapter will conclude with some ideas on how to support biodiversity on green roofs.

Species Diversity Case Studies

Recommendations regarding the best plant selection for green roofs to support biodiversity can only be provided within a regional context. In most cases, green roofs are man-made habitats planted with horticultural species in only a single thickness of media and, as such, do not contribute a great deal to local biodiversity. Although all around the world there are only a few common species that are specified on roofs, many more species are possible candidates (Olivares Esquivel, 2014). Increasing species richness should include a wide range of native and nonnative species and cultivars well-adapted for the particular climate region to enhance biodiversity on a broader level (Tan and Sia, 2008). Green roofs can contribute to ecosystem diversity by creating more microhabitats. This can increase beta-diversity when different habitat aspects such as dry, "hilly" structures contrast shallow structures, including ponds with higher moisture. Genetic diversity is the most difficult aspect of biodiversity to support, as it is a challenge to select and incorporate the right balance of genetically diverse target organisms into a green roof project.

When developed with all three aspects of biodiversity in mind, green roofs can contribute to novel urban ecosystems (Kowarik, 2011; Perrig et al., 2013). This requires high anthropogenic influence in construction, and later on requires minimal intervention and little maintenance. As of now however, the majority of existing green roofs do not match these targets.

Case studies highlight the potential of supporting biodiversity by starting with a variety of plant species on typical extensive green roofs. While increasing plant species diversity does not address all aspects of biodiversity, it provides a framework from which other experiments can be developed. The case studies presented here include vegetation surveys from the past few decades (Table 1).

- *Ufa, Ufafabrik, Berlin, Germany, (link: http://www.ufafabrik.de/de) (Köhler, 2006).* The oldest green roof at the Ufa complex was installed in the 1930s, and since the late 1980s, green roofs have been a typical element in all restoration work in this area. Ufa grows several types of extensive green roofs with a 10 cm depth of soil. Vegetation is both seeded and manually planted. There is mainly a wide range of *Sedum* species and cultivars, different perennial herbs, and small shrubs. Some species include garden cultivars of typical alpine plants. In the 1990s, planting indigenous species was not the main focus of such innercity projects. During the initial establishment years, these extensive green roofs were irrigated.
- *PLU, Paul—Lincke Ufer Restoration Project, Berlin, Germany (Fig. 1A).* This is a typical rental house from the beginning 20th century, retrofitted in 1985 with new apartments in the roof zone. The green roofs installed at this time have been monitored every year since that time (Köhler, 2006; Köhler and Poll, 2010). The same type of turf mats were installed on this roof with nine different roof pitches. Since installation, each year a minimum of one survey took place to count the number of plant species and their cover values (Köhler, 2006). At that time, the technology of

TABLE 1 Green Roof Case Study Sites Used to Explain Patterns of Plant Diversity Over Time and Possible Differences Between Green Roof Communities in Big and Small Cities

Acronym	Address/Location	Description of the Research Site	Established/Research, Since	Av. Number of Vascular Plants/Year	Remarkables
PLU	Berlin—Inner City, 4th story	Extensive, 10 subroofs, sloped roofs, about 900 m², 10 cm growing media	1985/since 1986	41, seeded and planted species together	No maintenance, annual observation since the beginning
Ufa	Berlin—Inner City, 2nd story	Extensive, nearly flat, more than 3000 m², 10 cm growing media	1985/since 1992	93, wide range of seeded species	Irrigation in the first years (1985−1991), use of seed from some attractive alpine vascular plants
HS 2	Neubrandenburg, suburban, 4th story	Extensive, flat, about 1000 m², 9 cm, 3 types of growing media	1998/ since 1999	28 seeded and planted species together	Seeded and spontaneous establishing on three different growing mediums, annual observation, and continual climate observation
HS 3	Neubrandenburg, suburban, 4th story	Extensive, flat, about 1000 m², 10 cm, 2 types of growing media, turf mats, and seeds	2001/ since 2001	55, small selection in the turf mats	Annual observation of plant development in relation to growing media, type of establishing, growing media, and run off measurements.
Torn	Tornesch, suburban, 3rd story	Semi-intensive, flat, 330 m², 12 and 18 cm, planted	1999 and 2015	82 species	92 species planted in 1999: no maintenance since 2002, a lot of dead biomass since that time

FIGURE 1   (A) Paul-Lincke Ufer green roofs from 1985; a view on three of the ten different subroofs in 2016. In the foreground, the dominant *Allium schoenoprasum* blooms. A steep-sloped roof is pictured on the left and dominated by *Sedum* species. (B) Tornesch intensive green roof in 2015, with 18 cm growing media depth. Rose and juniper species can be identified. (C, D) Two images of biodiverse green roofs with different structures and growing materials, seen in 2015 in Bunn, Czech Republic.

ready-made turf mats for extensive green roofs was developed. This was one of the first projects in which such thin vegetation layers were used to protect the growing media on a building against erosion. Today this is a typical technology in Germany.

- *HS2, HS3, Neubrandenburg, Germany.* These are two buildings at the University of Applied Sciences, Neubrandenburg, that were erected in 1999 and 2001. On both of these buildings, green roof research test plots have been installed and are continuously observed (Kohler et al., 2012). They are models of typical extensive green roofs constructed everywhere in Germany within the last 30 years.

- *Torn, Former Optima Headquarters, Tornesch, Germany (Fig. 1B).* In contrast to the majority of researched extensive roofs, in 1999 an intensive roof was installed to test plant performance on depths of 14 and 18 cm growing media that was both irrigated and maintained. The aim was to find robust plant species fit for the climate of northwestern Germany. Since a change in building ownership 10 years ago, this roof has received no maintenance. This change opened up the opportunity to compare the survival of plant species from the beginning to now and see how various plants survived without additional support. This research was conducted in the summer of 2015.

Species Richness Results From Selected Case Studies

The data collection methods used in these case studies were not completely similar, but they allow one to reflect on the typical patterns of succession on the green roofs of Germany since the 1980s.

The roof at PLU is similar to hundreds of extensive green roofs constructed throughout Germany between 1980 and 2000. At this time, it was typical for nonirrigated extensive green roofs to receive nearly no maintenance. Besides the dominant grass *Lolium perenne*, *Poa pratensis* sp. *angustifolia*, *Koeleria pyramidata*, *Festuca rubra*, *Poa compressa*, *Festuca ovina* s.str., *Allium schoenoprasum*, and *Sedum acre* were also included in the initial planting. The *Poa pratensis*, *Koeleria* sp. and *Festuca rubra* remained on the roof for a few years while *Poa compressa* has survived through today. Since 1991, the most abundant species has been *Allium schoenoprasum*, with more than 60% cover. These plant species provide resources to many insects that are adapted to urban environments. The vegetation that covers the roof today includes a small number of drought-, heat- and wind-resistant plants. The data reveal four distinct groups of plant species. The first group is made up of species exhibiting typical plant behavior of the early stages of establishment. The second group includes seeded species that died out after a few years. The third group is characterized by seeded species present in all years (i.e., *Poa compressa*, *Festuca ovina*, *Allium schoenoprasum*, *Sedum acre*, and *Bromus tectorum*). Finally, the fourth group is made up of a wide range of temporary species that colonize based on climate and other factors.

The data show that after about 28 years, there is a slight decline in cover of *Allium schoenoprasum*. All spontaneous colonizing plant species vary from year to year. *Bromus tectorum*, e.g., spontaneously colonized and became dominant from the first year on. Turf mat vegetation closes surface gaps from the beginning and prevents erosion. Half of the initially-planted species became the basic cover over the subsequent years. These results are in contrast to various other projects in which only the seed is used, such as on parts of HS3. In such cases, the plant community development still includes open spaces even after several years. This can provide opportunities for colonization by spontaneous plants, but has the negative consequence of potential wind erosion.

The data from PLU were supplemented with long-term surveys from Ufafabrik and roofs at the University of Applied Sciences in Neubrandenburg to document differences in plant species diversity in various growing media and in planted vs. seeded plant species (Table 1, Fig. 2).

Climate data sets for PLU from 1986−2011 were compared with climate data from Berlin-Dahlem to address the question of how monthly summer temperatures influence the plant species. The data from these 15 years show that a higher number of annual grasses were present with increasing hot summer conditions.

FIGURE 2 Comparison of the number of all vascular plants on four research sites throughout several years (horizontal lines: average species diversity for each site).

Of the four extensive green roofs, UFA had high mean species diversity at 93 species with a range of 49−131 species. The mean species diversity at HS3 was 54 species within a range of 43−72 species. PLU had between 25 and 56 species, with a mean diversity of 41 species. Species diversity ranged from 21 to 36 species at HS2 with the lowest mean diversity of the four roofs surveyed at 28 species. The data from these surveys reveals that there is oscillation in the number of plant species present at any site as a function of time. The reasons for these variations differ by site. Only in one case (PLU) was there a significant relationship between species diversity and high summer temperatures. This suggests that such plant selection at this location was appropriate for harsh and varied climatic conditions. Overall, the surveys show a wide range of species diversity. Despite differences between sites, the species richness remains relatively steady over time within a site, and climatic factors do not explain these differences overall. This can be interpreted optimistically; many plants species can survive on extensive green roofs under various climate conditions, including cold winters and dry or wet summers.

On the intensive roof "Torn" 92 plant species were monitored, with an average of 82 species present per year. Over the observation time period of 15 years, 66 species survived. An additional 16 species arrived spontaneously. Of the planted species, 17 moved out from their original test plots to other areas of the roof. The nearly full–coverage of the roof's surface by plant species can be considered an industry success and demonstrates that plant diversity can be achieved by incorporating the right plant species into the selection process (Köhler et al., in press). The plants that survived over time did so with very low maintenance. With the ability to support species with low competitive ability and increase urban niches and microhabitats, such intensive green roofs can offer habitat for local and perhaps even rare species of both plants and the animals that depend on them, thus supporting biodiversity.

To summarize these case studies (Table 1), the effects of soil depths and structures are important for biodiversity (i.e., at Torn), but (i.e., at Ufa), planting many species at the beginning and having high biodiversity in the surroundings also influences which species will be present. Maintenance is also important to support some of the slower-growing target species.

Connections Between Plant Species Richness and Faunal Benefits

Recently, Pille and Säumel (2017) conducted a meta-analysis of plant species diversity and habitat richness on typical extensive green roofs. They concluded that green roofs harbor a wide range of species and that richness varies between roofs. Some green roofs have fewer than five species, while others contain up to 60 species. The analysis also revealed that unique structures on green roofs could host up to ten microhabitats, from dry to wet, deep to shallow media, and local to nonlocal plants. Taken together, the diversity in plant species and habitats could attract local insects and other wildlife.

In contrast to simple extensive green roofs with only a single depth of soil, roofs intentionally built to support biodiversity are known by different names around the world. For example, the term "bioroofs" comes from Switzerland. This term describes the use of a wide range of soils from fine to rocky textures, the use of hilly structures to create microhabitats, and the use of seeds from local plant species (Fig. 3). In the United Kingdom, similar concepts include the idea of living roofs, e.g., that promoted by Gedge et al. (2010); including a high diversity of vegetation is one aspect of this concept. On the other hand, some roofs have extremely poor soils but are rich in microstructures to support other species. These projects are called brown roofs. Brown roofs are used by roof-nesting birds, including the black redstart. Although plant species are the most visible aspect of biodiverse roofs, there is a wide range of species beyond plants that can be supported. Green roofs can incorporate plants and habitats that support birds, wild bees, butterflies, beetles, fungi, and many other beneficial organisms.

Biodiverse roofs also play a part in city-wide green corridors and have been developed to do so throughout some European cities (Fig. 1C and D). Green roofs and facades with high plant diversity are used in migration

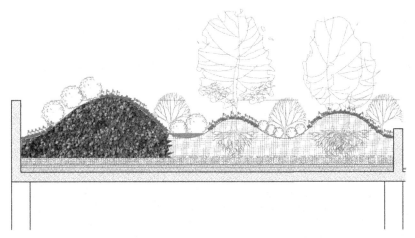

FIGURE 3 Principle scheme for a highly-biodiverse green roof includes different media depths, various growing mediums, and multiple layers of structural elements hosting floral and faunal opportunities.

pathways by species from other urban green open spaces. At the moment, questions remain about the usefulness of such green structures due to their relative small size. But in general, each habitat helps the individual organisms that use it. Long-term observations of established green roofs are needed to clarify details on how fauna use green roofs in urban corridors.

Exact mechanisms for establishing biodiverse green roofs are currently being researched by many institutions. Some basic concepts are similar across locations but others require different materials and design selections (Briz et al., 2015). To support biodiversity in the long term, it is important that a well-educated roof gardener performs annual maintenance and makes the necessary adjustments over the years. Long-term monitoring programs can deliver data on the degree of success for each project.

CONCLUSION

In conclusion, urbanization and human activities have caused fragmented natural habitats and declines in biodiversity. When incorporated into urban green corridors, green roofs can provide a habitat to reduce the effects of fragmentation for some plant and animal species. Biodiversity support means paying more attention to the details on green roofs. It means realizing that evolutionary processes occur within the roof as an ecosystem (Lundholm, 2015; Yang et al., 2015). Over time, these processes shape the plant and animal communities both on the roofs, and in the surrounding cities.

ACKNOWLEDGMENTS

Thanks to Daniel Kaiser for his support on several research projects included in this chapter and to Claire Diebel, who edited earlier versions.

REFERENCES

Briz, J., Felipe, I.D., Köhler, M., 2015. Green Cities in the World, second ed. Editorial Agricola, Espanola, MD, p. 358, ISBN 978-84-92928-30-9.

Francis, R.A., Lorimer, J., 2011. Urban reconciliation ecology: the potential of living roofs and walls. J. Environ. Manage. 92, 1429–1437.

Gedge, D., Grant, G., Kadas, G., Dinham, C., 2010. Buglife. Creating Green Roofs for Invertebrates. A Best Practice. Guide. 29p. https://www.buglife.org.uk/

Gyllin, M., 2004. Biological Diversity in Urban Environments Positions, values and estimation methods. Department of Landscape Planning. Alnarp. Phd.

Harper, J.L., Hawksworth, D.L., 1994. Biodiversity: measurements and estimation. Philosop. Trans. Royal Soc. London 345, 5–12.

Kadas, G., 2006. Rare invertebrates colonizing green roofs in London. Urban Habitats 4 (1), ISSN 1541-7115. http://www.urbanhabitats.org.

Köhler, M., Poll, P., 2010. Long-term performance of selected old Berlin greenroofs in comparison to younger extensive greenroofs in Berlin. Ecol. Eng. Elsevier B.V, ISSN: 0925-8574, doi:10.1016/j.ecoleng.2009.12.019.

Köhler, M., 2006. Long term vegetation research on two extensive green roofs in Berlin. Urban Habitats, Brooklyn Bot. Garden (USA) Vol 4 (1), Dec: 3–26.ISSN 1541-7115. http://www.urbanhabitats.org/v04n01/berlin_full.html.

Köhler, M., Ansel, W., Appl, R., Betzler, F., Mann, G., Ottelé, M., Wünschmann, S., 2012. Handbuch Bauwerksbegrünung. R. Müller Verlag, Köln, 250 S., ISBN 978-3-481-02968-5.

Köhler, M., Kaiser, D., Marrett-Fossen, M., in press. Plant developement on a semi-intensive roof Garden test installation – a 15 year survey in Northwest-Germany. UFUG.

Kowarik, I., 2011. Novel urban ecosystems, biodiversity and conservation. Environ. Pollut. 159, 1974–1983.

Ksiazek, K., 2015. The Potential of green roofs to support urban biodiversity. 103-126. In: Briz, J., Koehler, M., de Felipe, I. (Eds.), 2015. Green Cities in the World. Ediorial Agricola, Madrid, p. 478.

Larson, D.W., Matthes, U., Kelly, P., Lundholm, J.T., Gerrath, J.A., 2004. The Urban Cliff Revolution. Fitzhenry & Whiteside, Toronto, p. 198.

Lundholm, J., 2015. The ecology and evolution of constructed ecosystems as green infrastructure. Front. Ecol. Evol. 08 September.

MacIvor, S.J., Lundholm, J.T., 2011. Insect species composition and diversity on intensive green roofs and adjacent level-ground habitats. Urban Ecosyst. 14 (2), 225–241. Available from: https://doi.org/10.1007/s11252-010-0149-0.

Oberndorfer, E.C., Lundholm, J.T., Bass, B., Coffman, R., Doshi, H., Dunnett, N., Gaffin, S., Köhler, M., Liu, K., Rowe, B., 2007. Green roofs as urban ecosystems: ecological structures, functions and services. BioScience 57 (10), 823–833. www.biosciencemag.org http://www.aibs.org/bioscience-press-releases/resources/11-07.pdf.

Olivares Esquivel, E., 2014. Exploring the Potential of Mexican Crassulaceae Species on Green Roofs. PhD thesis, University of Sheffield.

Perrig, M.P., Manning, P., Hobbs, R.J., Lugo, A.E., Ramallo, C.E., Standish, R.J., 2013. Novel urban ecosystem and services. In: Hobbs, R.J., Higgs, E.C., Hall, C.M. (Eds.), Novel Ecosystems: Intervening in the New Ecological World Order. Wiley, New York, pp. 310–325.

Pille, L., Säumel, I., 2017: Arbeitsstand Biodiversitätseffekte. http://www.kuras-projekt.de/downloads/praesentationen-abschlussveranstaltung/ Potenzial von Regenwasserbewirtschaftung für Biodiversität in der Stadt, Technische Universität Berlin.

Tan, P.Y., Sia, A., 2008. A Selection of Plants for Green Roofs (second ed.). CUGE, Singapore. http://www.academia.edu/8893197/A_Selection_of_Plants_for_Green_Roof_in_Singapore.

Yang, Z., Liu, X., Zhou, M., Ai, D., Wang, G., Wang, Y., Chu, C., Lundholm, J., 2015. The effect of environmental heterogeneity on species richness depends on community position along the environmental gradient. Sci. Rep. 5, 15723. Available from: http://www.nature.com/articles/srep15723.

FURTHER READING

Ksiazek-Mikenas, K., 2017. The potential of green roofs to provide habitat for native plant conservation. PhD, Chicago.

Chapter 3.15

Green Streets as Habitat for Biodiversity

Eleanor Atkins

Chapter Outline

INTRODUCTION

Cities are defined by the buildings they contain and the roads that link them. These streets are essential to our modern transport infrastructure but represent sealed surfaces, designed to shed rainwater and surface run-off as quickly as possible. Surface water flooding is becoming an increasing issue worldwide due to climate change altering the intensity and frequency of rainfall events. However, with careful planning and design, streets present opportunities to retain and treat stormwater close to its source. If improved with elements of green infrastructure these bioengineered streets are generally termed as green-streets (Graham et al., 2015; Lukes and Kloss, 2008).

Green streets integrate landscaping, engineering, and biological processes to achieve multiple benefits. The term covers a wider remit than just the roadway itself, including: green roofs, green walls, and rain gardens. Other components include increased tree canopy, permeable pavements, bioretention pools and swales (Ahern, 2011; Barbosa et al., 2007; Biggs, 2003; Graham et al., 2015; Lukes and Kloss, 2008; Runfola and Hughes, 2014). The best schemes incorporate a range of habitats beneficial for wildlife and

Nature Based Strategies for Urban and Building Sustainability.
DOI: https://doi.org/10.1016/B978-0-12-812150-4.00023-9

251

for water management (Ahern, 2011) bringing urban wetlands and other wildlife friendly green spaces into our towns and cities (Fig. 1). With careful siting, they can link existing habitats to create green corridors with improved habitat connectivity (Graham et al., 2015).

This chapter examines the biodiversity value of elements of green streets and seeks to discuss their benefits to urban wildlife and the importance of their inclusion within city infrastructure.

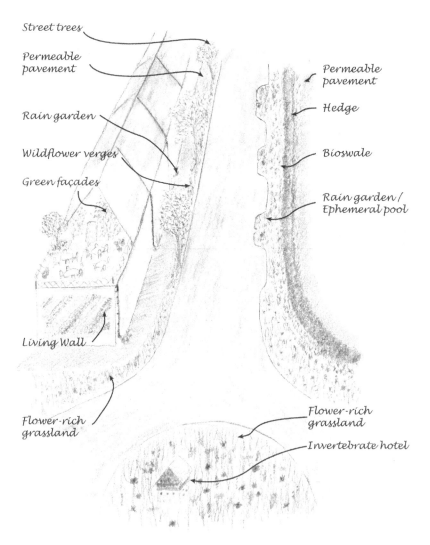

FIGURE 1 Incorporating a diverse range of habitat elements within green streets can increase aesthetic value and biodiversity value together with enhanced ecosystem service delivery.

GREEN ROOFS & GREEN WALLS

The potential for vegetated green roofs to provide space for wildlife is enormous, (Buglife, 2013; Grant, 2006) providing increased wildlife habitat, particularly for invertebrates and birds; some of which are rare or endangered (Anon n.d.[1]; Baldock et al., 2015; Biggs, 2003; Baumann, 2006; Brenneisen, 2005, 2006; Buglife, 2013; Graham et al., 2015; Kadas, 2006; Kadas and Gedge, 2016; Oberndorfer et al., 2007). For more information on biodiversity of green roofs see Chapter 3.14, Green Roofs as Habitat for Biodiversity. The presence of green walls benefits urban biodiversity, particularly for birds and invertebrates, and offers many other ecological services (Chiquet et al., 2012; Dover, 2015; Graham et al., 2015; Loh, 2008). For more information on biodiversity values of green walls see Chapter 3.13, Vertical Greening Systems as Habitat for Biodiversity.

RAIN GARDENS

Rain gardens are vegetated, permeable areas designed to retain rainwater following a rainfall event (Fig. 1). They provide a habitat for wetland plants, invertebrate cover, nectar for insects when suitable plants are included, foraging areas for birds and other wildlife, and wet vegetated areas provide habitats for amphibians. A range of species require ephemeral water features for part of their life cycle. If integrated as part of a green street scheme they offer potential for movement of species between other suitable habitats and support larger diversity and populations of species. Additional features such as invertebrate hotels (Fig. 1) can be included (Graham et al., 2015).

Bumblebees are in decline across the United Kingdom and rain gardens planted with native, nectar rich plant species provide food for insects such as butterflies, hoverflies, and bumblebees (Anon n.d.[2]; Graham et al., 2015) and can act as stepping stones between other urban habitats. Larger rain gardens are better for wildlife and can often offer areas of permanent water or ephemeral pools, which provide habitats for a range of species from invertebrates to herpetiles. All sizes offer benefits in terms of runoff reduction and wildlife value.

STREET TREES/THE URBAN FOREST

On a landscape scale, urban forests can be more biodiverse than the surrounding areas and, if well-managed, woody species provide benefits for a variety of wildlife including song birds (Alvey, 2006). Trees provide shelter, and a place to breed and forage for seeds, nuts, and berries (Kadir and Othman, 2012). The importance of planting native tree species is being recognized for reducing biotic homogenization between cities. Native species with local provenance should be favored as they offer greater value to wildlife, but planting a range of species is also important (Alvey, 2006; Snep et al., 2016); species that maximize fruiting and seed production for invertebrates and birds offer obvious benefits (Graham et al., 2015).

Urban forests can be important in promoting wider urban biodiversity (Alvey, 2006). Dead and decaying wood provide a habitat for mosses, fungi, and lichens and is particularly important for invertebrates, providing essential places to complete their life cycles. Veteran trees can provide cavities in which birds and bats can breed. A diverse urban forest offers resilience to disease and climatic change (Graham et al., 2015).

HEDGES

Hedges are human made structures composed of managed rows of trees or shrubs and are well known to support biodiversity in rural environments. Their composition, management and structure influence their suitability to support wildlife (Bellamy et al., 2000; Gelling et al., 2007; Hinsley and Bellamy, 2000; Kotzageorgis and Mason, 1997; Michel et al., 2006) and were designated as a priority habitat in the 1994 UK Biodiversity Action Plan (Biodiversity Reporting and Information Group, 2007). Little is known, however, of urban hedges in terms of species make-up, distribution and management, or their value to wildlife.

The linear nature of hedges helps to slow and intercept runoff on slopes. Inclusion of hedges within green streets design can assist in reduction of runoff but also increase biodiversity. Studies undertaken on urban hedges within the city of Stoke-on-Trent, UK found evidence of use by mammals in 63% of all study hedges. 87% of all hedges surveyed were used by birds with hawthorn (*Crataegus*) being particularly important for both mammal abundance and bird abundance and diversity (Atkins, in prep.). Faiers and Bailey (2005) studied canal-side hedges and noted that the quality and biodiversity value of urban hedges was lower than those of rural areas. Gosling et al. (2016) found no significant difference in the mean number of woody species found in urban and rural hedges but a difference in the component species with urban hedges scoring significantly lower in their wildlife and structural score. The mean number of invertebrate groups was not significantly different between rural and urban hedges, but different species assemblages inhabited urban hedges rather than rural. Adjacent hard surfaces had a detrimental impact on hedge biodiversity.

These studies suggest that urban hedges, even when poorly managed, do provide a habitat for wildlife. If such hedges were more sympathetically managed and utilized as a component of a larger green infrastructure scheme, they may offer high wildlife value and assist in runoff reduction as well as offer multiple ecosystem services.

PERMEABLE PAVEMENTS

Permeable pavements are designed to allow water to infiltrate through them to an aggregate base which provides runoff storage and pollution removal via filtering and adsorption. There are four types of permeable pavement: concrete, asphalt, permeable concrete pavers, and grid pavers (Lukes and Kloss, 2008). Potentially diverse microbial communities can develop in the geotextile below permeable pavements; which breakdown the polluting hydrocarbons from oils and convert them to sugars which they utilize as a food source. Providing a well aeriated soil is maintained, the oil conversion rate can be extremely high (Newman et al., 2006) and reduce water treatment costs. This contamination removal will benefit surrounding biodiversity and reduce the contamination of aquatic systems. Interplanting with nectar-rich, trample-tolerant species particularly in gravel and cellular blocks with a locally-sourced, appropriate wildflower mix (Graham et al., 2015) provides obvious benefits for nectar-feeding insects. If located adjacent to street trees permeable pavements can provide additional area for root growth and provide roots with additional air and water (Lukes and Kloss, 2008).

SWALES OR SEASONALLY WET DITCHES

Swales and ditches may provide a valuable habitat to both terrestrial and aquatic species (Biggs, 2003) and were found by Kazemi et al. (2011) to be much more biodiverse than mown grassland areas. They are likely to be more exposed to contaminants so they are likely to support only the more tolerant species, but they prevent damage to wildlife by removing pollutants from the water thus providing a benefit to biodiversity in downstream wetland areas. By definition, swales provide connectivity for flow of water and thus habitats, facilitating the movement of species such as amphibians and small mammals. Filter strips should be used in conjunction with swales and other elements creating a controlled flow of treated water with high wildlife value downstream from polluted water, which has low habitat value; mixing should be prevented. Filter strips planted with native wildflowers, tussocky grassland areas or native wetland species provide habitat, shelter, and a place to forage and breed for invertebrates, birds, mammals, and herpetiles. Shallow pools developing in undulating areas can offer wetland habitat for plants, invertebrates, reptiles, and amphibians (Biggs, 2003; Graham et al., 2015; Anon, n.d.[3]).

Kazemi et al. (2011) found that the invertebrate number, richness, and diversity of the bioretention swales they studied was significantly higher than conventional green spaces (gardens and lawned areas). An average of 64.5 species was recorded for bioretention swales and 23.13 for conventional green spaces. Their study indicates that maximum biodiversity can be provided for with a higher number of flowering plants and more diverse species assemblages particularly of mid stratum vegetation.

PONDS

Ponds and bioretention basins can vary in size or shape and this flexibility means they can easily be incorporated into street designs, carparks, roundabouts or traffic islands (Kazemi et al., 2009).

Studies by Kazemi et al. (2009) of the terrestrial invertebrates of bioretention basins in Melbourne, Australia suggested they have the potential to offer suitable habitat for invertebrate species, more so than other green spaces. They can provide habitat for wetland plants, invertebrate cover, nectar for insects when suitable plants are included, foraging areas for birds and other wildlife; and some species require ephemeral water features for part of their life cycle (Graham et al., 2015). More value to invertebrate species was offered in basins that had deeper plant/leaf litter. Maximum biodiversity could be achieved by including diverse microhabitats (such as varied ground topography, substrate type, and litter depths), planting more diverse taxa and managing sympathetically to facilitate greater leaf/plant litter coverage and depth (Kazemi et al., 2011, 2009). Including shallow water habitat is important as this is less affected by pollution than deep water habitat (Biggs, 2003).

In the United Kingdom, it is now thought that frogs are more common in garden ponds than in the countryside. Over half of the UK's ponds have been destroyed causing a 50% decrease in great crested newt populations and the extinction of 10% of breeding dragonfly species, therefore including ponds would create valuable habitat (Biggs, 2003). O'Brien (2015) studied twelve urban ponds over three years and found amphibians in seven of these ponds in 2010 and in eight ponds in 2011 and 2012. None of the ponds were found to have pollutants at levels harmful to amphibians. The study ponds were also found to have high habitat suitability index scores for great crested newts, and high habitat diversity suggesting that they would support a range of species.

VERGES & GRASSLAND/VEGETATED AREAS

Grassland is particularly important for wildlife. Pollinators have been declining, primarily due to the loss of flower-rich habitats, and urban areas contain much unused land including road verges and roundabouts prime for conversion to a flower-rich habitat (Blackmore and Goulson, 2014; Graham et al., 2015) (Fig. 1). It's conceivable that if more flower-rich areas are planted within cities, rarer species of bees may enter or reenter areas now urbanized. Blackmore and Goulson (2014) surveyed 30 urban wildflower plots and their results suggest that sowing wildflower seeds into areas such as grass verges can significantly increase the floral and pollinator diversity.

Vegetated areas in Glasgow, UK were investigated by Humphreys et al. (2012) who found that bird, moth, and butterfly abundance was high, including rare and protected species. Areas containing rank or unmown grassland showed a higher species richness than those without. The presence of a waterbody or wetland area within a green space increased the bird species richness.

Birds and mammals will forage in different lengths of grass for food. Long swards provide areas for egg laying, for pupae or larvae to over winter, and for bumblebees to nest. Flying insects may shelter during rain or extreme temperatures and roost overnight. Reptiles and amphibians will forage for insects and use it for cover while moving from site to site. Long swards beneath trees allow tree-feeding invertebrates a place to pupate and complete their life cycle (Graham et al., 2015). The vegetated areas may be cut in patches allowing for variation in structure and habitats along the length of the street if it is not possible to achieve this with distance away from the carriageway.

BENEFITS OF INTERCONNECTED GREENERY TO WILDLIFE

Amphibian populations are declining globally, and the importance of urban habitats for their survival is increasing for several species (O'Brien, 2015). Green streets can create both wet and dry habitat connectivity in the built environment permitting movement of reptiles and amphibians through this otherwise hostile landscape. Traditional drainage methods often contain barriers and traps such as drains and kerbs which are problematic for amphibians (Anon n.d.[3]), these types of barriers tend not to be as prevalent in green street designs.

Connectivity is a critical factor in the functionality of many green infrastructure initiatives. Urban streets need to link one place to another and stormwater management needs to flow eventually to the river network. Presently, the connectivity of built systems is well established but is lacking in the green elements which tend to be fragmented, with obvious impacts on the ecological processes such as movement of species and resilience to change (Ahern, 2011). Greening our already interconnected network of streets provides an answer to both of these issues.

There is evidence of the movement of species from sites in close proximity demonstrated by natural colonization of green roofs and derelict sites (Baumann, 2006; Buglife, 2013; Kadas and Gedge, 2016; Oberndorfer et al., 2007). Proximity to other such habitats increased the distinctiveness of their flora suggesting that there was dispersal between sites (Angold et al., 2006) and between wetland schemes situated close to existing wetland and freshwater habitats. The proximity allows movement between sites; this could be by flying, in flood water, by wind, or carried by other animals and will allow for more rapid, natural colonization and the strengthening of existing populations due to a greater complexity of habitats (Biggs, 2003). Dispersal distances are different for different species and a closer proximity often creates higher species diversity (Graham et al.,

2015). Larger sites were found to support more bird species; if larger sites cannot be created, a connected landscape of smaller sites is recommended (Humphreys et al., 2012).

Green streets offer the potential to create a real biological corridor and to be as natural as possible to facilitate citywide species movement (Savard et al., 2000), reduce fragmentation, allow migration, and support ecosystem resilience. Wetland systems should be designed into a future network of interconnected habitats. They can then act as link habitats, stepping stones or part of a corridor and, particularly important in urban environments, facilitate the movement of species into and from rural environments and between other urban habitats thus increasing adaptability to climate change (Ahern, 2011; Angold et al., 2006; Baldock et al., 2015; Biggs, 2003; Graham et al., 2015). Dispersal corridors may be utilized by small mammals but they also provide habitat for other species as well, even if these species are not proven to disperse along them (Angold et al., 2006).

CONCLUSION

Although there is an increase in the number of studies being published on the biodiversity value of urban environments (Dover, 2015) there are still a lot of gaps and we have little research available on many urban habitats. Green streets and their elements have the potential to offer a diverse, connected array of microhabitats (such as ephemeral pools, rough tussocks, and undulations) providing opportunities for birds, invertebrates, amphibians and mammals whilst offering essential flood defense. The general consensus is that the more habitat types and the greater proximity or connectivity included, the more potential is offered to increase biodiversity. Keeping habitats as natural as possible, incorporating local materials and creating habitats in proximity to, or linked to, other similar habitats leads to better colonization and higher biodiversity. Increasing the habitat heterogeneity by well-planned and varied management, sympathetic cutting regimes and the inclusion of pools or wetland areas can also increase biodiversity values. To increase pollinator abundance on any urban habitat the inclusion of wildflowers and wildlife orientated management appears to be key. Green streets should be included in our towns and cities much more widely than they are today and policy and legislation needs to become more explicit in this respect. These should be more clearly based on the evidence emerging from scientific studies to ensure our cities become greener for the benefit of both humans and biodiversity.

ACKNOWLEDGMENTS

With many thanks to Prof. John Dover, Dr. Ruth Swetnam, and Dr. Paul Mitchell for commenting on the draft of this manuscript.

REFERENCES

Ahern, J., 2011. From fail-safe to safe-to-fail: sustainability and resilience in the new urban world. Landscape Urban Plan. 100 (4), 341—343.

Alvey, A.A., 2006. Promoting and preserving biodiversity in the urban forest. Urban For. Urban Gree. 5 (4), 195—201.

Angold, P.G., Sadler, J.P., Hill, M.O., Pullin, A., Rushton, S., Austin, K., et al., 2006. Biodiversity in urban habitat patches. Sci. Total Environ. 360 (1—3), 196—204.

Anon, n.d.[1]. Green Roof and Brownfield Diversity. [Online]. Black Redstarts.org.uk. Available from: http://www.blackredstarts.org.uk/pages/greenroof.html. (accessed 1.11.2016a).

Anon, n.d.[2]. Sustainable Drainage Systems (SuDs) for Wildlife. [Online]. The Wildlife Trust of South and West Wales. Available from: https://www.welshwildlife.org/things-to-do/wildlife-gardening/sustainable-drainage-systems-suds-for-wildlife/. (accessed 1.012017c).

Anon, n.d.[3]. Sustainable Drainage Schemes (SuDS). [Online]. Amphibian and reptile conservation. Available from: http://www.arc-trust.org/suds. (accessed: 20.06.2001b).

Atkins, in prep. Biodiversity Value of Urban Hedges. Manuscript in preparation.

Baldock, K.C.R., Goddard, M.A., Hicks, D.M., Kunin, E., Mitschunas, N., Osgathorpe, L.M., et al., 2015. Where is the UK's pollinator biodiversity? The importance of urban areas for flower-visiting insects. In: Proceedings of the Royal Society Biologcal Sciences. 2015, p. 20142849.

Barbosa, O., Tratalos, J. a, Armsworth, P.R., Davies, R.G., Fuller, R. a, Johnson, P., et al., 2007. Who benefits from access to green space? A case study from Sheffield, UK. Landscape Urban Plan. 83 (2—3), 187—195.

Baumann, N., 2006. Ground-nesting birds on green roofs in Switzerland: preliminary observations. Urban Habit. 4, 37—50.

Bellamy, P.E., Shore, R., Ardeshir, D., Treweek, J.R., 2000. Road verges as habitat for small mammals in Britain. Mam. Rev. 30 (2), 131—139.

Biggs, J., 2003. Maximising the Ecological Benefits of Sustainable Drainage Schemes SR 625, December 2003. Department of Trade and Industry (December).

Biodiversity Reporting and Information Group, 2007. Report on the Species and Habitat Review.

Blackmore, L.M., Goulson, D., 2014. Evaluating the effectiveness of wildflower seed mixes for boosting floral diversity and bumblebee and hoverfly abundance in urban areas. Insect Conserv. Diversity 7 (5), 480—484.

Brenneisen, S., 2005. The Natural Roof (NADA)-Research Report on the Use of Extensive Green Roofs by Wild Bees. Hochschule Wadenswil. (November). p. 22.

Brenneisen, S., 2006. Space for urban wildlife: designing green roofs as habitats in Switzerland. Urban Habit. 4 (1), 27—36.

Buglife, 2013. Creating green roofs for invertebrates. A Best Practice Guide. 1—29.

Chiquet, C., Dover, J.W., Mitchell, P., 2012. Birds and the urban environment: the value of green walls. Urban Ecosyst. 16 (3), 453—462.

Dover, J.W., 2015. Green Infrastructure Incorporating plants and enhancing biodiversity in buildings and urban environments. routledge, Oxon.

Faiers, A., Bailey, A., 2005. Evaluating canalside hedgerows to determine future interventions. J. Environ. Manage. 74 (1), 71—78.

Gelling, M., Macdonald, D.W., Mathews, F., 2007. Are hedgerows the route to increased farmland small mammal density? Use of hedgerows in British pastoral habitats. Landscape Ecol. 22 (7), 1019—1032.

Gosling, L., Sparks, T.H., Araya, Y., Harvey, M., Ansine, J., 2016. Differences between urban and rural hedges in England revealed by a citizen science project. BMC Ecol. 16 (Supplement 1), 45—55.

Graham, A., Day, J., Bray, B., Mackenzie, S., 2015. Sustainable Drainage Systems Maximising the potential for people and wildlife. A guide for local authorities and developers.

Grant, G., 2006. Extensive green roofs in London. Urban Habit. 4 (1), 51−65.

Hinsley, S.A., Bellamy, P.E., 2000. The influence of hedge structure, management and landscape context on the value of hedgerows to birds: a review. J. Environ. Manage. 60, 33−49.

Humphreys, E., Kirkland, P., Russell, S., Sutcliffe, R., Coyle, J., Chamberlain, D., 2012. Urban Biodiversity: successes and challenges: the biodiversity in glasgow (BIG) project: the value of volunteer participation in promoting and conserving urban biodiversity. Glasgow Naturalist. 25, 4.

Kadas, G., 2006. Rare invertebrates colonizing green roofs in London. Urban Habit. 4, 66−86.

Kadas, G., Gedge, D., 2016. Bees and Green Roofs − studies in London and Switzerland. [Online]. 2016. Livingroofs.org. Available from: http://livingroofs.org/bees-green-roofs/. (accessed 1.12.2016).

Kadir, M.A.A., Othman, N., 2012. Towards a better tomorrow: street trees and their values in urban areas. Proc. Soc. Behav. Sci. 35, 267−274.

Kazemi, F., Beecham, S., Gibbs, J., Clay, R., 2009. Factors affecting terrestrial invertebrate diversity in bioretention basins in an Australian urban environment. Landscape Urban Plan. 92 (3−4), 304−313.

Kazemi, F., Beecham, S., Gibbs, J., 2011. Streetscape biodiversity and the role of bioretention swales in an Australian urban environment. Landscape Urban Plan. 101 (2), 139−148.

Kotzageorgis, G.C., Mason, C.F., 1997. Small mammal populations in relation to hedgerow structure in an arable landscape. Proc. Zool. Soc. London (A). 1974, 425−434.

Loh, S., 2008. Living Walls − a way to green the built environment. Aust. Inst. Arch. 1, 1−7.

Lukes, R., Kloss, C., 2008. Managing wet weather with green infrastructure action strategy. Municipal Handbook 1−17.

Michel, N., Burel, F., Butet, A., 2006. How does landscape use influence small mammal diversity, abundance and biomass in hedgerow networks of farming landscapes? Acta Oecologica 30 (1), 11−20.

Newman, A.P., Coupe, S.J., Smith, H.G., Puehmeir, T., Bond, P.C., 2006. The microbiology of permeable pavements. 8th International Conference on Concrete Block Paving, November 6-8, 2006 San Francisco, California USA, pp. 181−191.

Oberndorfer, E., Lundholm, J., Bass, B., Coffman, R.R., Doshi, H., Dunnett, N., et al., 2007. Green roofs as urban ecosystems: ecological structures, functions, and services. BioScience 57, 823−833.

O'Brien, C.D., 2015. Sustainable drainage system (SuDS) ponds in Inverness, UK and the favourable conservation status of amphibians. Urban Ecosyst. Ecosyst. 18 (1), 321−331.

Runfola, D., Hughes, S., 2014. What makes green cities unique? Examining the economic and political characteristics of the grey-to-green continuum. Land 3 (1), 131−147.

Savard, J.P.L., Clergeau, P., Mennechez, G., 2000. Biodiversity concepts and urban ecosystems. Landscape Urban Plan. 48 (3−4), 131−142.

Snep, R.P.H., Kooijmans, J.L., Kwak, R.G.M., Foppen, R.P.B., Parsons, H., Awasthy, M., et al., 2016. Urban bird conservation: presenting stakeholder-specific arguments for the development of bird-friendly cities. Urban Ecosyst. 19 (4), 1535−1550.

FURTHER READING

Dwyer, J.F., Mcpherson, E.G., Schroeder, H.W., Rowntree, R.A., 1992. Assessing the benefits and costs of the urban forest. J. Arboricult. 18, 227−234 (September).

Nature Based Strategies: Social, Economic and Environmental Sustainability

Chapter 4.1

Vertical Greening Systems: Social and Aesthetic Aspects

Adriano Magliocco

Chapter Outline

INTRODUCTION

There are few studies and publications on the observation of vertical greening systems, especially if we compare them with the many studies undertaken on urban green areas and, recently, on green roofs. This is probably because the best-known green walls—for instance, the living walls designed by the landscape architect Patrick Blanc (2012)—were constructed focusing on aesthetic aspects rather than other performances. These facades, characterized as they are by lush vegetation, appeared as expensive architects' toys, destined to embellish special-use buildings (museums, etc.).

Nevertheless, the progress of studies on the effectiveness of green facades, in terms of environmental quality improvement, led to the consideration of perception aspects. Perception is often, erroneously, considered to be a synonym of sight. Sight is clearly the sense that prevails in a perceptual experience, but perception is the interpretation of what we see and feel through our senses, in a synesthetic way. Synesthesia, as a perceptual experience, is clearly explained by Merleau-Ponty (2003) in the 1945 publication, *Phénoménologie de la Perception*. If the senses are the means through which we receive information, our personal history, i.e., our life experience, is the filter through which we interpret and understand it. So our perception may change depending on the level of knowledge that we have of what we are

Nature Based Strategies for Urban and Building Sustainability.
DOI: https://doi.org/10.1016/B978-0-12-812150-4.00024-0

observing. Benjamin, in 1936 (2011), reminded us that while art is understood through contemplation, architecture is perceived via habit. This is the starting point for what now follows.

INNOVATIVE GREEN

The practice of covering building facades with ivy, or creating porches from vines, is part of the European tradition, with climatic differences dictating some variations between north and south. However, the introduction of green facades on buildings in our chaotic cities may appear to be the introduction of extraneous elements. The introduction of new technologies always goes through an acceptance phase that can present some difficulties, as has indeed happened in the past with special inventions or innovations. To make these solutions acceptable, electric lightbulbs (Banham, 1969) were shaped as candles, and cast iron used in structural pillars was in the form of classical columns. No such mimicry is possible for a green facade, except for the previously mentioned vegetation cover solutions on traditional buildings.

Playing in favor of the introduction of integrated vegetation in architecture, are communication strategies that are using the word "green" as an evocative element of our society's lost naturalness that needs to be recovered (Biraghi, 2010). This mechanism is not without flaws and deceptions, but it is very effective. The presence of green elements in urban spaces plays a great role in the psychological well-being of society, as demonstrated by several studies (Chilla, 2004). People feel better in a natural environment; mainly for psychological reasons, according to the hypothesis of Wilson (1984) on the phenomenon of biophilia, which asserts that contact with nature and the complex geometry of natural forms is needed by humans as much as metabolism needs nutrients and oxygen.

"Concentrating on the environmental movement left students with the idea that the environment was the domain of activists and didn't much concern them" (Corbett, 2006). Using technologies to reduce the environmental impact of human activities and the improvement of environmental quality can therefore also have an "educational" effect if it is well-managed and communicated. The adoption of green facades, e.g., on buildings that house educational and training activities, such as schools and universities, can accustom future citizens to appreciate solutions that are normally considered unusual. Zui Ling and Ghaffarian Hoseini (2012) propose a wider adoption of green facades in Limkokiwing University Campus in Malaysia, noting that such a solution can improve the aesthetic value of buildings that may not always be of good quality, improve the microclimate of the campus and, consequently, the users' thermal comfort, reducing the carbon footprint and improving air quality. Finally, the extensive application of green facades can help provide a good "social perspective" through the integration of natural elements in the architecture, with the advantage of not being prone to graffiti.

Although, as we shall see shortly, a green facade is mostly accepted and positively evaluated, it comes with all the contradictions of a man-nature relationship. It is a loving mother, and a source of food and health benefits for our species (those that are called ecosystemic services), but it is also a cruel mother, a source of diseases, a home to unwelcome creatures, and a bearer of doom. While the value of a green facade is immediately perceived, with its ability to improve air quality, it also raises concerns regarding maintenance, the fear of more insects, and the distress of pollen allergies that, for some people, are prevalent.

VERTICAL GREEN PERCEPTION IN SINGAPORE

Hien Wong et al. (2010) tell of an extended campaign of interviews conducted in Singapore. The investigation involves various categories of people: Singapore architectural and landscape consultants; real estate consultants; people working at some government agencies such as the National Parks Board (NParks), the Ministry of National Development, the Building and Construction Authority, the Housing and Development Board; a group of residents, selected from neighborhoods where there are buildings with green roofs. The questionnaire was sent to a very large number of representatives of the first four targets, but only 220 forms were filled out and returned, and only 194 of those contained answers that could be considered. It was then decided to directly survey 400 residents instead.

It is evident from the responses that the level of knowledge of the interviewee influences the perception of different qualities of green facades. For example, the ability to retain and filter rainwater is usually recognized by landscapers but hardly at all by other groups of respondents, while almost everyone recognizes the ability of vegetation to mitigate the microclimate and clean the air. Inhabitants do not seem to know how to connect these benefits with a reduction in building cooling operating costs. The limited knowledge of how green facades are installed means that a majority of respondents—in all groups—considers that green facades could damage the walls on which they are installed, because of plant roots and moisture from irrigation. Real estate consultants and employees of government agencies, unlike the respondents in the other groups, do not believe that green facades are an element of aesthetic improvement.

The conclusions of these monitoring campaigns seem to be twofold. In general, the ability of green facades to reduce electricity demand for air conditioning is recognized, as is their capacity to reduce air pollution and noise, and the ability to improve the aesthetics of a city by bringing natural elements more closely into contact with humans. Resistance to their adoption persists because of a lack of knowledge in maintenance procedures, which are feared to be too expensive, and the belief that plants can soil public spaces with falling leaves.

A CASE OF GREEN FACADE PERCEPTION MONITORING IN GENOA, ITALY

An extended case of monitoring was conducted by the author (Magliocco et al., 2015). This entailed an analysis campaign of the social perception, by building users and citizens, of the green facade built in Genoa Sestri Ponente. This green facade was installed during the months of October and November 2014 on the south wall of a building owned by INPS (Istituto Nazionale di Previdenza Sociale, National Insurance Institute), in a high-density urban area (13000 inhabitants/km^2). The building, built early last century, was renovated in the 80s as an office building. The green facade was subjected to different types of monitoring: fine dust retention on plant leaves (Perini et al., submitted), energy performance (Perini et al., submitted), evapotranspiration of some plant species, and social investigation.

The social investigation was carried out in order to understand the perception impact that the vertical green system, capable of greatly modifying the aesthetics of a building, has on citizens. The aim was to see how and if answers change when opinions are requested in different moments (before, during and after facade construction). The assumption was that, after directly verifying a substantial absence of feared drawbacks, and being in the habit of seeing an undeniably unusual facade, people would have modified their subsequent judgment.

The first survey (Magliocco and Perini, 2015) was carried out in July 2014, 3 months before the installation of the green facade. Sixty citizens were involved in the survey, some working in the same office building (59.9%), others working (25.4%), or living (23.7%) in the same neighborhood. The survey was developed considering the results of previous researches about green envelopes and urban green (Alessandro et al.,1987; Perussia, 1990).

The conclusions of the first survey showed that, although vegetation was usually positively perceived as being a good element by improving the urban environment, green integrated architecture was seen as something abnormal, expensive, and problematic (especially relating to the possibility of an increased insect population). One relevant study (Chilla, 2004) had already noted several negative opinions about the integration of vegetation in dense urban areas in the case of German citizens not living in buildings covered with vegetation, fearing the presence of insects and possible damage to walls due to plant roots and irrigation water. In the Genoa perception campaign, 45% of the people interviewed had never seen a green facade and 32% had never even heard of green facades. While positive effects were in any case recognized—air quality improvement, more nature in cities—evaluation of the aesthetic value was controversial, with a positive ranking of the "visually enhanced cityscape" parameter, while most people didn't recognize the "building aesthetic parameter" at the same level. Many of them considered the "increase of biodiversity" (small animals) as a negative effect. In general, positive effects were rated with lower values by INPS employees than by people living and working in that area. Consistently, negative effect ranking was higher for INPS employees. The most voted negative effect by both groups was "more insects," while the most voted positive effect by both groups was "air quality improvement." The conclusion was that the average values given to positive effects were in the range of 3.0–3.6 out of 5.0, which is a sufficiency rating but not much more.

After 6 months (Magliocco et al., 2015), 21 semistructured interviews were administered again to citizens both living or working nearby the building and to INPS building workers, focusing on three main themes: vertical green on the INPS building, urban green in their neighborhood, the possibility of using vertical green systems on other Genoa city buildings. The inquiry results were that views on the application of green facades were definitely more positive than those found in the first survey. Only 3 out of 21 people interviewed expressed negative opinions. In general, young people seemed more favorable and the employees who work in the office appeared very involved, giving fewer but much more articulated answers. However, once again, the fear of insects and other animals climbing on the green facade was expressed. Referring to the controversy that had arisen during construction—due mainly to concerns about the use of public money on a solution for which they did not understand the usefulness—many respondents said they had changed their minds after talking with other people and with university researchers who had explained the reasons behind construction of the green facade. The perception is, therefore, also interpersonal; it is transformed through communication (Gazzola, 2011).

A SUSTAINABLE METROPOLIS

The adoption of integrated green infrastructures in architecture seems to be one of the strategies that big cities are trying to implement with the intention of mitigating the heat island effect and to improve air quality. Widespread application of these solutions obviously needs not only technologically adequate performance but a high acceptance level is necessary, indirectly by people who have to live with them or directly by clients.

Sydney, Australia, is committed to increasing the number of green roofs and walls. In April 2014, after 2 years of research activity, the city adopted the green roofs and walls policy, the first of its kind in Australia. In 2012, in order to commence research (v.a. 2012), a communication company was commissioned to carry out a perception study on greening infrastructure. There were several research objectives: understanding the level of awareness on green walls and roofs among participants, pointing out the limits to green wall and roof installation, determining satisfaction levels with already installed green walls and roofs, defining the role of the city in promoting green walls and roofs. A qualitative research involved, through a focus group and interviews, 22 industry stakeholders. Moreover, an online community survey and a street intercept survey of random respondents were conducted, collecting 416 responses. Although the investigation jointly considered vertical and horizontal green infrastructures—not permitting possible differences between them to be distinguished—the assessments are reasonably comparable to those already found in other surveys. Major concerns were maintenance costs and the impermeability of the layers separating walls from green systems, as well as the effectiveness of the irrigation system. Interviewed citizens also perceived several advantages in terms of environmental quality. Unlike the respondents of other case studies, they also seemed sensitive to the idea that an increase of biodiversity might improve an urban habitat. The city of Sydney used the inquiry results to draft a Green Roofs and Walls Strategy, as well as a good number of guidelines to encourage citizens and entrepreneurs to adopt green infrastructures.

London is also working on the adoption of urban resilience strategies, and supports the implementation of green roofs and facades. In 2008, Design for London (2008) and the Greater London Authority's London Plan and Environment Teams, along with the London Climate Change Partnership, prepared the Living Roofs and Walls Technical Report. To support the London Plan Policy, it was considered useful to draft guidelines for living roofs and walls as the key to providing living space; adapting the city to more extreme climatic conditions and reducing energy use and CO_2 emissions along the roads (Fig. 1). Perceived barriers to the implementation of living roofs in London were considered to be lack of policies encouraging the installation of living roofs and walls, lack of a common

FIGURE 1 A living wall on the side of an old gas building in Goods Way, King's Cross, designed by Biotecture.

FIGURE 2 The huge green facade of the Palace Hotel in Victoria, designed by Green Roof Consultancy.

technological standard, the fear of fire hazard, maintenance and structural costs, leakage and damage to waterproofing. Unlike other cases, the aesthetic value of green facades has been widely investigated in London. This is due to the presence of buildings of great value—and customers with great economic capacity—enabling the creation of case studies such as the huge green wall, designed by Gary Grant of Green Roof Consultancy, covering an entire 350 square meter facade of the Palace Hotel in Victoria (Fig. 2) (Andrews, 2013).

CONCLUSION

If green spaces can benefit, at the perceptual level, from the effects that they can induce in sensory terms, especially under heat stress conditions in the hot season (Lafortezza et al., 2009), this cannot be said for green facades, which almost never have a "wrap-around effect," as they are visible from one direction only. Although the green facade has a remarkable effect on the reduction of air temperature (Magliocco et al., 2016), this effect is perceived only when there are large extensions of facade as a source of relief. This means that few people are able to connect the effects of shading and evapotranspiration with possible energy and air conditioning cost savings.

For some authors the aesthetic value of urban green is indubitable and creating green wall elements, integrated into architectural elements, "could take advantage of the measurable improvements to the human condition that plants can provide" (Shaikh et al., 2015). As it has been verified during perceptive inquiries, to which reference is made in this text, although the aesthetic value of vegetation seems to be commonly recognized, the integration of plants isn't always an accepted solution because "plants should be on the ground" (as some interviewees said).

Since we can consider perception as an evaluation of what our senses detect, it was shown that a positive perception often goes with the experience and the knowledge of what we see. It is therefore possible to state that it is necessary to support the application of vertical greening systems with information campaigns to make people understand their qualities.

ACKNOWLEDGMENTS

The Genoa green wall social perception research was funded by INPS (National Insurance Institute) Liguria as part of the monitoring activity carried out to quantify the environmental and energy performance of vertical greening systems. Thanks are given to Dr. E. Cattaneo and U. Valle (INPS Liguria) for their important support.

REFERENCES

Alessandro, S., Barbera, G., Silvestrini, G., 1987. Quaderno Consiglio Nazionale delle Ricerche, in Stato dell'arte delle ricerche concernenti l'interazione energetica tra vegetazione e ambiente costruito, Istituto per l'edilizia ed il risparmio energetico, Palermo.

Andrews, K., 2013. London's largest living wall will "combat flooding." http://www.dezeen. com/2013/08/21/londons-largest-living-wall-will-combat-flooding/.

Banham, R., 1969. Architecture of the Well-Tempered Environment. Architectural Press, Chicago.

Biraghi, M., 2010. Green is the colour. In: Biraghi, M., Lo Ricco, G., Michelis, S. (Eds.), MMX Architettura zona critica, Gizmo, Emanuela Zandonai Editore, Rovereto.

Blanc, P., 2012. The Vertical Garden: From Nature to the City (Revised and Updated). Norton & Company, New York.

Chilla, T., 2004. "Natur" – Elemente in der Stadtgestaltung. Institute of Geography of the University of Köln, Köln.

Corbett, J.B., 2006. Communicating Nature. Island Press, Washington, Covelo, London.

Design for London, GLA, 2008. Living Roofs and Walls. Technical Report: Supporting London Plan Policy. Greater London Authority, London.

Gazzola, A., 2011. Uno sguardo diverso. La percezione sociale dello spazio naturale e costruito. Franco Angeli, Milano.

Lafortezza, R., Carrus, G., Sanesi, G., Davies, C., 2009. Benefits and well-being perceived by people visiting green spaces in periods of heat stress. Urban For. Urban Gree. 8 (2009), 97–108.

Magliocco, A., Perini, K., 2015. The perception of green integrated into architecture: installation of a green façade in Genoa, Italy. AIMS Environ. Sci. Available from: http://dx.doi.org/ 19.3934/environsci.2015.4.899.

Magliocco, A., Perini, K., Prampolini, R., 2015. Qualità ambientale e percezione dei sistemi di verde verticale: un caso studio. In Abitare insieme / Living together. Proceedings of the 3° Edition of "Inhabiting The Future", Naple 1-2 October 2015, Clean Edizioni, Naple.

Magliocco, A., Perini, K., Giulini, S., 2016. Vertical greening systems evaporation measurements: does plant species influence cooling performances? Special Issue on "Breakthrough of natural and hybrid ventilative cooling technologies". Int. J. Ventil. September 2016. https://doi.org/10.1080/14733315.2016.1214388.

Merleau-Ponty, M., 2003. Fenomenologia della percezione. Bompiani, Milano (original work published 1945).

Nyuk Hien Wong, N., Yong Kwang Tan, A., Yok Tan, P., Sia, A., Chung Wong, N., 2010. Perception Studies of Vertical Greenery Systems in Singapore. J. Urban Plan. Dev. 136 (4), 330–338.

Perini, K., Bazzocchi, F., Croci, L., Magliocco,A., Cattaneo, E., submitted. The use of vertical greening systems to reduce the energy demand for air conditioning. Field monitoring in Mediterranean climate for consideration to Energy and Buildings. Energy and Buildings.

Perini, K., Ottelé, M., Giulini, S., Magliocco, A., submitted. Vertical greening systems and air quality: quantifying the performances of different plant species in capturing fine dust particles.

Perussia, F., 1990. Immagini di natura: contribuiti di ricerca. Guerini, Milano.

Shaikh, A.F., Gunjal, P.K., Chaple, N.V., 2015. A review on green walls technology benefits & design. IJESRT Int. J. Eng. Sci. Res. Technol. 4 (4), April 2015.

Wilson, E.O., 1984. Biophilia. Harvard University Press, Cambridge, MA.

Zui Ling, C., Ghaffarian Hoseini, A., 2012. Greenscaping buildings: amplification of vertical greening towards approaching sustainable urban structures. J. Creative Sustain. Arch. Built Environ. vol. 2, December.

FURTHER READING

Benjamin, W., Valagussa, M., 2011. L'opera d'arte nell'epoca della sua riproducibilità tecnica. Einaudi, Torino (Original work published 1936).

Osmond, P., 2012. Green Roofs and green walls perception study final research report. City of Sidney.

Chapter 4.2

Green Roofs Social and Aesthetic Aspects

Benz Kotzen

Chapter Outline

INTRODUCTION

Tom Turner, (University of Greenwich), coined the acronym and title MOER (Multiobjective Environmental Roof) in 2012. Neither the acronym nor the title slip off the tongue, but the idea that all roofs should be environmentally multiobjective, in other words *multifunctional* is correct. In the future, it is not too outrageous to predict that the vast majority of all new-build roofs will have several functions and be environmentally useful. Not only will roofs protect buildings from weather, which, traditionally, is their main function, but every roof will be biodiverse, be planted and/or provide power or hot water through solar panels, and/or be used for recreation and other purposes. Additionally, when land values are so high in urban areas and developers and authorities are under pressure to maximize the footprints of properties for indoor use, it makes sense to locate some of the landscape functions at roof level. Change is coming, it is coming fast, and this is a good thing.

This chapter focuses particularly on the social and aesthetic aspects of green roofs, but in order to understand the benefits of green roofs as well as all green infrastructure it is important to locate these benefits within the context of ecosystem services provision. Out of the four main ecosystem services categories which include supporting, provisioning, regulating, and cultural services (Fig. 1), the social and aesthetic functions of green roofs

Nature Based Strategies for Urban and Building Sustainability.
DOI: https://doi.org/10.1016/B978-0-12-812150-4.00025-2

273

and other green infrastructure fall within the cultural service provision. Whilst it may be argued that some ecosystem services may be more important than others with regard to green roofs, such as providing water control, regulating the urban heat island effect, the social and aesthetic aspects are particularly important in our cities and indeed in more rural areas as well. The cultural services provision is often noted for providing the following nonmaterial benefits: ethical values, spiritual and religious, recreation and tourism, aesthetic, inspirational, educational, sense of place and cultural heritage (Millennium Ecosystem Assessment) (Fig. 1).

The Millennium Ecosystem Assessment (2005), notes that these cultural services are nonmaterial benefits obtained by people through spiritual enrichment, cognitive development, reflection, recreation, and aesthetic experiences, including:

- cultural diversity,
- spiritual and religious values,
- knowledge systems (traditional and formal),
- educational values,
- inspiration,
- aesthetic values,
- social relations,
- sense of place,
- cultural heritage values, and
- recreation and ecotourism.

It is possible to see most of these social values as a service provided by green roofs, but the most important aspects are that green roofs can provide a social space at rooftop level which can be enjoyed and that can improve the health and well-being of people. The Millennium Ecosystem Assessment

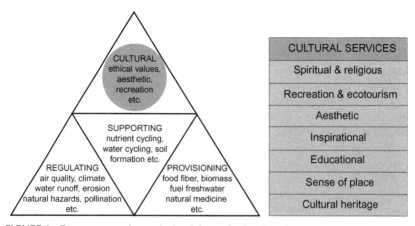

FIGURE 1 Ecosystems services and a breakdown of cultural services.

(2005), also notes a number of well-being constituents offered by cultural services and this can be seen with green roofs. These constituents include: (1) good social relations, (2) health, (3) basic materials for a good life, and (4) security. For green roofs the strongest provision is in health and well-being, good social relations, and social cohesion.

Urban real estate at ground level is so expensive in our cities that it is understandable that providing gardens and other public spaces is being rationalized and every square meter is going to be contested. Some of the social aspects of our city lives will thus have to move upwards. Although we decry any loss of green space at ground level, creating a balance of green space at a number of levels is possible and having green roofs has also been assessed to be economically advantageous. The 2011 publication *Benefits and Challenges of Green Roofs on Public and Commercial Buildings*, by the GSA (United States General Services Administration) and Arup in the United States found that "over ... 50 years an extensive green roof would generate the equivalent of $38/square foot" ($409/square m), "of public benefit" (GSA, 2011).

CONCISE HISTORY OF THE SOCIAL AND AESTHETIC ASPECTS OF GREEN ROOFS

The Hanging Gardens of Babylon (if they existed), incorporated planting on a series of stepped arches and elevated walkways. Osmundson (1999) notes that the first known references to "manmade gardens above grade" were Mesopotamian Ziggurats built from the 4th century BC. Apparently the landings of these stepped towers were planted with trees and shrubs. This may have been for microclimatic purposes, but undoubtedly both for the Hanging Gardens of Babylon and the Ziggurats aesthetics were as important. The roof terrace of the Villa of Mysteries, Pompei has been planted with grass, trees, shrubs and hedging as part of the restoration following detailed investigations that uncovered empty spaces occupied by plant roots within the roof. In the Middle Ages, gardens were located on top of vaulted chambers, e.g., at the cloister garden at Mont-Saint-Michel. There are a number of examples of Renaissance roof gardens in Italy, such as the Palazzo Piccolomino, Pienza, "one of the first and best preserved roof gardens," the Tower at Guinigis, Lucca, and the Medici roof garden at Careggi (Osmundson, 1999). The Kremlin hosts large scale roof gardens and terraces "laid out in 1822–1823" above a canalized river (Osmundson, 1999). All of the above examples were planted for aesthetic and social purposes in which people over the ages would have sat out amongst the plants and admired the integration of plants and architecture, delighting in the smells, textures, and wildlife that verdancy and nature brings. Over 170 species of plants thrive on the roof of the "Moos" filtration plant, Zurich, Switzerland built in 1914. "Most impressive are the 6000 individual specimens of *Orchis morio*, an orchid species that is thought to be now extinct in the Zurich area except on the roof" (Earth Pledge, 2005). This Swiss example illustrates the ecosystem service provision of aesthetic integration, whereby planting this large reservoir structure, the green planted roof fits into the surrounding green landscape. In the United Kingdom, the most famous roof garden was designed by Ralph Hancock on top of the six-story Derry and Toms department store, now commonly known as the Roof Gardens, owned by the Virgin Group. The 0.4 ha gardens comprise a Spanish Garden (Fig. 2), a Tudor Garden and an English Woodland Garden with flamingos and water which attract a large number of wild birds. The gardens have been classified as being of specific historical interest and are Grade II listed by English Heritage. The character of these roof gardens is rather surprising especially when one sees flamingos wading in the water and the porticos and the Granada-esque facades (Fig. 2). In the United States, the most famous roof gardens consist of the five installed on top of the Rockefeller Center in New York between 1933 and 1936. The landscape architect was also Ralph Hancock, but the visionary architect was Raymond Hood who "appalled at the ugliness and disorder of the hundreds of neighboring roofs that would be visible from the towers, he determined that higher rents could be charged for views of gardens and proposed that the lower roofs be designed and planted as foreground viewscapes" (Osmundson, 1999). Le Corbusier designed a number of buildings with green roofs. He initiated a set of five architectural points, the second of which notes that roofs can be used for domestic purposes or as a roof terrace or roof garden with the additional benefit that the concrete roof slab will be protected. He said "in this way the roof garden will become the most favored

FIGURE 2 The Spanish Garden on top of the old Derry and Toms department store building.

place in the building. In general, roof gardens mean to a city the recovery of all the built-up area." His scheme for the apartment of M. Charles de Beistegui in Paris, 1929 shows an arrangement of lawn, hedging cypress trees and topiary as well as a number of objects including a stucco fireplace, curlicue metal chairs, and cut off views to the Arc De Triomphe and the Eiffel Tower that create a distinct surrealist impression.

Friedensreich Hundertwasser, the Austrian/New Zealand national (1928–2000), the pioneer environmentalist, artist, and architect said "grass and vegetation in the city should grow on all horizontal spaces—i.e., to say, wherever rain and snow falls vegetation should grow, on the roads and on the roofs. The horizontal is the domain of nature..." (Gardenvisit.com [2]). Furthermore, he said "I've worked a great deal with grass roofs, putting soil on top and having things grow, but there is something strange in this, more than ecological. It is a religious act to have soil on your roof and trees growing on top of you; the act reconciles you with God, with nature, maybe not Christian or Jewish monotheism, but something wider, older—a very ancient wisdom" (Gardenvisit.com [2]). Whether planting green roofs reconciles one with God or not, the reintroduction of vegetation in cities on most surfaces can be seen as a reconciliation with nature and the environment in general. It fulfils mankind's biophilic urge to live with nature and to be close to nature. The biophilia hypothesis, is that humans possess an innate tendency to seek connections with nature and other forms of life (Rogers). According to Rogers' writing for Encyclopedia Brittanica, "the term biophilia was used by German-born American psychoanalyst Erich Fromm in *The Anatomy of Human Destructiveness* (1973)." The term was later popularized by the American biologist Edward O. Wilson in his work *Biophilia* (1984), "which proposed that the tendency of humans to focus on and to affiliate with nature and other life-forms has, in part, a genetic basis" (Rogers).

THE SOCIAL AND AESTHETIC BENEFITS OF GREEN ROOFS

While the environmental benefits of green roofs are becoming well recognized, the social and aesthetic values may be overlooked, but are likely to become more important as ground level space is squeezed. The type of provision of social space as well as aesthetic pleasure provided by green roofs is not dissimilar to those provisions provided at ground level and are largely determined whether the space is in private or public ownership or it is lodged somewhere in between. While most green roofs are in private ownership with restricted access, there is the potential for green roofs to become public parks and used in much the same way that ground level parks are used. In most cases it is likely that these parks would have time limited access, during the hours of daylight and locked at night to prevent antisocial behavior. This has been the case for many years at the Jardin Atlantique. Opened in

FIGURE 3 Jardin Atlantique, above Gare Monparnasse, Paris.

1994, the roof garden that was built over Gare Monparnesse in Paris. The garden which was designed by François Brun and Michel Péna, is open from dawn to dusk every day (Fig. 3).

As our cities become more populated and busier, and people, (authorities as well ordinary citizens), realize that the quality and character of the environment affects the health and well-being of populations; there will be an increased drive to provide facilities that will increase the health and well-being of urban populations. Green roofs can play their part in this provision by providing the following types of places:

Garden space: These gardens can be publicly accessible or private in much the same way that gardens at ground level are organized and used. They can be simple or complex designs and be ornamental or nature based. One great advantage of the elevated gardens is the ability to (1) provide visual and physical delight within the gardens as well as through elevated views which may be expansive or more contained by local roof-tops, (2) be quieter and more tranquil than many ground level gardens that are affected by traffic noise, and (3) be above much of the pollution caused by vehicular traffic.

Growing space: Growing food and flowers for harvest is an important part of advocating health and well-being in communities. Rooftop gardens can readily be used for this purpose, and the same fruits and vegetables can be grown at rooftop level as at ground level (Fig. 4). Greater care may need to be taken in terms of wind and the use of dwarf rooting stock for trees as well as espaliered trees to reduce space. Most vegetables are nutrient-demanding so these will need to be replenished continuously as with any productive garden. One advantage is that slugs and snails and many pests may be reduced at rooftop level, although slugs started to appear on the

FIGURE 4 Sweetcorn, tomatoes, cosmos, and calendula growing in the urban agriculture gardens on The Stockwell Street Roofs, University of Greenwich.

green roofs at the University of Greenwich after about 2 years. It is presumed that the eggs are transported on the feet of birds. Rooftop urban agriculture has real potential for producing locally grown, niche market products destined for local farmers' markets and restaurants, outdoors as well as in greenhouses.

Amenity space: The word amenity is most often used to describe a feature or facility provided by a building or space. In terms of green roofs, a great number of social amenities can be provided including cafés, bars, restaurants, and swimming pools. These can be pop-up or permanent. These types of amenities are usually privately owned and managed and can be simple or as sophisticated as the Marina Bay Sands Hotel in Singapore. The Sands Sky Park is located 57 stories from ground level and spans across three separate buildings. The garden includes an infinity pool as well restaurants and bars (Fig. 5).

THE MOER (MULTIOBJECTIVE ENVIRONMENTAL ROOF)

The number of green roofs is expanding and this will promote more green roofs as people, clients, authorities, workers, and dwellers become more familiar with their benefits and the opportunities they provide. The multiobjective environmental roof provides the way forward. Although each roof may not be able to provide every single function that a green roof can potentially provide, the larger structures should be able to provide multiple functions. The 0.4 ha roofs at No. 10 Stockwell Street, University of Greenwich do this by providing amenity spaces as well as wildlife roofs with ponds, urban agriculture with field grown crops as well as fruit bearing trees

FIGURE 5 "Sands Sky Park"—Marina Bay Sands Hotel, Singapore. *Photo Andrew Thomas, Shrewsbury, UK.*

(apricots and quince). The roofs also have beehives and photovoltaic solar panels. Moreover the roofs incorporate two greenhouses, one for horticulture and the other for aquaponics, which grow fish and vegetables in a recirculating water system.

CONCLUSION

Green roofs have the ability to moderate some of the environmental consequences of the last and present centuries, improve biodiversity in our cities as well as create remarkable spaces where people can work, relax, and grow food. They can provide real, measurable socially important ecosystem services. However, they should also be seen as part and parcel of a biophilic urban realm, as part of green infrastructure where the sum human and environmental benefits is greater than the individual parts. We should celebrate every green roof that is installed, either as a new build project or as retrofit. Clients and owners who specify green roofs should be encouraged and praised as they are not only providing benefits for themselves but their roofs provide tangible health and well-being benefits for the wider community.

REFERENCES

Earth Pledge, 2005. Green Roofs – Ecological Design and Construction. Schiffer Publishing Ltd, Pennsylvania.

GSA (United States General Services Administration), 2011. The Benefits and Challenges of Green Roofs on Public and Commercial Buildings - A Report of the United States General Services Administration. Available at https://www.gsa.gov/portal/getMediaData?mediaId = 158783.

Osmundson, T., 1999. Roof Gardens: History, Design and Construction Web Sources Le Corbusier and Pierre Jeanneret, 'Five Points Towards a New Architecture', Art Humanities Primary Source Reading 52. http://www.learn.columbia.edu/courses/arch20/pdf/art_hum_ reading_52.pdf (accessed 06.01.2017).

Gardenvisit.com [2], Friedenreich Hundertwasser – Quaotations from Hundertwasser. Gardenvisit. com. http://gardenvisit.com/garden_products/sustainable_design/friedenreich_hundertwasser_ landscapeThe Garden Guide (accessed 29.01.2017).

Millennium Ecosystem Assessment, 2005. Ecosystems and Human Well-being: Synthesis. Island Press, Washington, DC. http://www.millenniumassessment.org/documents/document.356. aspx.pdf (accessed 23.07.2017).

Rogers, K., Biophilia hypothesis. Encyclopaedia Britannica, https://www.britannica.com/science/ biophilia-hypothesis (accessed 23.07.2017).

FURTHER READING

Caroun.com, The Moscow Kremlin – Around Kremlin: Kremlin Gardens. http://www.caroun.com/ Countries/Europe/Russia/Kremlin/22-Kremlin-KremlinGardens.html (Accessed 06.01.2017).

Gardenvisit.com [1], 'Mont St Michel Cloister Garden', Gardenvisit.com The Garden Guide, http://www.gardenvisit.com/gardens/mont_st_michel_cloister_garden (accessed 08.01.2017).

Chapter 4.3

Green Streets Social
and Aesthetic Aspects

Paola Sabbion

Chapter Outline

INTRODUCTION

Urban ecosystem services can be grouped into four categories: provisioning services, regulating services, habitat (or supporting services), and cultural services (TEEB, 2011). Cultural services include the nonmaterial benefits people receive from ecosystems: aesthetic, spiritual, physical, and psychological. As a result of interaction with natural features, cultural benefits enhance social and educational experiences, thus promoting recreation and tourism. Because green is a key feature for a salutogenic transformation of our living environment—as it will be described in the following pages—this chapter focuses on the social and aesthetic aspects of vegetated areas.

With increasing levels of obesity, diabetes, and poor cardiovascular health, but also mental disorders (WHO, 2009), Western lifestyle has been associated with emergent health issues in cities, like the constantly rising circulatory diseases (i.e., coronary ischemia) which are the leading causes of mortality in Western countries (Eurostat, 2013). The World Health Organization (1948) has defined health as "a state of complete physical, mental, and social well-being," and not merely the absence of disease. This definition was subsequently accomplished, coming to define the state of

Nature Based Strategies for Urban and Building Sustainability.
DOI: https://doi.org/10.1016/B978-0-12-812150-4.00026-4

health as "a state of balance, that an individual establishes within himself and between himself and his social and physical environment" (Sartorius, 2006). This is an extension of the concept of health, which identifies the environment as a key factor that may support or worsen a person's ability to cope with everyday life (Ward Thompson, 2016).

Nowadays, most people live in urban areas worldwide. As a consequence, it is crucial to take action to increase the quality of urban development, not only to trigger virtuous processes globally (considering the global impact of cities on the environment), but also because the living environment has important repercussions on the health of citizens, as well as on their social relations, working capacity and productivity, with important effects on health and public spending.

IMPACTS OF CITIES GREEN STREETS ON HUMAN HEALTH

As stated in previous chapters, vegetation in cities acts on water and air quality improvement, with important impacts on human environment, i.e., contrasting atmospheric pollution and leading to a reduction of respiratory disorders. Nevertheless, parks, greenways and green streets—as pathways that connect larger green areas—are also essential infrastructures for the practice of physical activity, consequently acting on the reduction of diseases caused by a sedentary lifestyle, such as overweight and obesity (which is rising globally, having increased from 15% to 34% among adults between 1976 and 2008; CDC, 2009).

Physical activity performed within natural environments have resulted more effective compared to activity in synthetic environments. The effects of short-term exposure to a physical context during a walk performed in a natural environment (such as green streets or parks) are stronger than the same activity performed in synthetic environments such as indoor or outdoor built spaces, regarding to better self-reported states of mind (Bowler et al., 2010). A number of studies suggest that the outdoor physical activity improves self-esteem, mental, and psychological health in general, as a result of natural environments' positive impacts on well-being (Barton and Pretty, 2010; Hansmann et al., 2007).

In a broader sense, research on physician-assessed morbidity refers evidence for a positive relation between living in green environments and self-reported indicators of physical and mental health. In particular, the annual prevalence rate of most disease clusters was lower in living environments with more green space in a 1 km radius (Maas et al., 2009). Green streets can thus stimulate a greater involvement of city-dwellers and encourage the performance of healthy physical activities.

Furthermore, the opportunity of daily walking along green streets and verdant paths increases life expectancy. A study conducted in Tokyo, lasting 5 years and conducted on 3000 participants aged between 74 and 91 years, states that those living near walkable green spaces were more long-lived, independently from gender, socioeconomic, and civil state (Takano et al., 2002).

BENEFITS OF VEGETATION AND NATURAL ENVIRONMENTS ON PSYCHOLOGICAL HEALTH

The psychological effects of natural environments are recognized by well-established researches. As stated by Almo Farina et al. (2007), if the appearance of landscape influences the human psyche through perception, therefore a landscape that reproduces the signs of nature can be defined as a therapeutic environment. Beyond promoting physical activity, in fact, the mere perception of green and vegetation enhances psychological health. Access to green spaces has a proven beneficial effect on stress reduction and numerous studies have shown the correlation between the availability of green and disorders such as depression and anxiety (Maas et al., 2009; Bowler et al., 2010; White et al., 2013).

A study based on the Center for Disease Control and Prevention data on 60,000 people in the United States, considering stress levels, has confirmed that people living in urban areas were significantly more stressed than those who lived in rural areas (Austin, 2014).

Even just a walk in a more natural environment is a beneficial exercise: performing this simple activity in green settings can reduce attention deficit-hyperactivity disorder symptoms (AD/HD) in children. A study on 16 children aged between 7 and 12 years suffering from AD/HD reported the effects of a 20 minutes walk in three environments with different levels of greenery by measuring the ability to perform tasks that required concentration. The three environments were an urban park, a suburban neighborhood, and a central area of the city. Despite all the paths being in good condition and with minimum levels of traffic, the walk in the park resulted as the most effective to achieve a significant increase in the ability to concentrate, an outcome similar to usual pharmacological therapy (Taylor et al., 2001). This also highlights the importance of the availability of greenways and green areas close to schools for children.

IMPORTANCE OF EVERYDAY PERCEIVED LIVING LANDSCAPE

The positive effect of green on psychological health is commonly recognized not only by specialists, but also by a huge number of people who regularly visit national parks and protected natural areas. In spite of the relative rarity of pristine natural environments, it is mandatory to focus on the proximity green that ultimately offers even greater benefits to city-dwellers. In fact, if some exceptional places can be visited only a few times in life, urban landscape is an everyday environment with which to cope, directly affecting the health (regarding stress-related illnesses) and everyday quality of life of town-dwellers (Wen et al., 2006).

According to Grahn and Stigsdotter (2003), planning and realizing more green spaces and making these more accessible, produces more restorative and salutogenic environments, considering that "people do not usually compensate for lack of green environments in their own residential area with more visits to public parks." This leads to an important consideration: significant positive effects on the health of town-dwellers is possible only if green streets and parks are widespread and available as part of everyday life.

Furthermore, aesthetically, structurally, and ecologically altered everyday-environments, although not polluted, can cause serious problems to human health. Even in the absence of specific pollutants, harmful health conditions can be produced simply because of the structural and spatial dysfunctions of ecological systems (Ingegnoli, 2015). The spatial deconstruction and the lack of aesthetic quality of the landscape can, in fact, trigger tadrenergic stress, the alarm system of an organism activated by environmental conditions. Recovery from stress, ordinary matter in a more natural environment, can be far away and relatively expensive in cities, with an increase of the environmental alarm status. As a consequence, the spontaneous rebalancing of stress, feasible in a more natural environment, becomes difficult: in cities the syndromes have increased from episodic to epidemic, and morbidity has become high (Ingegnoli, 2015; Fig. 1).

It is therefore compelling for city planning to be oriented towards the restoration of human habitat, able to reduce the environmental alarm, or the adrenergic stress, and not merely focused on air or water pollution.

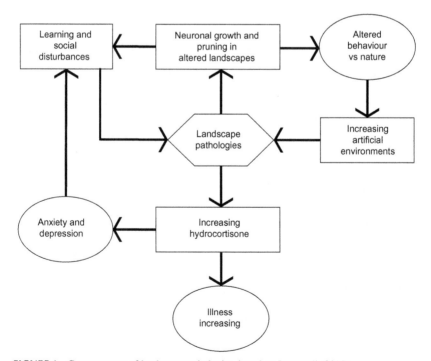

FIGURE 1 Consequences of landscape pathologies, based on Ingegnoli (2015).

It is necessary not only to promote green spaces in urban areas, but "to put back biodiverse, functioning ecosystems" (Connop et al., 2016; Hostetler et al., 2011).

ENVIRONMENTAL JUSTICE AND GREEN STREETS

Pursuing equity of access to green environments is also a key element of social sustainability. In fact, green areas in urban context promote social cohesion, increasing contact between community members and reducing crime rates (Dillen et al., 2012; Tarran, 2009). According to recent studies, green streets and vegetated areas have up to 50% lower crime levels than areas with no vegetation and a 10% increase in the amount of tree cover is associated with a 12% decrease in crime rates (Mullaney et al., 2015; Troy et al., 2012), probably as a result of a greater sense of community spread among residents. Moreover, green space is an important key to support social justice; according to Mitchell and Popham (2008), green environments are important to reduce socioeconomic health inequalities: health inequalities related to income deprivation result lower in populations living in the greenest areas. Equity of access to health-supportive environments is a key feature to address health inequalities, and, broadly, environmental justice in cities (Pearce et al., 2010; Ward Thompson, 2013).

Furthermore, green street programs can increase social cohesion involving local communities participation and sustainable political agendas. For example, in New York City there are numerous programs that are being currently implemented to increase citizens' participation. *GreeNYC* is aimed at encouraging residents to adopt sustainable practices in their daily environments and *MillionTreesNYC* is a public-private program intending to plant and care for one million new trees across the city's five boroughs over the next decade (Perini and Sabbion, 2017).

GREEN STREETS' RECREATIONAL VALUE IN SPATIAL PLANNING

Street greening at the moment is a growing practice especially in North America, Australia, and Europe to counteract soil erosion and to improve stormwater management, increasing soil water storage potential based on vegetation potential (Thompson and Sorvig, 2007). The design of green streets has been recently implemented in cities such as Portland, Seattle, New York City (Fig. 2), and Chicago, where a great number of permeable pavements and natural drainage projects with a subsoil infiltration system is currently being planned and implemented.

This action also supports a rise of recreational value of entire boroughs and areas. This is due to the fact that green streets are often planned together with other sustainable design practices, such as pedestrian and cycle paths,

FIGURE 2 Hudson River Park Greenway, New York City.

traffic moderation, and scenic amelioration of streetscapes. Moreover, green streets are often realized transforming unused road space, traffic islands, and former industrial areas into green assets, as in the cases of NYC and Portland.

To enhance their recreational value, green streets should be flexible and adaptive for multiple and sociable uses. Attributes of green streets should improve neighborhood walkability, reducing traffic levels and speed, and promoting pedestrian and cyclist use. Delivering opportunities for resting, providing a number of different options of seating (in sun or shade, sheltered from the wind, out of a café, or informally on steps or benches) is a good way to increase recreational value of green streets (Ward Thompson, 2013).

Mixing natural features with human activities (cafés, shopping opportunities, events-spaces) can increase aesthetic values, as a lively sidewalk is a prime attraction for street users. The highest level of recreational value in pedestrian streets is reached when natural features are combined with the opportunity to see, hear, and meet other people, allowing to access from larger spaces to smaller and more intimate ones (Gehl, 2011).

CONCLUSION

Virtuous processes generated in the most advanced cities define the current trend of promoting the reappropriation of abandoned urban areas, through requalification, walkability, and forestation, supporting new ecologies, to ensure a better quality of streetscape. As a result, the most advanced countries in planning have "high environmental metastability and greater longevity" (Ingegnoli, 1986).

A number of studies have shown that increasing the percentage of green space in people's living environment has a positive effect on the perceived health of residents. This confirms that urban landscape is "more than just a luxury and consequently the development of green space should be allocated a more central position in spatial planning policy" (Maas et al., 2006). In fact, the establishment of such a virtuous system not only works to the advantage of the environmental and ecological balances (both locally and globally), but it is essential for the well-being of future generations, which will choose the city as a primary place of opportunity, relationship, and sustainable development.

REFERENCES

Austin, G., 2014. Green Infrastructure for Landscape Planning: Integrating Human and Natural Systems. Routledge, Abingdon, Oxon.

Barton, J., Pretty, J., 2010. What is the best dose of nature and green exercise for improving mental health - a multi-study analysis. Environ. Sci. Technol. 44, 3947–3955.

Bowler, D.E., Buyung-Ali, L.M., Knight, T.M., Pullin, A.S., 2010. A systematic review of evidence for the added benefits to health of exposure to natural environments. BMC Public Health 10, 456.

CDC, National Center for Health Statistics (NCHS), 2009. National Health and Nutrition Examination Survey [WWW Document]. US Department of Health and Human Services, CDC, NCHS, Hyattsville, MD, http://www.cdc.gov/nchs/nhanes/nhanes_questionnaires.htm (accessed 10.23.16).

Connop, S., Vandergert, P., Eisenberg, B., Collier, M.J., Nash, C., Clough, J., et al., 2016. Renaturing cities using a regionally-focused biodiversity-led multifunctional benefits approach to urban green infrastructure. Environ. Sci. Policy 62, 99–111.

Dillen, S.M.E., van, Vries, S., de, Groenewegen, P.P., Spreeuwenberg, P., 2012. Greenspace in urban neighbourhoods and residents' health: adding quality to quantity. J. Epidemiol. Community Health 66, e8–e8.

Eurostat, 2013. Causes of death—standardised death rate, 2013 [WWW Document]. http://ec.europa.eu/eurostat/statistics-explained/index.php/File:Causes_of_death_%E2%80%94_standardised_death_rate,_2013.png (accessed 10.23.16).

Farina, A., Scozzafava, S., Napoletano, B., 2007. Paesaggi terapeutici: basi paradigmatiche e potenzialità applicative in Paesaggi terapeutici, a cura di A. Ghersi, Alinea, Firenze.

Gehl, J., 2011. Life Between Buildings: Using Public Space, 6 edizione. ed. Island Press, Washington, DC.

Grahn, P., Stigsdotter, U.A., 2003. Landscape planning and stress. Urban For. Urban Gree. 2, 1–18.

Hansmann, R., Hug, S.-M., Seeland, K., 2007. Restoration and stress relief through physical activities in forests and parks. Urban For. Urban Gree. 6, 213–225.

Hostetler, M., Allen, W., Meurk, C., 2011. Conserving urban biodiversity? Creating green infrastructure is only the first step. Landscape Urban Plan. 100, 369–371.

Ingegnoli, V., 1986. Considerazioni sul rapporto fra transizione demografica e crisi ecologica. In: Ecologia e Longevità, Atti Conv. Naz. diEcologia Umana, Firenze, Boll. SIEU:53-63.

Ingegnoli, V., 2015. Landscape Bionomics Biological-Integrated Landscape Ecology. Springer, Milano.

Maas, J., Verheij, R.A., Groenewegen, P.P., De, V., Spreeuwenberg, P., 2006. Green space, urbanity, and health: how strong is the relation? J. Epidemiol. Community Health 60, 587–592.

Maas, J., Verheij, R.A., De, V., Spreeuwenberg, P., Schellevis, F.G., Groenewegen, P.P., 2009. Morbidity is related to a green living environment. J. Epidemiol. Community Health 63, 967–973.

Mitchell, R., Popham, F., 2008. Effect of exposure to natural environment on health inequalities: an observational population study. Lancet 372, 1655–1660.

Mullaney, J., Lucke, T., Trueman, S.J., 2015. A review of benefits and challenges in growing street trees in paved urban environments. Landscape Urban Plan. 134, 157–166.

Pearce, J.R., Richardson, E.A., Mitchell, R.J., Shortt, N.K., 2010. Environmental justice and health: the implications of the socio-spatial distribution of multiple environmental deprivation for health inequalities in the United Kingdom. Trans. Instit. Bri. Geogr. 35, 522–539.

Perini, K., Sabbion, P., 2017. Urban Sustainability and River Restoration: Green and Blue Infrastructure. Wiley Blackwell, Chichester.

Sartorius, N., 2006. The meanings of health and its promotion. Croat. Med. J. 47, 662–664.

Takano, T., Nakamura, K., Watanabe, M., 2002. Urban residential environments and senior citizens' longevity in megacity areas: the importance of walkable green spaces. J. Epidemiol. Community Health 56, 913–918.

Tarran, J., 2009. People and trees, providing benefits, overcoming impediments. Proceedings of the 10th National Street Tree Symposium, pp. 63–82.

Taylor, A.F., Kuo, F.E., Sullivan, W.C., 2001. Coping with ADD The surprising connection to green play settings. Environ. Behavior 33, 54–77.

TEEB, 2011. TEEB Manual for cities: ecosystem services in urban management. In: UNEP and the European Union (ed.), The Economics of Ecosystems and Biodiversity. Manual for Cities: Ecosystem Services in Urban Management.

Thompson, J.W., Sorvig, K., 2007. Sustainable Landscape Construction: A Guide to Green Building Outdoors, second ed. Island Press, Washington.

Troy, A., Morgan, G., O'Neil-Dunne, J., 2012. The relationship between tree canopy and crime rates across an urban-rural gradient in the greater Baltimore region. Landscape Urban Plan. 106, 262–270.

Ward Thompson, C., 2013. Activity, exercise and the planning and design of outdoor spaces. J. Environ. Psychol. 34, 79–96.

Ward Thompson, C., 2016. Editorial: landscape and health special issue. Landscape Res. 41, 591–597.

Wen, M., Hawkley, L.C., Cacioppo, J.T., 2006. Objective and perceived neighborhood environment, individual SES and psychosocial factors, and self-rated health: an analysis of older adults in Cook County, Illinois. Soc. Sci. Med. 63, 2575–2590.

White, M.P., Pahl, S., Ashbullby, K., Herbert, S., Depledge, M.H., 2013. Feelings of restoration from recent nature visits. J. Environ. Psychol. 35, 40–51.

World Health Organization, 1948. Preamble to the constitution of the World Health Organization as adopted by the International Health Conference in New York, 22 July 1946 and entered into force on 7 April 1948.

WHO (Ed.), 2009. Global Health Risks: Mortality and Burden of Disease Attributable to Selected Major Risks. World Health Organization, Geneva.

Chapter 4.4

Economic Benefits and Costs of Vertical Greening Systems

Paolo Rosasco

Chapter Outline

INTRODUCTION

Like other green building envelope solutions, most of vertical greening systems (VGS) are characterized by high installation and maintenance costs but also by a large range of individual and social benefits, which, most of them, can be evaluated in economic terms.

The main personal economic benefits are related to the energy savings for winter heating (Krusche et al., 1982; Mc Pherson et al., 1998), summer cooling (Mazzali et al., 2012; Alexandri and Jones, 2008; Dunnet and Kingsbury, 2008), and with the increase of the durability of some building components, like plaster facade (Perini and Rosasco, 2013).

Other economic benefits are related to the increase of real estate value because the solutions improve the aesthetic quality of the building and the internal comforts for residential (due to the lower sound level environments)

Nature Based Strategies for Urban and Building Sustainability.
DOI: https://doi.org/10.1016/B978-0-12-812150-4.00027-6

(Veisten et al., 2012). The social benefits are mainly related with the increase of environmental quality nearby dense urban areas where these solutions are applied: greenhouse gases output reduction, heat island phenomena reduction, better air quality due to the abatement of pollution, indoor and outdoor comfort conditions improvement, urban wildlife (biodiversity) (Collins et al., 2017).

Unlike green roofs, widespread in the countries of northern Europe and America and characterized by a several studies which investigate their economic cost and benefits (Clark et al., 2008; Bianchini and Hewage, 2012; Mahdiyar et al., 2016), in many countries green facades are less widespread because they are still today considered expensive architectural solutions; this is also related to the economic evolution of some costs and benefits because of the scarcity of data related to: the lifespan of systems and components, the maintenance and management costs (pruning, irrigation, repairs, etc.) within the lifespan, and their frequency.

This chapter will provide information about the costs and benefits of vertical greenery systems, both in a private and social perspective and also describe their importance on the main economic sustainability indicators (net present value—NPV).

PRIVATE COSTS OF VERTICAL GREENING SYSTEMS

Installation Costs

The installation and maintenance costs related to a VGS depend on some factors like: location (urban or peripheral area, local climate, etc.), type of plants and system (direct, indirect, living wall) (Table 1); they can be evaluated with a bill of quantities. Many costs are supplied by manufacturing companies (farmers, etc.).

An indirect green facade made by high-density polyethylene (HDPE) costs about 125 €/m^2 of facade (plant species, supporting system, and installation); if made with steel mesh the cost is about 240 € m^{-2}. For the indirect green facades combined with planter boxes the costs are higher: about 165 € m^{-2} if made with HDPE and 330 € m^{-2} if made with steel; besides these systems, as well as the living wall, require an irrigation system which costs about 10 € m^{-2}. The living wall system is the most expensive system; the panels and plant species cost from a minimum of 185 € m^{-2} to a maximum of 500 € m^{-2}, also the cost of irrigation system is much higher: about 30 € m^{-2}.

Transportation costs of VGS depend on the distance between building and the manufacturing companies' location. The design cost depends on the complexity of the green facade solution; it can vary from a minimum of 6% to a maximum of 10% of the total cost. The disposal cost at the end of lifespan includes the greening systems disposal (removal of plants and structure,

TABLE 1 Private and Social Costs Related to a Vertical Greening System

Type of Green Facade	Perspective	Category	Cost	Cost Range ($€\ m^{-2}$ Facade)	Time Frame
Direct green facade	Private and social	Initial	Design	6%–10% of installation cost	One time
			Dig + pot	450–550 ($€\ m^{-2}$)	One time
			Plant species and installation	18–28	One time
			Irrigation system	4–7	One time
		Maintenance	Pruning	2.5–3.5	Annual
			Irrigation (H_2O)	0.2–0.5	Annual
			Pipes replacement (irrigation system)	0.3–0.4	Annual
		Disposal	Green layer disposal	30–45	One time (end of lifespan)
	Social	Tax incentives	Tax reduction	Depends on local regulation	One time or annual
Indirect green facade	Private and social	Initial	Design	6%–10% of installation cost	One time
			Dig + pot	400–550 ($€\ m^{-2}$)	One time
			Supporting system and transportation	35–110	One time
			Installation of supporting system	70–100	One time

(Continued)

TABLE 1 (Continued)

Type of Green Facade	Perspective	Category	Cost	Cost Range (€ m^{-2} Facade)	Time Frame
(HDPE or steel)			Plant species and installation	18–28	One time
			Irrigation system	4–7	One time
		Maintenance	Pruning	2.5–3.5	Annual
			Irrigation (H$_2$O)	0.2–0.5	Annual
			Pipes replacement (irrigation system)	0.3–0.4	Annual
		Disposal	Green layer disposal	160–220	One time (end of lifespan)
	Social	Tax incentives	Tax reduction	Depends on local regulation	One time or annual
Indirect green facade	Private and social	Initial	Design	6%–10% of installation cost	One time
with planter boxes			Supporting system and transportation	35–110 (€ m^{-2})	One time
			Installation of supporting system	70–100	One time
(HDPE or steel)			Plant species	25–50	One time
			Planter boxes	35–70	One time
			Irrigation system	25–35	One time
		Maintenance	Pruning	4–6	Annual
			Irrigation (H$_2$O)	1.0–1.5	Annual

		Plant species replacement (5%)	3.5–5.0	Annual
		Pipes replacement (irrigation system)	2–3	Annual
Social	Disposal	Green layer disposal	160–220	One time (end of lifespan)
	Tax incentives	Tax reduction	Depends on local regulation	One time or annual
Living wall system	Private and social	Initial	Design	6%–10% of installation cost	One time
		Panels, installation, and transportation	160–450 (€ m^{-2})	One time
		Plant species	25–50	One time
		Irrigation system	25–35	One time
	Maintenance	Pruning and panels adjustment	12–18	Annual
		Irrigation (H2O)	1.0–1.5	Annual
		Panels replacement (5%)	5–8	Annual
		Plant species replacement (10%)	3.5–5.0	Annual
		Pipes replacement (15%)	2–3	Annual
	Disposal	Green layer disposal	180–240	One time (end of lifespan)
Social	Tax incentives	Tax reduction	Depends on local regulation	One time or annual

transport to landfill, and dump taxes); they vary from 30 to 45 € m^{-2} for a direct green facade and from 180 to 240 € m^{-2} for a living wall system (Table 1).

Maintenance Costs

The maintenance of plants and vertical systems play an important role in the global cost of the system within lifespan. The cost and frequency of pruning and plant species replacement are indicated by companies and it depends on the vertical green system, plant species, and local climate (solar irradiation, local temperature, and humidity, etc.).

The annual pruning cost can vary from 2.5 € m^{-2} for a direct and indirect green facade with *Hedera helix* and without planter boxes (with planter boxes from 4.0 € m^{-2}) to 18 € m^{-2} for a living wall system (including panels adjustment) (Table 1).

The annual plants species replacement vary widely reaching up 50% annually in some cases (Pérez-Urrestarazu et al., 2015); normally, it can vary between 5% and 15% (Perini and Rosasco, 2013; Ottelè et al., 2011; Pérez-Urrestarazu et al., 2015).

Also, the annual irrigation cost depends on local climate (air temperature and humidity, incoming solar energy, speed of air flow, vegetation type) (Pérez-Urrestarazu et al., 2014).

The maintenance of the irrigation system (annual pipes replacement) depends on the length, diameter, and materials of pipes; it can reach the 15% of the total length pipes annually. For vertical green facades, it can vary from a minimum of 0.3 € m^{-2} for an indirect green facade (with a plot at the base of the facade) to a maximum of 3 € m^{-2} for a living wall system (Table 1).

SOCIAL COSTS OF A VERTICAL GREENING SYSTEM

The tax reduction can be considered a social cost because it reduces the public tax revenue. Their amounts and modality depend on the national and local law; in Italy they can be applied when the VGS (like other building systems) increase the insulation performance of the building and, in 2017, can reach up the 65% of the installation and design costs.

PRIVATE BENEFITS OF A VERTICAL GREENING SYSTEM

Also the private benefits are related to the type of vertical greening system; they can all be evaluated in monetary terms (Table 2).

TABLE 2 Private and Social Benefits Related to a Vertical Greening System

Perspective	Benefit	Time Frame	Evaluation Modality
	Energy saving for cooling	Annual	Quantitative
	Energy saving for heating	Annual	Quantitative
Private	Longevity facade	One time	Quantitative
	Aesthetic (real estate value)	Annual/one time	Quantitative
	Noise isolation	Annul	Quantitative
	Air quality improvement (No_2; PM_{10})	Annual	Quantitative
	Carbon reduction	Annual	Quantitative
Social	Urban heat island mitigation		Qualitative
	Urban hydrology		Qualitative
	Habitat creation (biodiversity)		Qualitative

Building Insulation

Some recent studies (Cameron et al., 2015; Hunter et al., 2014; Pérez et al., 2011) have shown that green facade may modulate indoor temperature, thank a combination of different actions: interception of solar radiation by vegetation, thermal insulation provided by the vegetation and substrate (especially for living wall systems), evaporative cooling due to evapotranspiration, modification of wind impact on the building facade. Susorova et al. (2013) show that a plant layer added to the facade can improve its thermal resistance.

Comparing temperature measures on a plant covered wall and a bare wall during summer, Bartfelder and Köhler (1987a,b) show a temperature reduction at the green facade in a range of $2-6°C$.

Pérez et al. (2017) show the high capacity of leaves to intercept the direct solar radiation, which implies representative reductions on the external surface wall temperatures up to $10.1°C$ on the south orientation and indoor temperature reductions around $2.5°C$.

For a living wall system located at Wuban (China) characterized by hot and humid climate, Chen et al. (2013) discovered that the system has notable cooling effect; the temperature of exterior wall surface is reduced by a maximum of $20.8°C$, the temperature of interior wall surface by a maximum of $7.7°C$ while the temperature of interior space is reduced by a maximum of $1.1°C$; moreover, the living wall system does not bring more humidity into the indoor space.

Eumorfopoulou and Kontoleon (2009) have reported a temperature cooling potential of plant covered walls in Mediterranean climate; the reduction can reach up 10.8°C. Davis et al. (2016a) demonstrate that a vertical garden can cool the external air near the building facade by approximately 6°C while Mazzali et al. (2013) show that the presence of a living wall system outside a building can change the external temperature of a wall up to 20°C.

During the summer, a VGS significantly reduce the cooling energy demands (Meier, 2010).

Hasan et al. (2012) showed that the green facade can save 9.5%−18.0% of the cooling energy consumption in commercial buildings.

In a case study developed for the city of Toronto (Canada), Bass and Baskaran (2001) discover that the shading and cooling effect of a VGS reduce the energy needed for cooling by approximately 20%. For a north facade with a living wall system the power consumption for heating is reduced up to 37% (Carlos, 2015).

Annual energy saving for cooling and heating can be evaluated by technical software which allows comparing the insulation performance of building before and after the vertical green installation.

Plaster Renovation

The adoption of VGS increase the building plaster facade durability reducing the frequency of intervention thanks to a protective action given by the leaves (reduces the quantity of UV light, the rain-off along the facade, etc.); this effect is evident in the case of a living wall system in which an addition protective action on plaster is provided by continuous supporting layers.

The savings come from the minor plaster reconstruction works and the postponement of the intervention over time; it can be evaluated comparing the discounted cost of facade plaster renovation without the vertical green system at time "t_1" (usually varying from 25 to 30 years) and the same cost time at "$t_1 + n$" in which "n" is the increase of plaster lifespan due to the presence of the vertical green system (about 10−15 years).

In a case study located in Genoa (north Italy), Perini and Rosasco (2013) estimate a total savings of 91 € m^{-2} for an indirect green facade with planter boxes and 108 € m^{-2} for a living wall system within 50 years of lifespan.

Real Estate Value

Vertical greenery also increases the real estate value of the building according to specific areas of local real estate market. Several authors evaluate the value property increase due to the presence of a green wall; Hunt (2008) does not differentiate between green walls and green roofs and states that the premium of a house value can vary from 3.9% to 15.0%. Peck et al. (1999) assumed that a green roof or a green wall would yield the same property increase as a "good tree cover" and assume a value increase interval for a property of 6%−15%. In a cost-benefitaAnalysis of a green wall affecting courtyards Veisten et al. (2012) assume an aesthetic appreciation of 5.8 € per household converted to 2.4 €/person per year. The presence of a vertical greening system is relevant to three aesthetic real estate features: the typology of building, the presence of a green surface, the lower acoustic and air pollution within the building due to the presence of green layer.

The real estate value can be evaluated considering the different economic impacts of the three features recognized in different areas of the real estate market; for the real estate market of Genoa (Italy), Sdino (1994) indicate— for the three real estate futures—a total percentage increase value between 2.0% (for a location in a central urban area) and 5.0% (for a location in a peripheral urban area).

Noise Isolation

The benefits of the vertical green system also comprise the noise attenuation, especially for living wall systems, as the substrate where plants grow has a sound absorbing effect (Azkorra et al., 2015; Davis et al., 2016b). Veisten et al. (2012) estimated that a 9.2 m high green wall facade reduces the noise level by 4.1 while a 3 m high green wall facade reduces the noise level by 4.5 dB; in a CBA of a green wall affecting courtyards they assume a noise attenuation benefits 10.095 € per person (living inside the building) per dB(A) per year (valid for noise levels below 71 dB(A)).

SOCIAL BENEFITS OF VERTICAL GREENING SYSTEMS

The social benefits of VGS relate to air quality improvement, carbon reduction, habitat creation, and urban heat island (UHI) mitigation. As shown in Table 2 some of these benefits can't be quantified in an economical way due to a lack of reliable data from literature or difficulty to estimate their effect connected to a single VGS.

Air Quality Improvement

Air quality depends on the amount of dust, particulates, and nitrates in the air (Carter and Keeler, 2007; Peck et al., 1999).

Several studies have demonstrated the important role of green against air pollution; in an urban street with trees there are only 10%−15% of the total dust particles of similar streets without trees (Johnston and Newton, 1996). Gaseous pollutants can be dissolved or sequestrated through stomata on plants and leaves (McPherson et al., 1994; McPherson et al., 1998).

The oxygen-carbon dioxide exchange rate differs between plant types; Ottelé et al. (2011) demonstrated that *H. helix* collects particulate matter, resulting in air quality improvement (although difficult to quantify). Yang et al. (2008), Currie and Bass (2005) estimate that 1 ha of green roofs remove between 72 and 85 kg of pollutants ($7.2 \bullet 10^{-3}$ kg m^{-2}–$8.5 \bullet 10^{-3}$ kg m^{-2}).

According to Currie and Bass (2008) a green wall absorbs slightly less than an extensive roof (-61% of NO_2, -65% of O_3, -62% of SO_2) except for PM_{10} (-37%).

In a cost-benefit analysis (CBA) applied on six vertical green systems located in Genoa (Italy), Perini and Rosasco (2013) assumed the 50% of the removal value assumed by Currie and Bass (2008) for green roofs, and estimated an annual social benefit for air quality improvement of $9.4 \bullet 10^{-3}$ to $10.9 \bullet 10^{-3}$ € m^{-2} year^{-1}.

Carbon Reduction

VGS also reduce greenhouse gas emissions as a result of reducing the demand for energy. A study developed in two Chicago neighborhoods quantified the effects of shade trees around buildings (shading, evapotranspiration, and wind speed reduction) and found a decrease in carbon emission by 3.2%–3.9% per year for building types in two residential blocks located in north-west Chicago where tree cover was respectively 33% and 11% per residential unit (Jo and McPerson, 2001).

The oxygen-carbon dioxide exchange rate differs between plants types (Bianchini and Hewage, 2012) installed on a VGS; for the evaluation of carbon reduction due to a VGS installation, the same air pollution removal range considered by Currie and Bass (2005) ($7.2 \bullet 10^{-3}$ kg m^{-2}–$8.5 \bullet 10^{-3}$ kg m^{-2}) for a green roof can be assumed.

Urban Hydrology

A vertical green system can also improve the urban hydrology, reducing the storm water to the public sewers during rainfall water (Carter and Jackson, 2007). The effect is evident if the vertical green system is combined with roof technologies that act as a water buffer and a water collection system even if is difficult to quantify the benefit of VGS.

Habitat Creation

The integration of vegetation with VGS in built spaces can improve biodiversity creating a habitat for microorganisms and also for smaller animals (bees, bats, birds, etc.). A study conducted by Köhler (1993) shows that mainly sparrows, blackbirds, and green finches were found between the climbers of green facades in Berlin; green fac,ades, green roofs, and vegetation in general, function hereby as a food source (insets) and as a nesting or breeding opportunity. Green systems also enhance urban diversity by allowing spontaneous vegetation to colonize these systems (Dunnet and Kingsbury, 2008).

A recent study developed by Collins et al. (2017) in Southampton (south England) reveal a positive value associated with green infrastructure (like living walls) that increase biodiversity.

According to Bianchini and Hewage (2012) this can be considered as a social benefit, however habitat creation is not a common investment in many cities. Due to the difficulties to evaluate this benefit in an economic way, for the economic evaluation of VGS a qualitative assessment instead of a quantitative one can be done.

Urban Heat Island

Due to the capability of plants to absorb a considerable quantity of solar radiation for their biological function, the green facade plays an important role in shaping the urban microclimate (Pérez et al., 2017; Wong et al., 2010a); at district and urban scale, if combined with green roofs they can significantly reduce the UHI effect (Gomez et al., 1998) by reducing the amount of reradiated heat (Bass and Baskaran, 2001).

The UHI phenomenon can cause air temperature in the cities to be 2−5 degrees Celsius higher than those in the surrounding rural areas, mainly due to the amount of artificial surfaces (high albedo) compared with natural land cover and by atrophic activities (Taha, 1997; Oke, 1989; Roth et al., 1989; Quattrochi et al., 2010).

Alexandri and Jones (2006) observed a 8−9°C temperature reduction in an urban canyon (with tall buildings on both sides of the street) when walls and roofs were covered with vegetation. Another positive effect is the fact that the air introduced from outside in the air conditioning system has a lower temperature, allowing an electrical energy savings for its cooling (Wong et al., 2010b).

The benefit is quantified in a qualitative way, due to the impossibility to evaluate the effect of a single green facade on the UHI mitigation.

THE USE OF SENSITIVITY ANALYSIS FOR VGS ECONOMIC ASSESSMENT

A sensitivity analysis of two VGS (225 m^2 on a south facade in a building office located in Genoa,Italy) shows how the variables influence the net present unitary value (NPV) calculated considering only the personal costs and benefits (Perini and Rosasco, 2013, 2016). System 1 is an indirect green facade (with a well grown *H. helix* supported by steel mesh) while System 2 is an indirect green facade (steel planter boxes with a well grown *H. helix* supported by a steel mesh). For System 1, the main economic factors which negatively influence the NPV are the installation and maintenance costs (Fig. 1).

The same trend is show by System 2 even if the maintenance costs are more relevant due to the presence of planter boxes and the irrigation system (Fig. 2).

For both systems, the longevity facade positively influence the NPV: an increase of 5 years (+ 15% of lifespan) increases the NPV about 15% due to the postponement of costs restoration.

Economic savings for installation and maintenance cost, which can be increased with careful design and material choice, are relevant for both two systems while discount and inflation rate are less relevant factors.

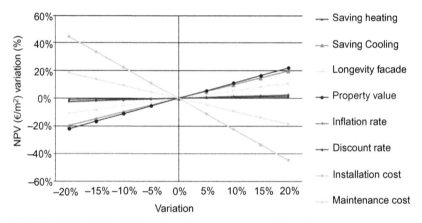

FIGURE 1 System 1: Sensitivity analysis for personal costs and benefits.

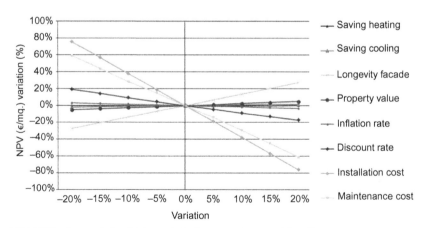

FIGURE 2 System 2: Sensitivity analysis for personal costs and benefits.

CONCLUSION

External VGS are technical and architectural solutions that have many personal and social benefits; the high installation and maintenance costs within the lifespan are balanced with economic private benefits such as energy savings for cooling and heating, lifespan increase of building plaster, and higher real estate value.

There are also many social benefits such as carbon reduction, better urban hydrology, biodiversity and habitat creation, and reduction of UHI; although these benefits are less relevant in economic terms, they are very important in a social point of view because they improve the quality of urban environments and alleviate climate change.

In these last few years, new technical solutions are developed in order to improve the performance and reduce costs; some studies have also demonstrated that green facades are also economically sustainable (Perini and Rosasco, 2013, 2016).

As shown by a sensitivity analysis of two VGS installed in a south facade of a building office located in Genoa, the economic sustainability of vertical greening system is strongly influenced by the installation and maintenance costs and energy savings for heating. For the dissemination of these systems a careful assessment of costs and benefits is essential, not only at private, but also at social scale.

A key role is played by the public administration that can, by means of tax incentives, promote the implementation of these systems at urban scale, providing a better city environment and reducing some of the negative effects generated by climate change.

REFERENCES

Alexandri, E., Jones, P., 2006. Temperature decreases in an urban canyon due to green walls and green roofs in diverse climates. Build. Environ. 43 (4), 480–493.

Alexandri, E., Jones, P., 2008. Temperature decrease in a urban canyon due to green walls and green roofs in diverse climates. Build. Environ. 43, 480–493.

Azkorra, Z., Pérez, G., Coma, J., Cabeza, L.F., Bures, S., Alvaro, J.E., et al., 2015. Evaluation of green walls as a passive acoustic insulation system for buildings. Appl. Acoust. 89, 46–56.

Bartfelder, F., Köhler, M., 1987a. Experimentelle untersuchungen zur function von fassadenbegrünungen, Berlin.

Bartfelder, F., Köhler, M., 1987b. Experimentelle untersuchungen zur function von fassadenbegrünungen, Abbildungen, tabellen und literaturverzeichnis, Berlin.

Bass, B., Baskaran, B., 2001. Evaluating Rooftop and Vertical Gardens as an Adaptation Strategy for Urban Areas. Institute for Research and Construction NRCC-46737, Project number A020 http://www.roofmeadow.com/technical/publications/BBass_GreenRoofs_2001.pdf.

Bianchini, F., Hewage, K., 2012. Probabilistic social cost e benefit analysis for green roofs: a lifecycle approach. Build. Environ. 58, 152–162.

Cameron, R.W.F., Taylor, J., Emmett, M., 2015. A Hedera green façade e Energy performance and saving under different maritime-temperate, winter weather conditions. Build. Environ. 92, 111–121.

Carlos, J.S., 2015. Simulation assessment of living wall thermal performance in winter in the climate of Portugal. Build. Simulat. 8, 3–11.

Carter, T., Jackson, C.R., 2007. Vegetated roofs for stormwater management at multiple spatial scales. Landscape Urban Plan. 80, 84–94.

Carter, T., Keeler, A., 2007. Life-cycle cost benefit analysis of extensive vegetated roof systems. J. Environ. Manage. 87 (3), 350–363.

Chen, Q., Li, B., Liu, X., 2013. An experimental evaluation of the living wall system in hot and humid climate. Energy Build. 6, 298–3017.

Clark, C., Adriaens, P., Talbot, F.B., 2008. Green roof valuation: a probabilistic economic analysis of environmental benefits. Environ. Sci. Technol. 42, 2155–2161.

Collins, R., Schaafsma, M., Hudson, M.D., 2017. The value of green walls to urban biodiversity. Land Use Policy 114–123.

Currie, B.A., Bass, B., 2005. Estimates of Air Pollution with Green Plants and Green Roofs Using the UFORE Model. In: Proc. of 3rd North American Green Roof Conference Greening Rooftops for Sustainable Communities, Washington, 4–6 May, 495–511.

Currie, B.A., Bass, B., 2008. Estimate of air pollution mitigation with green plants and green roofs using the UFORE model. Urban Ecosyst. 11, 409–422. Available from: https://doi.org/10.1007/s11252-008-0054-y.

Davis, M.J.M., Ramirez, F., Pérez, M.E., 2016a. More than just a Green Façade: vertical gardens as active air conditioning units. Proc. Eng. 145, 1250–1257.

Davis, M.J.M., Tenpierik, M.J., Ramírez, F.R., Pérez, M.E., 2016b. More than just a Green Facade: the sound absorption properties of a vertical garden with and without plants. Build. Environ. Accepted Manuscript 10 January 2017.

Dunnet, N., Kingsbury, N., 2008. Planting Green Roofs and Living Walls. Timber Press, Oregon.

Eumorfopoulou, E.A., Kontoleon, K.J., 2009. Experimental approach to the contribution of plantcovered walls to the thermal behaviour of building envelopes. Build. Environ. 44 (5), 1024–1038.

Gomez, F., Gaja, E., Reig, A., 1998. Vegetation and climatic changes in a city. Ecol. Eng. 10, 355–360.

Hasan, M.M., Karim, A., Brown, R.J., Perkins, M., Joyce, D., 2012. Estimation of energy saving of commercial building by living wall and green facade in sub-tropical climate of Australia.

Hunt, H.D., 2008. Green house values. Tierra Grande 15, 2.

Hunter, A.M., Williams, N.S.G., Rayner, J.P., Aye, L., Hes, D., Livesley, S.J., 2014. Quantifying the thermal performance of green facades: a critical review. Ecol. Eng. 63, 102–113.

Jo, H.K., McPerson, E.G., 2001. Indirect carbon reduction by residential vegetation and planting strategies in Chicago, USA. J. Heat Fluid Flow 29, 364–373.

Johnston, J., Newton, J., 1996. Building green. A guide for using plants on roofs, wall and pavements. London, The London Ecology Unit.

Köhler, M., 1993. Fassaden und Dachbergrunung. Stuttgart: Ulmer Fachbuch Landschafts- und Grunplanung.

Krusche, P., Krusche, M., Althaus, D., Gabriel, I., 1982. Ökologisches bauen, Herausgegeben vom umweltbundesamt, Bauverlag.

Mahdiyar, A., Tabatabaee, S., Nobahar Sadeghifam, A., Reza Mohandes, S., Abdullah, A., Moharrami Meynagh, M., 2016. Probabilistic private cost-benefit analysis for green roof installation: a Monte Carlo simulation approach. Urban For. Urban Gree. 20, 317–327.

Mazzali, U., Peron, F., Scarpa, M., 2012. Thermo-physical performances of living walls via field measurements and numerical analysis. In: Eco-architecture IV. Harmonization between architecture and nature WIT transactions on ecology and the environment, vol. 165 ISBN 978-1-84564.

Mazzali, U., Peron, F., Romagnoni, P., Pulselli, R.M., Bastianoni, S., 2013. Experimental investigation on the energy performance of living walls in a temperate climate. Build. Environ. 64, 57−66.

McPherson, E.G., Nowak, D. J., Rowntree, R.A., 1994. Chicago's Urban Forest Ecosystem: Results of the Chicago Urban Forest Climate Project. (includes Executive Summary). Forest Service General Technical Report (final) (No.PB-94-203221/XAB; FSGTR-NE−186). Forest Service. Northeastern Forest Experiment Station., Delaware, OH (United States).

McPherson, E.G., Scott, K.I., Simpson, J.R., 1998. Estimating cost effectiveness of residential yard trees for improving air quality in Sacramento, California, using existing models. Atmosp. Environ. 32 (1), 75−84.

Meier, K., 2010. Strategic landscaping and air-conditioning savings: a literature review. Energy Build. 15−16, 479−486.

Oke, T.R., 1989. The micrometeorology of the urban forest. Philos. Trans. R. Soc. Lond. 324, 335−349.

Ottelé, M., Fraaij, A.L.A., Haas, E.M., Perini, K., Raiteri, R., 2011. Vertical greening systems and the effect on air flow and temperature on the building envelope. Build. Environ. 46 (11), 2287−2294.

Peck, S., Callaghan, C., Kuhn, M., Bass, B., 1999. Greenbacks from green roofs: forging anew industry in Canada, Status report on benefits, barriers and opportunities for green roof and vertical garden technology diffusion, prepared for Canada Mortgage and Housing Corporation.

Pérez, G., Rincon, L., Vila, A., Gonzalez, J.M., Cabeza, L.F., 2011. Green vertical systems for buildings as passive systems for energy savings. Appl. Energy 12, 4854 e 4859.

Pérez, G., Coma, J., Sol, S., Cabeza, L.F., 2017. Green facade for energy savings in buildings: the influence of leaf area index and facade orientation on the shadow effect. Appl. Energy 187, 424−437.

Perez-Urrestarazu, L., Egea, G., Franco-Salas, A., Fernandez-Canero, R., 2014. Irrigation systems evaluation for living walls. J. Irrigat. Drainage Eng. Available from: https://doi.org/10.1061/(ASCE)IR.1943-4774.0000702.

Pérez-Urrestarazu, L., Fernández-Cañero, R., Franco-Salas, A., Egea, G., 2015. Vertical greening systems and sustainable cities. J. Urban Technol. 22 (4), 65−85. Available from: https://doi.org/10.1080/10630732.2015.1073900.

Perini, K., Rosasco, P., 2013. Cost-benefit analysis for green facades and living wall systems. Build. Environ. 70, 110−121.

Perini, K., Rosasco, P., 2016. Is greening the building envelope economically sustainable? An analysis to evaluate the advantages of economy of scope of vertical greening systems and green roofs. Urban For. Urban Gree. 20, 328−337.

Quattrochi, D.A., Al-Hamdan, M., Estes, M. Jr. - NASA Marshall Space Flight Center, 2010. Heat Island Mitigation Measures in Response to Climate Change Impacts. In Urban Climate: Local, Regional & Global Impacts, 2010. Association of American Geographers (AAG) Meeting, 14-18 Apr. 2010, Washington, DC, United States.

Roth, M., Oke, T.R., Emery, W.J., 1989. Satellite-derived urban heat islands from three coastal cities and the utilization of such data in urban climatology. Int. J. Remote Sensing 10, 1699−1720.

Sdino, L., 1994. Valore e caratteristiche immobiliari. Contributi e riflessioni economiche, estimative, finanziarie per le professioni immobiliari. Tecnocopy, Genova, pp. 117–160.

Susorova, I., Angulo, M., Bahrami, P., Stephens, B., 2013. A model of vegetated exterior facades for evaluation of wall thermal performance. Build. Environ. 67, 1–13.

Taha, H., 1997. Urban climates and heat islands: albedo, evapotranspiration, and anthropogenic heat. Energy Buildings 25, 99–103.

Veisten, K., Smyrnova, Y., Klæboe, R., Hornikx, M., Mosslemi, M., Kang, J., 2012. Valuation of green walls and green roofs as soundscape measures: including monetised amenity values together with noise-attenuation values in a cost-benefit analysis of a green wall affecting courtyards. Int. J. Environ. Res. Public Health 9, 3770–3788.

Wong, N.H., Kwang Tan, A.Y., Chen, Y., Sekar, K., Tan, P.Y., Chan, D., et al., 2010a. Thermal evaluation of vertical greenery systems for building walls. Build. Environ. 45 (3), 663–672.

Wong, N.H., Kwang Tan, A.Y., Tan, P.Y., Sia, A., Wong, N.C., 2010b. Perception studies of vertical greenery systems in Singapore. J. Urban Plan. Dev. 136 (4), 330–338.

Yang, J., Yu, Q., Gong, P., 2008. Quantifying air pollution removal by green roofs in Chicago. Atmosp. Environ. 42, 7266–7273.

FURTHER READING

Francis, R.A., Lorimer, J., 2011. Urban reconciliation ecology: the potential of living roof and walls. J. Environ. Manage. 92, 1429–1437.

Landsberg, H.E., 1981. The Urban Climate. Academic Press, New York.

Nakamatsu, R., Tsutsumi, J.G., Arakawa, R., 2003. Relations of energy consumptionand local climate in a subtropical region. Atti del Fifth International Conference on Urban Climate, IAUC and WMO, Lodz, Poland, 1–5 September.

Oke, T.R., 1982. The energetics basis of urban heat island. Meteorol. Soc. 108, 1–24.

Santamouris, M., 2013. Using cool pavements as a mitigationstrategy to fight urban heat island-a review of the actualdevelopments. Renew. Sustain. Energy Rev. 26, 224–240.

Tereshchenko, I.E., Filonov, A.E., 2001. Air temperature fluctuations in Guadalajara, Mexico, from 1926 to 1994 in relation to urban growth. Int. J. Climatol. 21, 483–494.

Chapter 4.5

Economic Benefits and Costs of Green Roofs

Haibo Feng and Kasun N. Hewage

Chapter Outline

INTRODUCTION

Green roofs are well-suited for urban areas, as they provide excellent value for money at both individual and public levels in comparison with other currently available green or gray infrastructure. However, the high initial investment required for green roofs acts as a barrier to their market penetration. In general, individual benefits of green roof include reduction in energy use for heating and cooling, membrane longevity, acoustic insulation, aesthetic benefits, and LEED certification bonus (Berardi, 2016; Clark et al., 2008; Nurmi et al., 2013; USGBC, 2015; Bianchini and Hewage, 2012). Public benefits include reduction of stormwater runoff, improvement of air quality, mitigation of urban heat island effect, and increment of urban biodiversity, etc. (Driscoll et al., 2015; Connelly and Hodgson, 2008; Rosenzweig et al., 2006; Brenneisen, 2006). The costs in green roofs involve their initial construction, operations, maintenance, demolition, and disposal (Bianchini and Hewage,

Nature Based Strategies for Urban and Building Sustainability.
DOI: https://doi.org/10.1016/B978-0-12-812150-4.00028-8

2012). The objective of this chapter is to present the economic benefits of a green roof in terms of its public and individual benefits, summarize the total costs of a green roof throughout its lifecycle, and estimate the payback period based on the benefits and costs.

INDIVIDUAL BENEFITS OF GREEN ROOFS

Energy Reduction in Heating and Cooling

Green roofs reduce energy consumption in space heating through shading, evapotranspiration, insulation, increase in thermal mass, and reduction of heat loss through radiation. Green roofs can also be more efficient in preventing heat loss in the winter compared with conventional roofs (Liu and Baskaran, 2003; Berardi, 2016). The reduction in energy bills is usually the most convincing factor for building owners to install green roofs. For example, an experiment conducted in Ottawa found that a 6-inch extensive green roof reduced heat gains by 95%, and heat losses by 26% compared to a conventional roof (Liu, 2002). Another study on a two-story building was conducted by Florida Solar Energy Center. Its findings revealed that 18% of energy used for space cooling was saved by a green roof compared with the conventional roof, and 44% was saved when the plants were more established (Sonne and Parker, 2006). The economic benefit of reduction in space conditioning demand has been quantified by a previous study, which demonstrated that a green roof can save $0.18−0.68 m^{-2} in cooling, and 0.22 m^{-2} in heating annually (Bianchini and Hewage, 2012).

Membrane Longevity

Green roof technology increases the lifespan of a building's roof by protecting against diurnal fluctuations, UV radiation, and thermal stress. Studies have revealed that the lifetime of roofing membrane can be easily lengthened up to 40−50 years by green roofs (Clark et al., 2008), while a conventional roof's lifespan ranges from 10 to 30 years (Oberndorfer et al., 2007). The cost of replacing a conventional roof at the end of its lifespan is estimated at around $160 m^{-2} (Bianchini and Hewage, 2012). The benefit of installing a green roof is the cost of installing a conventional roof 20 years in the future, which is at $160 m^{-2}.

Acoustic Insulation

Green roofs improve the soundproofing of a building, and reduce the sound reflection by increasing absorption (Azkorra et al., 2015). For buildings located near very strong sources of noise such as night clubs, highways, or flight paths, the sound insulation created by green roofs can be especially useful. There are no reliable estimates in the literature about the economic value of the sound insulation benefit of green roofs. A commonly used technique to improve noise insulation is to apply an extra layer of plasterboard into the ceiling. The noise insulation benefits acquired due to green roofs are similar or higher than that gained by such an additional ceiling element, since green roofs have more than one layer (Connelly and Hodgson, 2008). Material and installation costs are approximately $29 m^{-2} (€20 m^{-2}) for plasterboard. Therefore, the noise insulation benefit of green roofs is also estimated to be around $29 m^{-2} in air noise zones (Nurmi et al., 2013).

Aesthetic Benefits

Aesthetics are the most intangible benefit, generally left out in cost-benefit analyses due to the difficulty in valuing aesthetics in monetary terms. An individual's willingness to pay a higher price can be used as a method to attribute a monetary value to qualitative characteristics such as aesthetics. Commission for Architecture and the Built Environment in London states that the price of buildings or houses will increase by 6% if there is a park nearby, and by 8% if the building has a direct view of the park. Green roofs, especially if spread over a larger area, has a similar function as a local park. Accordingly, 2%−5% and 5%−8% of property value increments for extensive and intensive green roof respectively have been assumed (Bianchini and Hewage, 2012). The extensive green roof may raise property value from $2.6 to $8.3 m^{-2}, while intensive green roofs may increase property value from $8.3 to $43.2 m^{-2}. Besides the aesthetic benefits, green roofs can also provide recreational spaces in urban areas if they are designed for public use similar to parks.

LEED Certification Bonus

LEED certified buildings are gaining in popularity because of their lower operating costs, better employee performance (in commercial and industrial buildings), improved public relations, better health standards, as well as other community benefits (CaGBC, 2014). The most attractive aspect for owners is that it increases access to capital. It is estimated that the return-on-interest of LEED certified buildings improved by 19.2% on average for green retrofit projects in existing buildings, and 9.9% on average for new green construction projects (USGBC, 2015). Under the Canada Green Building Council LEED program, buildings with green roof installations gain one point for stormwater management, and one point for reducing heat island effect if the roof covers at least 50% of the building. Another benefit of green roof technology is that the vegetation and soil media of green roof can be used as a filter for the storm runoff, so that the water from the green roof system can be used to irrigate other landscaping features without pretreatment (LEED

Canada, 2009). Under the LEED scheme, this may warrant an additional point for water efficient landscaping. The ability to reduce energy demand for cooling and heating, and increased energy efficiency may also garner additional points for optimized energy performance. Furthermore, potential points can be gained for reduced site disturbance, protection or restoration of open space, and innovation in design.

PUBLIC BENEFITS OF GREEN ROOFS

Reduction of Stormwater Runoff

Green roofs can impact the stormwater retention capacity of buildings. Most importantly, with the presence of green roofs, the rainwater that falls onto the roof surfaces flows into the sewers at a slower rate, as green roofs are able to retain water. Depending on regional climate, green roofs can lower the sewer system capacity requirement, by holding as much as 50%–95% of annual rainfall precipitation (Driscoll et al., 2015; Beecham and Razzaghmanesh, 2015). An investigation by the city of Portland revealed that $30 m^{-2} year^{-1} is needed to manage the stormwater falling on impervious areas that do not absorb rainwater (City of Portland, 2008). Based on the retention performance of green roofs listed above, green roofs will be able to create $15–28 m^{-2} savings per year by reducing the public infrastructure management fees.

Improvement of Air Quality

Green roofs are recognized as an air quality control technology. The vegetation reduces air pollution by actively absorbing many pollutants, and by passively filtering and directing airflows. It was estimated that eight metric tons of unclarified air pollutants can be removed per year by 109 ha of green roofs in Toronto, Canada (Currie and Bass, 2010). Another study conducted in Chicago estimated that the annual mass of air pollutants which can be removed by 19.8 ha of green roofs amounts to 1675 kg (Yang et al., 2008).

The cost estimate for the air quality benefit of a green roof is calculated by considering the negative effects of pollutant on health, environment, infrastructure, and climate change. The cost would be significantly higher in urban environments, due to the effect on a larger number of people. In North America, the NO_x emissions tax is $3375 ton^{-1} (Clark et al., 2008). In Europe, the SO_x cost in a populated area is $2500 ton^{-1}, and $500 ton^{-1} for NO_x cost (Nurmi et al., 2013). Based on the results from Yang et al. (2008) and Clark et al. (2008), the benefits from the improvement of air quality would be around $0.03 m^{-2} annually assuming all the air pollutants removed by green roof are NO_x.

Mitigation of Urban Heat Island Effect

In urban environments, vegetation has often been replaced by impervious and dark surfaces. Dark surfaces reflect less solar radiation and absorb more energy. Due to the lack of vegetation and the presence of dark surfaces, the urban heat island effect is created. A simulation study in New York showed that the average roof temperature can be reduced by as much as 0.8°C if 50% of the roof area is covered with vegetation (Rosenzweig et al., 2006). In Venice, the field observation and simulations results showed that the temperature of a green permeable surface could be 4°C lower than the existing paved roof (Peron et al., 2015). It was also estimated that the urban heat island effect can be reduced by 1−2 degrees Celsius if 6% of Toronto was covered with green vegetation (Peak, 2004). Another report on the Mediterranean region shows that 10%−14% of the electrical energy consumed in cooling residential buildings can be saved by green roofs (Zinzi and Agnoli, 2012). Green roof performance in reducing the urban heat island effect varies in different locations, due to the conditions in the surrounding environment, and changes in building density.

Increment of Urban Biodiversity

Green roofs can help to increase local biodiversity by providing habitats for different animal species such as birds and insects within a city. A study conducted in Switzerland found that 79 beetles and 40 spider species were supported by a single green roof, of which 20 species were endangered (Brenneisen, 2006). Another study conducted in England on green roofs which mimic conditions found in derelict sites discovered that these sites are favored by black redstart, a rare species of bird in the United Kingdom (Grant and Lane, 2006).

However, creation of a habitat for animals is treated only as a bonus compared with other quantifiable benefits. It is not easy to quantify the increase in biodiversity and estimate the corresponding costs and benefits using a common methodology. While it is difficult to directly quantify the economic benefits of habitat increase due to green roofs, the resulting environmental benefits may be translatable to economic terms based on environmental priorities.

LIFE CYCLE COST OF GREEN ROOFS

Initial Cost

There is a significant price variation among green roofs due to factors such as type and size, locations of green roofs, and country. The current cost in British Columbia, Canada for a standard extensive green roof varies from $130 to $165 m^{-2}, and the cost of a standard intensive green roof starts from

$540 m^{-2} (Bianchini and Hewage, 2011). Many factors such as labor and equipment costs affect the installation price. In Singapore, a green roof price ranges from \$40 to \$65 m^{-2} depending on the type of green roof and structure of the foundation (Wong et al., 2003). In China, the average price of a green roof investigated from three provinces is between \$48 and \$76 m^{-2} (Jia and Wang, 2011; Liu and Hong, 2012). In a mature market like Germany, the average green roof costs range from \$15 to \$45 m^{-2}. The lower green roof prices in Germany are a result of ongoing research and development as well as market penetration spanning two decades. In newer markets, no economies of scale exist and competition is scarce. Labor is also more expensive because of the lack of experience and the tendency to use custom design systems. One way to reduce the initial cost of green roofs is to adopt the low-cost techniques developed by mature markets. The cost of green roof generally decreases by 33%−50% once the industry has established itself (Toronto and Region Conservation, 2007).

Operation and Maintenance (O&M) Cost

Economic and environmental benefits of green roofs rely on their performance. Therefore, O&M of vegetative roofs are critical in securing their positive impacts. The maintenance cost also depends on the size of green roofs, the characteristics of the building, the complexity of the green roof system, the type of vegetation, as well as the market O&M price. It is estimated that annual O&M cost of green roofs in the United States is between \$0.7 and \$13.5 m^{-2} (Bianchini and Hewage, 2012).

Disposal Cost

There are different disposal options for green roofs at the end of life. Materials can be landfilled, reused, or recycled. Water retention layer, drainage layer, and root barrier layers of green roof can be recycled again at the end of the lifespan. However, many cities do not have the necessary facilities for the recycling process. Landfill costs depend on many factors such as technology, location, size of the facility, and available landfill capacity in a municipality.

A study indicated that the operations and maintenance cost in landfilling is on average \$56 per ton waste disposed without considering the energy recovery option (Chang and Wang, 1995). Another report compiled in Europe did a complete analysis on the green roof disposal cost, including inert material landfill, sanitary landfill, and incineration with energy recovery. The disposal cost for an entire green roof is estimated at \$1120 ton^{-1} (€784 ton^{-1}) (Peri et al., 2012). Bianchini and Hewage (2012) illustrated that the cost to dispose green roof materials is in the range between \$0.03 and \$0.2 m^{-2}.

GREEN ROOF COST BENEFIT ASSESSMENT

Net Present Value and Payback Period

In order to assess the total benefits and costs of green roof, the values need to be converted into a net present value (NPV) by the means of discounting. The lifespan of a green roof has been estimated as about 40 years minimum and 55 years maximum (Mahdiyar et al., 2016). In this analysis, 40 years is used to conduct the assessment. Based on the study from Gollier and Weitzman (2010), 3% of the discount factor was applied to this analysis. Based on the benefits and costs of green roofs introduced above, Table 1 summarized all the economic inputs for the analysis and NPVs as output.

There is a wide range in terms of the values in Table 1, especially the aesthetic benefits and stormwater runoff reduction benefits, and the lifecycle costs. One of the reasons is due to the different systems of green roofs. For example, extensive green roofs have shallow soil roofs with simple growing plants, and are usually not accessible. Therefore they have a lower lifecycle cost.

On the other hand, intensive green roofs are similar to a ground level garden with a deep growing medium and artificial irrigation (Kosareo and Ries, 2007). Therefore, the initial cost and O&M cost are higher. At the same time, intensive green roofs have higher benefits in stormwater runoff deduction due to its deep growing medium, and better aesthetic values because it acts like a garden. Another reason is the cost and technique variances between different markets. In the mature market like Germany, the costs are much lower than the new markets in Asia and North America, and the benefits generated from green roofs are more than the new markets because of its mature techniques and great popularity. Some other reasons are sizes of green roofs, weather conditions, and building features etc.

As shown in Table 1, the total NPV of individual benefits in 40 years is between \$135.9 and \$195.8 m^{-2}, and the total NPV of public benefits in 40 years is between \$478.7 and \$751.7 m^{-2}. Based on the result, it is obvious that the public benefits are over three times greater than the individual benefits, even though two of the public benefits are not counted in the calculation due to the unavailable data.

If the total NPV of lifecycle costs for green roofs in 40 years is close to \$42.3/m^2, which is at the lower side of the range (\$42.3−978.8 m^{-2}), it will only take 13 years of the individual benefits to balance the cost of green roofs. If the public benefits are considered, the payback period will be reduced to 3 years. If the total NPV of lifecycle cost for green roofs in 40 years is close to \$978.8 m^{-2}, this cost could still be paid back in its lifetime by the individual benefits and public benefits together.

TABLE 1 Economic Data Input and NPV Output ($ m^{-2}) for the Cost Benefit Assessment

		Value	Time Frame (Year)	NPV ($ m^{-2})
Economic Factor	Lifespan (year)	\	40	\
	Discount rate (%)	3	\	\
Individual Benefits ($ m^{-2})	Reduction of energy Use in heating and cooling	0.4–0.9	Annual	15.7–35
	Membrane longevity	160	At year 20	88.6
	Acoustic insulation	29	One time	29
	Aesthetic benefits	2.6–43.2	One time	2.6–43.2
	LEED certification bonus	n/a	n/a	n/a
Total NPV				135.9–195.8
Public Benefits ($ m^{-2})	Reduction in stormwater runoff	15 - 28	Annual	477.5–750.6
	Improvement of air quality	0.03	Annual	1.18
	Mitigation of urban heat island effect	n/a	n/a	n/a
	Increment of urban diversity	n/a	n/a	n/a
Total NPV				478.7–751.7
Lifecycle Costs ($ m^{-2})	Initial cost	15–540	One time	15–540
	Operation and maintenance cost	0.7–13.5	Annual	27.3–438.7
	Disposal cost	0.03–0.2	At year 40	0.01–0.06
Total NPV				42.3–978.8

Scale of Implementation

As shown in Table 1, the values created by the mitigation of urban heat island effect and increment of urban diversity are not available in this analysis, because the value would be very small if only one or a few green roofs were installed. However, the benefits of green roof will increase tremendously if implemented at a larger scale. Intangible benefits such as aesthetic appeal of green roofs and increased urban biodiversity can be gained with large scale of implementation (Niu et al., 2010; Nurmi et al., 2013) . The costs of green roofs will also be reduced with a higher implementation rate. Large scale of implementation would also reduce the volume of stormwater entering local waterways, which will lead to lower water temperatures, less in-stream scouring, and better water quality (Spengen, 2010).

Green Roof Policy Initiatives

Based on the analysis above, the public benefits of green roofs are over three time larger than the private benefits. Therefore, municipal authorities should play a key role in promoting green roofs in urban areas and residential neighborhoods through policy and regulatory measures.

In Toronto, Green roofs are required on all new institutional, commercial, and multiunit residential developments. The incentive offered for green roof is $75 m^{-2} up to an upper limit of $100,000 (City of Toronto, 2016). In New York, green roof tax abatement is implemented, so that each square foot of green roof can get a rebate of $5.23, up to $200,000 per project (NYC, 2014). In Singapore, the National Parks Board aims to increase greenery provision by funding up to 50% of the installation cost of rooftop greenery (National Parks, 2011). In Tokyo, it is mandatory for a new building to cover 25% of roof with greenery (Growing Green Guide, 2013).

In Munich, all building roofs with a surface area larger than $100 m^2$ should be landscaped. This policy was implemented around 20 years ago, and it makes the green roof a recognized construction standard in Munich (IGRA, 2011). As a world leader in green roof development, Germany's experience shows that it is necessary to introduce a green roof policy rather than rely solely on the goodwill of building owners (Ngan, 2004).

CONCLUSION

Green roofs have personal and social benefits. The cost benefit assessment showed that the lifecycle costs of green roofs can be retrieved in most of the markets around the world. The payback periods in the mature markets and markets with average initial costs are shorter than the lifespan of green roofs. With a larger implementation scale, the social benefits of green roofs will be increased tremendously. Governments should play a key role in promoting the green roof construction by providing incentives to transfer the social benefits into private investors, such as tax abatement, direct cash rebate, low interest loans, etc. These incentives will also expand the public benefits, and lower the lifecycle cost of green roofs.

REFERENCES

Azkorra, Z., Pérez, G., Coma, J., Cabeza, L.F., Bures, S., Álvaro, J.E., et al., 2015. Evaluation of green walls as a passive acoustic insulation system for buildings, Appl. Acoust., 89. pp. 46—56.

Beecham, S., Razzaghmanesh, M., 2015. Water quality and quantity investigation of green roofs in a dry climate., Water Res., 70. pp. 370—384.

Berardi, U., 2016. The outdoor microclimate benefits and energy saving resulting from green roofs retrofits, Energy Build., 121. pp. 217—229.

Bianchini, F., Hewage, K., 2011. How 'green' are the green roofs? Lifecycle analysis of green roof materials., Build. Environ., 48. pp. 57—65.

Bianchini, F., Hewage, K., 2012. Probabilistic social cost-benefit analysis for green roofs: a life-cycle approach., Build. Environ., 58. pp. 152—162.

Brenneisen, S., 2006. Space for urban wildlife: designing green roofs as habitats in Switzerland. Urban Habitats 4 (1), 27—36.

CaGBC, 2014. Canada Green Building Trends: Benefits driving the new and retrofit market. Available at https://www.cagbc.org/cagbcdocs/resources/CaGBC%20McGraw%20Hill%20Cdn%20Market%20Study.pdf (accessed at 15.11.16).

Chang, N.B., Wang, S.F., 1995. The development of material recovery facilities in the United States: status and cost structure analysis. Resour. Conserv. Recycling 13 (2), 115—128.

City of Portland, 2008. Oregon Cost benefit evaluation of Ecoroofs. Available at https://www.portlandoregon.gov/bes/article/261053 (accessed at 14.11.16).

City of Toronto, 2016. Eco-roof incentive program review. Available at http://www1.toronto.ca/City%20Of%20Toronto/Environment%20and%20Energy/Programs%20for%20Residents/PDFs/Eco-Roof/Eco-Roof%20Incentive%20Program%20Review%202016.pdf (accessed at 14.11.16).

Clark, C., Adriaens, P., Talbot, F.B., 2008. Green roof valuation: a probabilistic economic analysis of environmental benefits. *Environmental Science and Technology*, American Chemical Society, Department of Civil and Environmental Engineering, College of Engineering, University of Michigan, Ann Arbor, MI 48109-2125, United States, 42(6), 2155—2161.

Connelly, M., Hodgson, M., 2008. Thermal and acoustical performance of green roofs: sound transmission loss of green roofs. Green. Rooftops Sustain. Communities 1—11.

Currie, B.A., Bass, B., 2010. Using green roofs to enhance biodiversity in the city of toronto. (April).

Driscoll, C.T., Driscoll, C.T., Eger, C.G., Chandler, D.G., Roodsari, B.K., Davidson, C.I., et al., 2015. Green Infrastructure: Lessons from Science and Practice. (June).

Gollier, C., Weitzman, M.L., 2010. How should the distant future be discounted when discount rates are uncertain? Econ. Letters 107 (3), 350—353.

Grant, G., Lane, C., 2006. Extensive green roofs in London. Urban Habitats 4 (1), 51—65.

Growing Green Guide, 2013. Green roofs, walls & facades policy options background paper. Available at http://imap.vic.gov.au/uploads/Growing%20Green%20Guide/Policy%20Options%20Paper%20-%20Green%20Roofs,%20Walls%20and%20Facades.pdf (accessed at 14.11.16).

International Green Roof Association (IGRA), 2011. Green Roof News. Available at http://www.igra-world.com/links_and_downloads/images_dynamic/IGRA_Green_Roof_News_1_11.pdf (accessed at 14.11.16).

Jia, R., Wang, Y., 2011. Analysis of cost-benefit of green roof in Xi'an. 2011 2nd International Conference on Mechanic Automation and Control Engineering, MACE 2011 - Proceedings 5581—5583.

Kosareo, L., Ries, R., 2007. Comparative environmental life cycle assessment of green roofs. *Building and Environment*, Elsevier Ltd, Department of Civil and Environmental Engineering, University of Pittsburgh, 949 Benedum Hall, 3700 O'Hara Street, Pittsburgh, PA 15260, United States, 42(7), 2606−2613.

LEED Canada, 2009. LEED Canada for new construction and major renovation 2009 rating system. Available at http://www.cagbc.org/cagbcdocs/LEED_Canada_NC_CS_2009_Rating_System-En-Jun2010.pdf (accessed at 15.11.16).

Liu, K., Baskaran, B., 2003. Thermal performance of green roofs through field evaluation. In: Proceedings for the First North American Green Roof Infrastructure Conference, Awards, and Trade Show, pp. 1−10.

Liu, K.K.Y., 2002. Energy efficiency and environmental benefits of rooftop gardens NRCC-45345 energy efficiency and environmental benefits of rooftop gardens. Construct. Canada 44 (17), 20−23.

Liu, L.-P., Hong, G.-X., 2012. Popularizing path research on green roof project in China rural region: cost-effectiveness assessment. 2012 World Automation Congress, WAC 2012.

Mahdiyar, A., Tabatabaee, S., Sadeghifam, A.N., Mohandes, S.R., Abdullah, A., Meynagh, M. M., 2016. Probabilistic private cost-benefit analysis for green roof installation: a monte carlo simulation approach. Urban For. Urban Gree. 20, 317−327.

National Parks, 2011. New incentives to promote skyrise greenery in Singapore. Available at https://www.nparks.gov.sg/news/2011/3/new-incentives-to-promote-skyrise-greenery-in-singapore (accessed at 14.11.16).

New York City, 2014. Green roofs for stormwater management. Available at http://columbia-green.com/wp-content/uploads/2014/08/NYC-1-pager.pdf (accessed at 14.11.16).

Ngan, G., 2004. Green roof policies: tools for encouraging sustainable design. (December), 1−45.

Niu, H., Clark, C., Zhou, J., Adriaens, P., 2010. Scaling of economic benefits from green roof implementation in Washington, DC. Environ. Sci. Technol. 44 (11), 4302−4308.

Nurmi, V., Votsis, A., Perrels, A., Lehvävirta, S., 2013. Cost-benefit analysis of green roofs in urban areas: case study in Helsinki.

Oberndorfer, E., Lundholm, J., Bass, B., Coffman, R.R., Doshi, H., Dunnett, N., et al., 2007. Green roofs as urban ecosystems: ecological structures, functions, and services. BioScience 57 (10), 823.

Peak, S., 2004. The green roof infrastructure monitor. North 5 (May), 1−24.

Peri, G., Traverso, M., Finkbeiner, M., Rizzo, G., 2012. The cost of green roofs disposal in a life cycle perspective: covering the gap. *Energy*, 48(1), 406−414.

Peron, F., De Maria, M.M., Spinazz, F., Mazzali, U., 2015. An analysis of the urban heat island of Venice mainland. *Sustain. Cities Soc.* 19, 300−309.

Rosenzweig, C., Gaffin, S., Parshall, L., 2006. Green roofs in the New York metropolitan region research report. Columbia University Center for Climate Systems Research and NASA Goddard Institute for Space Studies. p. 59.

Sonne, J.K., Parker, D., 2006. Energy performance aspects of a florida green roof. Fifteenth Symposium on Improving Building Systems in Hot and Humid Climates.

Spengen, J. Van., 2010. The effects of large-scale green roof implementation on the rainfall-runoff in a tropical urbanized subcatchment, pp. 1−222.

Toronto and Region Conservation, 2007. An economic analysis of green roofs : evaluating the costs and savings to building owners in Toronto and surrounding regions. (July), 15.

USGBC, 2015. The business case for green building. Available at http://www.usgbc.org/articles/business-case-green-building (accessed at 15.11.16).

Wong, N.H., Tay, S.F., Wong, R., Ong, C.L., Sia, A., 2003. Life cycle cost analysis of rooftop gardens in Singapore. Build. Environ. 38 (3), 499–509.

Yang, J., Yu, Q., Gong, P., 2008. Quantifying air pollution removal by green roofs in Chicago. *Atmospheric Environment*, Elsevier Ltd, Department of Landscape Architecture and Horticulture, Temple University, 580 Meetinghouse Road, Ambler, PA 19002, United States, 42(31), 7266–7273.

Zinzi, M., Agnoli, S., 2012. Cool and green roofs. An energy and comfort comparison between passive cooling and mitigation urban heat island techniques for residential buildings in the Mediterranean region. Energy Build. 66–76. Available from: http://dx.doi.org/10.1016/j.enbuild.2011.09.024.

Chapter 4.6

Economic Benefits and Costs of Green Streets

Benz Kotzen

Chapter Outline

INTRODUCTION

This chapter is about green streets, and street "treeconomics." Treeconomics is a relatively new blend word, which combines the words tree and economics. The planting of street trees both historically and in the present is readily justified by various ecosystem services provisions namely aesthetic improvement, the mitigation of air pollution and the provision of shade, especially in areas with long hot summers. Today it is recognized that street trees provide a much broader set of ecosystem services such as flood control and energy savings etc. (EPA, 2017). These should help what defines an urban tree and an urban forest and provide a paradigm for these, which also includes street trees and community green space (Fig. 1). While they posit a rational placement of street trees in geo-spatial terms, it is just as important, if not more crucial, to locate street trees within a socioeconomic framework that relates to public and private space and at the same time relates to public and private goods. On the whole most streets, roads, avenues, lanes, highways etc. are part of city infrastructure and although the streets are managed and indeed owned by various city or municipal authorities, the streets themselves should be seen as part of public goods and this includes the street trees. Public

Nature Based Strategies for Urban and Building Sustainability.
DOI: https://doi.org/10.1016/B978-0-12-812150-4.00029-X
© 2018 Elsevier Inc. All rights reserved.

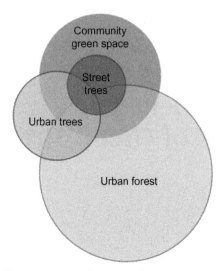

FIGURE 1 Conceptual paradigm explaining the domain of street trees after Roy et al. (2012).

goods are readily explained. If the benefit of an item or a system is not private and the benefit is for everyone, such as air and street lighting, then it is a public good.

In terms of public good or benefit there is no difference between a street light that provides luminance to anyone that passes and a street tree that provides beauty, shade, and reductions in solar radiation, cleaning the air and providing sanctuary for urban biodiversity to all who pass or live within the vicinity. An alternative paradigm for urban forests, including street trees can be characterized by a public vs. private division where the ecosystems' services provision is largely mutually beneficial as people continuously move between the two realms. The public and private realms thus need to be considered within the greater governing systems, which enclose, encompass, and connect all development. These systems (wind patterns, precipitation, air temperatures) are not affected by the ownership and boundaries of public and private space (Fig. 2).

The message is that all vegetation in urban areas should be considered as part of the urban forest and each part, including street trees provide ecosystem services benefits which can also be measured in economic terms.

The Economics of Municipal Street Trees (Stanley and Crossman, 1973) is a vintage document which illustrates some of the problems that arose and that still exist with street trees. However and more importantly, it provides a baseline view of the past and it illustrates how far we have come from considering trees as problematic, to recognizing the value of these trees to solve some of the issues that arise from urbanization as well as more global issues

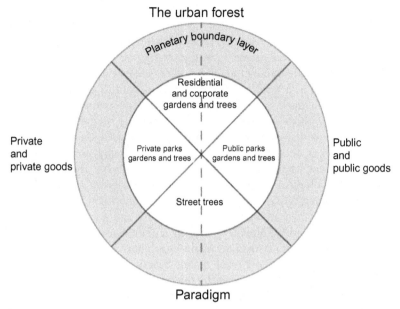

FIGURE 2 Conceptual paradigm of trees within urban areas as public or private goods.

such as global warming. The document which looks at street trees in Massachusetts, USA, points out some of the problems that still exist today including:

1. conflicts of space between trees and below and above ground utilities and infrastructure,
2. pest control,
3. lack of community support and,
4. adverse growing conditions. The article presents in some detail the difficulties involved in street tree planting in terms of the cost, maintenance, removal etc., but there is no mention at all of the ecosystem services benefits and the related economic benefits that street trees can bring.

THE ECONOMIC BENEFIT OF GREEN STREETS

The benefits of having green streets is best explained through ecosystem services provision. The system is well-known and understood by academics, environmental, and ecological consultants and by some local authorities but it is not a concept that is known to the larger public and it should be better understood and taught in schools. If people as voters do not know what is at stake, politicians and other decision-makers can ignore the issues and do nothing as it is cheaper, in the short term at least, and more expedient than doing something which requires capital and negotiations and very often social and political compromises. On the other hand, it is fortunate that many governments and international institutions (the UN, FAO, UNCCD, WHO) recognize that what is done (both positively and negatively), affects local conditions and local people as well as having an impact far beyond the local area. The benefits brought about by green streets can thus be calculated at a global as well as a local scale. However, it is hard to assess the "impact of street trees on human health and the urban environment" and this is because the ecosystem service provision is generally singled out "without considering the wider synergistic impacts of street trees on urban ecosystems" (Salmond et al., 2016). This is an issue that many proenvironmental developments have when including street trees, green roofs, living walls, etc. Whereas, the ecological benefits, e.g., of connected green spaces, such as green corridors connecting green areas is readily understood, green interventions in a broader area are calculated individually and not cumulatively. In other words, the broader cumulative benefits are not considered, even though they are likely to be greater than the sum of each individual part.

In order to understand how green streets benefit people, it is important to also understand ecosystem services in reference to natural capital. Natural capital can be defined as the "world's stocks of natural assets which include geology, soil, air, water and all living things" (World Forum on Natural Capital). "It is from this natural capital that humans derive a wide range of services, often called ecosystem services, which make human life possible" (World Forum on Natural Capital). Most people understand that mankind is having a large impact on the world's fauna and flora, on rainforests where millions of hectares are lost due to deforestation and on soils worldwide where overgrazing and other poor land management practices deplete soil structures and fertility, thus making farming impossible. The issues of maintaining and protecting natural capital, however, need to go hand-in-hand with proactive measures, such as tree planting. Street tree planting needs to be seen as part and parcel of the response to improving natural capital. While street tree planting and other vegetative interventions into the street environment will have little or no direct effect on what happens outside of the city, a large increase in urban vegetation can have considerable local benefits. Indeed, the introduction of vegetation by other means in streets is something we need to consider seriously through the introduction of living walls, urban agriculture, and other green and blue interventions. The need to intervene in our public areas in our cities is growing stronger as the world's urban populations expand. The World Health Organization (WHO) notes that the world's urban population grew from 34% in 1960 to 54% in 2014 (WHO [GHO] data). The Edible Bus Stop, by the Edible Bus Stop consultancy is a good example of introducing vegetation in an alternative way into the urban realm (Fig. 3). The economic benefits of this are hard to quantify, but the community well-being benefits are well-known.

FIGURE 3 The Edible Bus Stop, London by The Edible Bus Stop consultancy brings alternative methods of greening streets.

Thus far we have omitted the blue component(s) of green streets, but the sustainable management of water in our urban areas and particularly our streets can have positive impacts. This can be achieved through rain gardens and rills which are becoming ubiquitous in the northwest United States, particularly in Portland, Oregon and Seattle. The economic benefits of water capture can be readily calculated in the United States through The National Green Values™ Calculator which compares the performance, costs, and benefits of green infrastructure, or low impact development. In the United Kingdom, sustainable water management is termed SuDS (Sustainable Drainage Systems)' and it should be the case that this term is used and the methods grasped internationally.

Benefitting the World

Whether individuals or groups believe that global warming is anthropogenic or not, climatic studies have verifiably shown that the world is getting warmer and this is due mainly to the increase in carbon dioxide in the atmosphere. "Since 1906, the global average surface temperature has increased between 1.1°F and 1.6°F" which is 0.6−0.9°C (National Geographic). The effects which often lead to large financial losses include:

- rising sea levels with the consequent loss of land (agricultural, urban, etc.),
- increasing and stronger storms,
- floods and droughts becoming more common,

- less availability of fresh water and,
- disease continuing to spread such as malaria and the Zika virus (National Geographic).

There is a great need to reduce global warming, and the planting of trees can help with this mainly through carbon storage and carbon sequestration: during photosynthesis trees absorb carbon, part of which are stored as carbon compounds in the wood (Thompson and Matthews).

In order to mitigate the above looming crisis, urban forests (including street trees) have significant capacity to absorb greenhouse gases by capturing carbon as they grow (Kovacs et al., 2013). Additionally, trees produce oxygen which we need to breathe. It is estimated that a person breathes in approximately 9.5 tons of air per annum. Of this 23% is oxygen of which people extract one third when we breathe (Villazon, 2016). We thus inhale and use roughly 740 kg of oxygen per annum which is produced by about seven or eight trees. With an estimated population in 2050 of 9.7 billion people, this means we will require 77.6 billion trees just for our own oxygen needs. Additionally, there are the oxygen needs of domestic and wild animals, etc. This discussion has been raised not to say that we will run out of oxygen but to raise attention on an issue that is important albeit, not yet critical. As people have reduced the Earth's trees by half over the centuries and are continuing to reduce the tree population by 10 billion per annum −15 billion trees that are cut down very year and adding the 5 billion that are either planted or that have naturally sprouted (Pennisi, 2015)—this really could become a future issue.

Nowak et al. (2013) note that the total tree carbon storage in urban areas in the United States (c. 2005) is estimated at 643 million tons with a value of \$50.5 billion and annual sequestration is estimated at 25.6 million tons with a value of \$2.0 billion. It should be noted however that urban soils "play a significant role in the overall storage of carbon in urban landscapes" (Pouyat et al., 2006) and that urban soils are estimated to store approximately 1.9 billion tons of carbon in the United States (Pouyat et al., 2006) which is three times more than that of urban trees. It is essential that carbon sequestration is seen in its broadest terms in urban areas taking account of trees, other vegetation as well as soils.

In New York City (NYC) the trees store 1.35 million tons of carbon valued at \$24.9 million per annum and they remove 42,000 tons of carbon each year (Million Trees NYC). The trees intercept 890.6 million gallons (3371 million L) of storm water annually or 1525 gal (5773 L) per tree on average. The total annual value of this benefit to NYC is over \$35 million. (Million Trees NYC) It is calculated that \$5.60 of benefits is provided for each \$1 spent on tree planting and care. It has been noted that street trees have increased property values by \$52 million each year (Million Trees NYC) The New York City Street Tree Map (NYCSTM), allows web viewers to identify 684,040 street trees of 213 species. This wonderful resource illustrates the commitment there is to trees in the city (Fig. 4).

FIGURE 4 A composite of four screen grabs of New York City Tree Map showing districts and numbers of trees, which can be explored at individual tree level. The Google Map photo was added to define the area.

In Lisbon, using the i-Tree STRATUM software, Soares et al. (2011) report that there are 41,247 trees that provide services valued at $8.4 million annually, while $1.9 million is spent on their maintenance. They note that for every $1 invested in tree management, residents receive $8.48 in benefits. The benefits per tree per annum were as follows:

- energy savings—$6.20 (similar to the United States),
- CO_2 reduction—$0.33 (similar to the United States),
- air pollution mitigation—$5.40 (similar to the United States),
- storm water runoff reduction—$47.80 (greater than the United States),
- increased real estate values—$144.70 (greater than the United States).

Benefitting Local Areas

Apart from capturing and sequestering carbon, green streets can directly affect carbon emissions in built up areas. Nowak et al. (2013), note that planting trees "in energy-conserving locations around buildings ... can reduce building energy use and consequently emissions from power plants" and thus also the costs associated with running the building. The plants' leaves transpire and increase cooling and the beneficial change in albedo also alters urban microclimates that can also reduce carbon emissions from cities.

The TorbayI-Tree survey *Torbay's Urban Forest—Assessing Urban Forest Effects and Values* reveals the following based on a total of 818,000 trees, 11.8% tree cover, and an average of 11.5cm tree stem diameter (Treeco$_2$nomics, 2011):

- Pollution removal: 55 tons per annum valued at £281,495 (USEC) £1,330,000 (UKSDC);
- Carbon Storage: 98,100 tons per annum valued at £1,474,508 (USEC) and £5,101,200 (UKSDC);
- Carbon Sequestration: 3320 tons (net) per annum valued at £64,316 (USEC) and £172,640 (UKSDC) and;
- Structural Value: £280, 000,000.

Britt and Johnson (2008), *Trees in Towns II, A new survey of urban trees in England and their condition and management* indicate that across the 147 English towns and cities and 10 London boroughs, the average density of trees and shrubs recorded in the survey was 58.4 ha^{-1} and that approximately 50% of all surveyed sites had between 10 and 50 trees or shrubs/ha. Two thirds of trees and shrubs were located on private property. Approximately 20% were located in public parks and open space and approximately 12% were street or highway trees (Britt and Johnson, 2008). The most treed cities in the world include Singapore, Sydney, Sacramento, Cambridge (MA, USA), and Durban. Table 1 illustrates the cities with the most tree cover and their population densities.

Goodwin (2017) in *The Urban Tree* briefly discusses the positive economic effects trees can have on property prices. Past studies undertaken in the 1980s indicated that property value enhancement provided by medium to large front garden trees was in the order of 3.5%−4.5% (Goodwin, 2017 after Morales, 1980). A study by Donovan and Butry in Orgeon, USA, found that the combined effects of a number of trees fronting the property and overall crown area within 30.5 m of the house raised the sales price by an average of three percent (Goodwin, 2017, after Donovan and Butry, 2010). Trees also affect property and consumer behavior in retail areas. Kathleen Wolf, (Wolf, 2003) notes that trees are important components of a welcoming, appealing consumer environment. Surveys suggest shoppers are prepared to travel further to shopping areas with trees, visit more often, pay more for parking, and spend longer shopping. This disproves the argument that car parking lost to trees is economically negative.

VALUATION OF TREES

There have been numerous methods for determining the monetary value of trees but a globally accepted method has yet to be determined. However, the i-Tree method, is likely to become the standard as it has been used comprehensively in the United States and is started to be used elsewhere such as in

TABLE 1 A Comparison of Tree Canopy Percentages and Population Density for the Top 10 Canopied Cities and New York, London, and Paris after Treepedia (Treepedia)

City	Percentage Canopy Cover	Population Density/ Square Kilometer
Singapore	29.3	7797
Sydney, Australia	25.9	400
Vancouver, Canada	25.9	5249
Cambridge(MA), USA	25.3	6586
Durban, South Africa	23.7	2600
Johannesburg, South Africa	23.6	2900
Sacramento, USA	23.6	1800
Frankfurt, Germany	21.5	3000
Amsterdam, the Netherlands	20.6	4908
Geneva, Switzerland	21.4	12,000
New York, USA	13.5	10,831
London, UK	12.7	5518
Paris, France	8.8	21,000

the United Kingdom. The National Tree Benefit Calculator in the United States, is an online calculator based on i-Tree. This method works as follows: location of the tree according to zip code (post code), species of tree, land use type, and diameter of trunk at 4.5 ft (1.4 m) above ground level. The service then provides an overall calculation for the tree.

There have also been other numerous methods for valuing trees across the world. A comprehensive system has been devised by the city of Melbourne in Australia. The city uses the i-Tree method for evaluating the monetary costs of the ecosystem service provision, but they also provide a method for evaluating the amenity value (City of Melbourne) (These types of methods are quite common and advanced by arboricultural associations around the world. In the United Kingdom, the Arboricultural Association uses the Helliwell System to place a monetary value on the visual amenity provided by individual trees and/or woodland. It has been used extensively in legal cases. The Helliwell system is used to apply scores according to different factors including tree size, life expectancy, suitability to setting etc. These scores are added up and then a value is derived according to a monetary conversion factor). The amenity value is derived from the formula of the Maurer-Hoffman Formula (City of Melbourne). The basic monetary value of the tree was derived from the table of values devised by the American Council of Tree and Landscape Appraisers and the International Society of Arboriculture as follows:

$$\text{Value(V)} = \text{Basic Value(\$)} \times \text{Species}(S) \times \text{Aesthetics}(A) \times \text{Locality}(L) \times \text{Condition}(C)$$

An explanation is as follows. The Value (V) of the tree is made up of the:

1. Basic Value—This is a set value relating to the tree's diameter at 4.5 ft above the ground level, multiplied by;
2. Species Factor—This relates to the longevity of the species in a range from 1 to 6 where 1 is short life span, less than 50 years and trees of a long life span more than 150 years, multiplied by;
3. Aesthetics Factor—The aesthetic value of a tree is determined by the impact on the landscape if the tree were removed. This is determined on a scale from 0.5 to 1.0 in which 0.5 is a tree that contributes little to the landscape to 1.0 in which the tree is a solitary specimen tree, multiplied by;
4. Locality Factor—This relates to the position of the tree. Trees in a main street or boulevard score highest because of the stressful growing environment in which the tree has to survive. As the location becomes more rural, the significance of the tree diminishes (City of Melbourne). The score is from 0.5 in undeveloped bushland or open forest to 2.5 in a city center main street or boulevard, multiplied by;
5. Tree Condition—This is a complex assessment based on trunk condition, growth characteristics, structure, pests and diseases, canopy development, and life expectancy. A total score is derived from combining these aspects (City of Melbourne).

Economic Benefits of Other Greenery in Streets

There is no doubt that individuals, communities, and businesses use plants to attract people. These interventions vary considerably from a rose being planted in the sidewalk / pavement and trailed up the wall outside a shop to much more elaborate schemes which include small trees in planters, drainage channels with planting and living walls (green walls). One prime example of a major planted object that is used to attract visitors is the Guggenheim Puppy (Fig. 5) by Jeff Koons at the Guggenheim Museum in Bilbao, Spain. The Guggenheim Museum itself was created as a means to provide a renewal for the city, an attraction that would bring tourists and funds to an area not then well-known to visitors. Visitors brought in over €100 million over the first 3 years after it opened which more than paid for the construction costs (The Economist, 2014). A focal point of the landscape planning around the

FIGURE 5 Puppy by Jeff Koons outside the Guggenheim Museum, Bilbao. How much is this green sculpture worth to the region?

museum is the planted puppy, a giant dog at its entrance. It is impossible to calculate the economic benefit of this flowery dog. It would most likely be less attractive as just a big metal dog without the greenery and the museum ensures that it changes the flowers regularly to keep the experience fresh and attractive.

CONCLUSION

The argument for retaining, maintaining, and creating green streets is overwhelming. On the whole and in the right location, people prefer streets that have trees and other plants. Not only do the trees and other vegetation provide localized ecosystem services but they benefit our planet as a whole. These services are readily calculated via the i-Tree methodology as well as through other methods and it can be proven that trees increase monetary value of real estate as well as commercial activities. While street trees need to be protected and planted, other forms of street greening and "bluing" need to be encouraged. These interventions can be more sophisticated and high-tech in the use of proprietary living wall systems which include irrigation systems with dosatrons which supply the right amounts of water with the appropriate nutrients to low-tech and opportunistic guerilla gardening. Water is a precious commodity that should not only be used wisely but also brought to the surface for aesthetic delight and for cooling. Thus greening and bluing streets have aesthetic and financial benefits and should be actively encouraged. Town planners as well as individuals have a huge responsibility in securing the benefits vegetation brings to our built environment.

REFERENCES

Britt, C., Johnston, N., 2008. Trees in Towns II, A New Survey of Urban Trees in England and Their Condition and Management. ADAS UK Ltd., Myerscough College and Department for Communities and Local Government, London.

Donovan, G.H., Butry, D.T., 2010. Trees in the city: valuing street trees in Portland, Oregon. Landscape Urban Plan. 94, 77–83.

Goodwin, D., 2017. The Urban Tree. Routledge, London.

Kovacs, K.F., Haight, R.G., Jung, S., Locke, D.H., O'Neil-Dunne, J., 2013. The marginal cost of carbon abatement from tree planting in New York City. Ecol. Econ. 95, 1–10.

Morales, D.J., 1980. The contribution of trees to residential property value. J. Arboricul. 6 (11), 305–308.

Nowak, D.J., Greenfield, E.J., Hoehn, R.E., Lapoint, E., 2013. Carbon storage and sequestration by trees in urban and community areas of the United States. Environ. Pollut. 178, 229–236.

Pennisis, N. Earth home to 3 trillion trees, half as many as when human civilization arose. Science — AAAS (American Association for the Advancement of Science). 02 September 2015. http://www.sciencemag.org/news/2015/09/earth-home-3-trillion-trees-half-many-when-human-civilization-arose (accessed 10.07.2017).

Pouyat, R.V., Yesilonis, I.D., Nowak, D.J., 2006. Carbon storage by urban soils in the United States. J. Environ. Quality 35, 1566–1575.

Roy, S., Byrne, J., Pickering, C., 2012. A systematic quantitative review of urban tree benefits, costs and assessment methods across cities in different climatic zones. Urban For. Urban Gree. 11, 351–363.

Salmond, J.A., Tadaki, M., Vardoulakis, S., Arbuthnott, K., Coutts, A., Demuzere, M., et al., 2016. Health and climate related ecosystem services provided by street trees in the urban environment. Environ. Health 15 (Suppl. 1), 36.

Soares, A.L., Rego, F.C., McPherson, E.G., Simpson, J.R., Peper, P.J., Xiao, Q., 2011. Benefits of costs of street trees in Lisbon, Portugal. Urban For. Urban Gree. 10, 69–78.

Stanley, P.W., Crossmon, B.D., 1973. The economics of municipal street trees. J. Northeastern Agric. Econ. Council II, 1.

City of Melbourne, Tree Valuations in the City of Melbourne. https://www.melbourne.vic.gov.au/SiteCollectionDocuments/Tree-valuations.DOC (accessed 11.07.2017).

City of Melbourne, Tree Evaluations in the City of Mebourne, https://www.melbourne.vic.gov.au/SiteCollectionDocuments/Tree-valuations.DOC (accessed 29.11.2017).

The Economist, The Bilbao Effect. 6 January 2014. https://www.economist.com/news/special-report/21591708-if-you-build-it-will-they-come-bilbao-effect (accessed 10.07.2017).

EPA (United States Environmental Protection Agency), Benefits of Green Streets. https://www.epa.gov/G3/benefits-green-street (accessed 30.07.2017).

Thompson, D.A., Matthews, R.W., The storage of carbon in trees and timber. Research Information Note 160, Forestry Commission Research Division.

Million Trees NYC, NYC's Urban Forest — Benefits of NYC's Urban Forest. http://www.milliontreesnyc.org/html/urban_forest/urban_forest_benefits.shtml (accessed 04.01.2017).

National Geographic, Effects of Global Warming. http://www.nationalgeographic.com/environment/global-warming/global-warming-effects (accessed 07.05.207).

NYCSTM) New York City Street Tree Map. https://tree-map.nycgovparks.org (accessed 04.01.2017).

Treeco$_2$nomics, About i-Tree Eco. http://www.treeconomics.co.uk/about-treeconomics/i-tree-eco-uk (accessed 29.11.2017).

Treepedia, Exploring the Green Canopy in cities around the world. http://senseable.mit.edu/treepedia/cities/cambridge (accessed 11.07.2017).

Villazon, L., How many trees does it take to produce oxygen for one person? Science Focus, 6 September 2016, http://www.sciencefocus.com/qa/how-many-trees-are-needed-provide-enough-oxygen-one-person (accessed 10.07.2017).

WHO —(GHO) Data, World Health Organisation Global Health Observatory Data — Urban Population Growth. http://www.who.int/gho/urban_health/situation_trends/urban_population_growth_text/en (accessed 11.07.2017).

World Forum on Natural Capital, What is natural capital. http://naturalcapitalforum.com/about (accessed 07.05.2017).

Wolf, K., 2003. Public response to the urban forest in inner-city business districts. J. Arboricul. 29 (3), May.

FURTHER READING

Amos, J., Earths tree number 'three trillion. BBC News — Science and Environment, 3/09/2015, http://www.bbc.com/news/science-environment-34134366 (accessed 11.07.2017).

BBC2, Seven Man Made Wonders', BBC — Seven wonders. http://www.bbc.co.uk/england/sevenwonders/info (accessed 04.01.2017).

CNYP&R - City of New York Parks & Recreation, Calculating Tree Benefits for New York City. http://www.nycgovparks.org/sub_your_park/trees_greenstreets/images/treecount_report.pdf (accessed 04.07.2017).

Edible Bus Stop, http://theediblebusstop.org/thessaly-road (accessed 04.01.2017).

Fletcher, T.D., Shusterb, W., Hunt, W.F., Ashleyd, R., Butlere, D., Arthurf, S., et al., 2015. SUDS, LID, BMPs, WSUD and more — The evolution and application of terminology surrounding urban drainage. Urban Water J. 12 (7), 525–542.

Forestry.gov., Valuing London's Urban Forest. https://www.forestry.gov.uk/pdf/LONDONI-TREEECOSUMMARY160331.pdf/$FILE/LONDONI-TREEECOSUMMARY160331.pdf (accesssed 30.07.2017).

Gardner, C., Old Urbanist, We Are the 25%: Looking at Street Area Percentages and Surface Parking, 12 Dec. 2011. http://oldurbanist.blogspot.co.uk/2011/12/we-are-25-looking-at-street-area.html (accessed 04.07.2017).

Venables, B., The Secret History Of The London Plane Tree. The Londonist, http://londonist.com/2015/03/the-secret-history-of-the-london-plane-tree (accessed 03.01.2017).

Chapter 4.7

Life Cycle Assessment of Vertical Greening Systems

Katia Perini

Chapter Outline

INTRODUCTION

Vertical greening systems can improve the environmental quality of dense urban areas, improving air quality, mitigating the urban heat island effect, fostering biodiversity, and improving building envelope performances mainly due to plants cooling capacity (Chiquet et al., 2012; Coma et al., 2017; Dunnett and Kingsbury, 2008; Perini et al., 2017).

The different types of vertical greening systems—i.e., green facades based on climbing plants directly attached or with a supporting system and living wall systems, and panels for the growth of different species with irrigation system for water and nutrients—besides aesthetic characteristics, entail different maintenance needs, environmental and economic costs (Ottelé et al., 2011). Greening systems can be sustainable design strategies, when environmental benefits are higher than environmental costs. In this scenario, life cycle assessment (LCA) methodology can be used to compare environmental impacts, deriving from construction, installation, maintenance, and disposal, of a vertical greening system and the environmental benefits provided. Indeed, LCA is defined by the ISO14040:2006 as the compilation and evaluation of the inputs (e.g., raw materials, water, energy), outputs (e.g., emissions in the air, waste), and the potential environmental impacts of a product system throughout its life cycle. In the case of a building material/system, the life cycle mainly includes: raw materials extraction, transports,

Nature Based Strategies for Urban and Building Sustainability.
DOI: https://doi.org/10.1016/B978-0-12-812150-4.00030-6

production, construction, use, maintenance, disposal (Lavagna, 2008). In order to perform a LCA several tools and national or international databases can be used.

This chapter provides an overview on the most influencing aspects to be considered to evaluate the environmental impact of vertical greening systems. Since green envelopes can play an important role in reducing the environmental impact of a building, the main environmental benefits which can be considered in LCA are listed and discussed. Finally, the results of some researches, which quantify the environmental impact of green facades and living wall systems, are presented.

VERTICAL GREENING SYSTEMS ENVIRONMENTAL BURDEN

In order to calculate the environmental burden of a vertical greening system, LCA methodology can be used. LCA framework includes: goal and scope definition, inventory analysis, impact assessment and interpretation. Usually the aim of the study of a greening system LCA is the calculation of environmental burden and the comparison with the environmental benefits obtainable, during use (or operation) phase. The goal and scope definition also includes the definition of the system boundaries and of a functional unit— i.e., what is being studied and quantifies the service delivered—which, in the case of VGS, is $1 \, m^2$ or a whole facade. System boundaries are related to several aspects, e.g., boundaries between the technological system and nature, delimitations of the geographical area and time horizon considered, boundaries between the life cycle of the product studied, and related life cycles of other products (Tillman et al., 1994). In the case of vertical greening systems, e.g., plants cultivation in farms—the plants needed for the VGS construction—impacts could be excluded from an analysis.

The environmental burden of a system is calculated according to selected impact categories. Nowadays a material's environmental impact is often equated with its effect on greenhouse gas emissions and climate change (Van den Heede and De Belie, 2012). However, some studies consider, besides global warming, human toxicity and fresh water aquatic ecotoxicity (Ottelé et al., 2011).

In order to assess the environmental burden of a vertical greening system, an inventory analysis has to be created. The data inventory includes all the materials and components (and related dimensions and weight), depending on the greening system type. Table 1 shows the main materials used, which will all have different environmental burdens, for direct green facade, indirect green facade, indirect green facade with planter boxes and a living wall system. When materials are commonly used in the building sector these are usually available locally and therefore the impact of transport is not relevant.

Considering the whole life cycle of a system, the environmental burden derived from raw material extraction, transportation, production, and

TABLE 1 Components and Materials of Some Vertical Greening Systems

Components	Direct Green	Indirect Green	Indirect Green With Planter Boxes	Living Wall System
Structural support member	–	Stainless steel bolts, anchor, spacer brackets	Stainless steel bolts, anchor, spacer brackets—steel profile	Steel profile and anchors
Water proofing material	–	–	–	PVC foam plate, PP, etc.
Supporting system/panel		Steel mesh, wood trellis, plastic mesh, etc.	Steel mesh, wood trellis, plastic mesh, etc.	Felt, HDPE planter boxes, geotextile, etc.
Growing material	Substrate soil	Substrate soil	Potting soil	Potting soil or wool fleece or mineral wall, etc.
Irrigation system	–	–	PE pipes	PE pipes
Water demand	Ground water	Groundwater	Tapwater + nutrients	Tapwater + nutrients
Vegetation	Climbing plant	Climbing plant	Climbing plant	Climbing plant and shrubs

construction highly depends on the system type and related materials involved (Fig. 1). A study conducted by Pan and Chu (2015) shows that the material stage contributed to 43%−97% of all the environmental impact categories. Also Ottelé et al. (2011) show that the material choice plays a key role in the environmental burden of a system: stainless steel mesh to support climbing plants, when reuse and recycle is not assumed (due to a foliage weaving which can occur at the end of the service life separating plants from a support could be difficult), it has a much higher environmental burden compared to HDPE, coated steel or hard wood.

In order to reduce the environmental burden, when designing a VGS, the use of recycled materials can be considered. Recycled materials can be used for up to 50% of the total substrate volume, improving also the water retention, and facilitating root development (Serra et al., 2017).

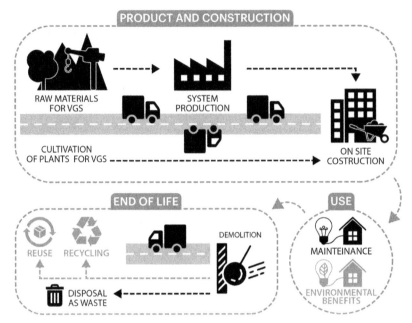

FIGURE 1 Main life cycle phases of vertical greening systems.

Vertical greening systems also demand different maintenance: pruning one or twice per year, replacement of dead plants or ruined panels; living wall systems also require different amounts of tap water for irrigation (around 3 L/m^2/day). The latter varies depending on the system, climate, facade orientation, plant species, influencing the total environmental burden especially in terms of the impact categories eutrophication potential and freshwater aquatic ecotoxicity (Pan and Chu, 2015).

According to Giordano et al. (2014), environmental information for each material that should be collected include the country of origin and the availability of the materials on the local market (in order to assess the transportation impact), the embodied energy and carbon dioxide equivalent emissions (to evaluate the depletion of the energy sources and the related climatic changes), the end of life scenarios (to assess the recycling potential), and environmental labeling (when available).

The durability of a system is very important as well. For example, replacing or fixing some panels during the assumed life span implies raw materials extraction, transport, production, construction, etc., with a consequent increase of the total environmental burden produced by a system. Ottelé et al. (2011) show that, due to the replacement of panels every 10 years, in a life span of 50 years the environmental burden of a living wall system based on felt layers can be higher, compared to other VGS, for both the impacts related to material production and waste. It's worth mentioning that, since the living wall system is a pretty recent technology—most of the systems were developed in the last 10 years—collecting real data on the durability of such systems is difficult and therefore LCA often relies on assumptions.

VERTICAL GREENING SYSTEMS ENVIRONMENTAL BENEFITS

A wide replication of green envelopes can be a good opportunity to improve the urban environment conditions, considering the many ecological and environmental imbalances faced by most of the urban areas (Dover, 2015). The environmental benefits provided by vertical greening systems over their lifetime can be taken into account to balance their environmental impact. As shown in Fig. 1, the potential savings are related to the phase of use.

The above mentioned benefits concern several fields, which are all related and operated on a scale range; some only work if a large surface in the same area is greened and their benefits are only apparent at a neighborhood or city scale, others operate directly on the building scale (Coma et al., 2017; Dunnett and Kingsbury, 2008; Perini and Magliocco, 2014). Some of these benefits, especially the ones related to the macro (urban) scale, are usually not taken into account in LCA studies. This happens mainly because there is a lack of data or simply because a benefit is unquantifiable for a single VGS and highly depends on the context/site where the VGS is installed. Among these air quality improvements, habitat creation and urban heat island mitigation can be hardly quantified due to the impossibility to estimate their effect connected to a single facade or roof (Perini and Rosasco, 2013).

At building scale the benefits of green facades and living wall systems are mainly related to their potential energy use reduction through summer cooling. In addition, the extension of the durability of facades, protected by a green layer, could be taken into account (Perini and Rosasco, 2013), as also the air cleaning potential (Feng and Hewage, 2014).

Studies demonstrate that a vertical green layer can contribute to the building envelope performances by creating an extra stagnant air layer, which has an insulating effect (Perini et al., 2011), and reducing the energy demand for air-conditioning, depending on the climate (Alexandri and Jones, 2008), on the greening system, facade orientation, etc. (as described in Chapter 3.1: Energy Performances and Outdoor Comfort). Indeed, under Mediterranean continental climate during cooling season, a real scale analysis of the thermal performance shows energy saving rates for the living wall, and the indirect green facade analyzed of respectively 58.9% and 33.8%, in comparison to a reference system. Yoshimi and Altan (2011) consider for Hong Kong climate an average daily electricity savings offered by the VGS of 16% in the hot summer months of August and September. The potential energy savings is often calculated by means of building energy simulation tools (e.g., EnergyPlus), applying energy savings percentage derived from specific studies on VGS energy performances (literature). Although this approach is common, it could reduce the accuracy of environmental benefits calculation due to the several influencing parameters involved in the thermal performance calculation.

ENVIRONMENTAL IMPACT ASSESSMENT

Few LCA studies have been conducted in the past few years on vertical greening systems, green facades, and living wall systems. Results show that in some cases the environmental burden is lower than the benefits obtainable during the life span of a system, while in others the durability and maintenance needs of greening systems play the most important role.

Ottelé et al. (2011) present a life cycle analysis of different vertical greening systems attached to a double brick wall with insulation material: two green facades—a direct system based on climbing plants and an indirect system in which climbing plants are supported by a stainless steel mesh—and two living wall systems, one based on HDPE planter boxes filled with potting soil and one based on felt layers. In order to include the energy savings as an environmental benefit in the analysis, the authors consider two locations: the Netherlands (temperate climate) and Italy (Mediterranean climate). Energy savings for cooling (only for Mediterranean area) and heating are calculated according to other studies, depending on the system and on the plants' growing rate. The life span assumed is 50 years. The results show that the direct greening systems have a very small influence on the total environmental burden, since they are made of vegetation only, with no other materials involved. For the other cases the material choice plays a key role. Living wall systems can either be a sustainable option or a system with a very high environmental burden (as in the case of the felt layer system), a climbing plant supported by plastic or wood mesh can have a low environmental impact, while the latter can be very high if made of stainless steel. The study shows that the cooling potential calculated for the Mediterranean area is fundamental to balance the environmental burden of the vertical greening system.

Feng and Hewage (2014) analyzed the same systems, assuming a life span of 50 years, by comparing air cleaning and energy savings (operation phase) to material production, construction, maintenance, and disposal stages. Results confirm what shown by Ottelé et al. (2011), i.e., that the felt layer system is not environmentally sustainable and that materials and plants determine the environmental performance of greening systems.

Pan and Chu (2015) quantify the environmental benefits and burdens of a commercially available vertical greenery system in Hong Kong, considering the electricity consumption for cooling in flats with and without a VGS (8.22 m^2) in a public housing estate, for a life span of 50 years. Results, thanks to a 16% energy savings, show that the environmental burden of the VGS in regard to abiotic depletion of fossil fuels could be paid back in 20 years.

CONCLUSION

LCA methodology can effectively be used to demonstrate if a vertical greening system can be considered environmentally sustainable or not. The studies discussed in the present chapter show that green facades and living wall systems can be a sustainable option for new construction and retrofitting projects, if the impact of the materials involved and the benefits obtainable are considered. Energy savings for air conditioning is the most influencing benefit to consider in order to balance the environmental burden produced by vertical greening systems. In this scenario, in order to increase the reliability of the results and to consider all the benefits, real and objective data related to each study specific location are needed.

LCA indicates that materials with high durability and low environmental burden should be used to design low impact greening systems, especially in the case of living wall systems. In general, in order to avoid the use of greening only with an aesthetic role, when a long (e.g., 50 years) life spam is assumed, durability aspects and maintenance need to be considered and vegetation cooling potential exploited.

REFERENCES

Alexandri, E., Jones, P., 2008. Temperature decreases in an urban canyon due to green walls and green roofs in diverse climates. Build. Environ. 43, 480–493. Available from: https://doi.org/10.1016/j.buildenv.2006.10.055.

Chiquet, C., Dover, J.W., Mitchell, P., 2012. Birds and the urban environment: the value of green walls. Urban Ecosyst. 16, 453–462. Available from: https://doi.org/10.1007/s11252-012-0277-9.

Coma, J., Pérez, G., de Gracia, A., Burés, S., Urrestarazu, M., Cabeza, L.F., 2017. Vertical greenery systems for energy savings in buildings: a comparative study between green walls and green facades. Build. Environ. 111, 228–237. Available from: https://doi.org/10.1016/j.buildenv.2016.11.014.

Dover, J.W., 2015. Green Infrastructure: Incorporating Plants and Enhancing Biodiversity in Buildings and Urban Environments, first ed. Routledge, London; New York.

Dunnett, N., Kingsbury, N., 2008. Planting Green Roofs and Living Walls. Timber Press, Portland, OR.

Feng, H., Hewage, K., 2014. Lifecycle assessment of living walls: air purification and energy performance. J. Clean. Prod. 69, 91–99. Available from: https://doi.org/10.1016/j.jclepro.2014.01.041.

Giordano, R., Montacchini, E.P., Tedesco, S., 2014. Eco-innovation based on Life Cycle Assessment and Green-Design. Strategies in manufacturing a Living Wall System, in: World Sustainable Building WSB14. Presented at the World Sustainable Building WSB14, Barcelona.

Lavagna, M., 2008. Life cycle assessment in edilizia. Progettare e costruire in una prospettiva di sostenibilità ambientale. Hoepli.

Ottelé, M., Perini, K., Fraaij, A.L.A., Haas, E.M., Raiteri, R., 2011. Comparative life cycle analysis for green façades and living wall systems. Energy Build. 43, 3419–3429. Available from: https://doi.org/10.1016/j.enbuild.2011.09.010.

Pan, L., Chu, L.M., 2015. Energy saving potential and life cycle environmental impacts of a vertical greenery system in Hong Kong: a case study. Build. Environ . Available from: https://doi.org/10.1016/j.buildenv.2015.06.033.

Perini, K., Magliocco, A., 2014. Effects of vegetation, urban density, building height, and atmospheric conditions on local temperatures and thermal comfort. Urban For. Urban Green. Available from: https://doi.org/10.1016/j.ufug.2014.03.003.

Perini, K., Ottelé, M., Fraaij, A.L.A., Haas, E.M., Raiteri, R., 2011. Vertical greening systems and the effect on air flow and temperature on the building envelope. Build. Environ. 46, 2287–2294. Available from: https://doi.org/10.1016/j.buildenv.2011.05.009.

Perini, K., Ottelé, M., Giulini, S., Magliocco, A., Roccotiello, E., 2017. Quantification of fine dust deposition on different plant species in a vertical greening system. Ecol. Eng. 100, 268–276. Available from: https://doi.org/10.1016/j.ecoleng.2016.12.032.

Perini, K., Rosasco, P., 2013. Cost–benefit analysis for green façades and living wall systems. Build. Environ. 70, 110–121. Available from: https://doi.org/10.1016/j.buildenv.2013.08.012.

Serra, V., Bianco, L., Candelari, E., Giordano, R., Montacchini, E., Tedesco, S., Larcher, F., Schiavi, A., 2017. A novel vertical greenery module system for building envelopes: the results and outcomes of a multidisciplinary research project. Energy Build. 146, 333–352. Available from: https://doi.org/10.1016/j.enbuild.2017.04.046.

Tillman, A.-M., Ekvall, T., Baumann, H., Rydberg, T., 1994. Choice of system boundaries in life cycle assessment. J. Clean. Prod. 2, 21–29.

Van den Heede, P., De Belie, N., 2012. Environmental impact and life cycle assessment (LCA) of traditional and "green" concretes: Literature review and theoretical calculations. Cem. Concr. Compos. 34, 431–442. Available from: https://doi.org/10.1016/j.cemconcomp.2012.01.004.

Yoshimi, J., Altan, H., 2011. Thermal simulations on the effects of vegetated walls on indoor building environments, in: Proceedings of Building Simulation, pp. 1438–1443.

Chapter 4.8

Life Cycle Assessment of Green Roofs

Julià Coma, Gabriel Pérez and Luisa F. Cabeza

Chapter Outline

INTRODUCTION

Green roof systems are successfully tested tools that strengthen the three main bases for a sustainable development: the environment, the economics, and the society by using natural solutions as well as help in reducing the dependence on gray standard solutions in urban areas (International Green Roof Association, 2017). This type of green infrastructures are included in the European green strategies with the aim to provide and ensure the restoration, protection, creation, and enhancement of green infrastructures to becoming integral parts of spatial planning development in cities whenever they offer a better alternative, or are complementary, to standard gray choices (Green Infrastructure and the Biodiversity Strategy; Environment. European Commission, 2013).

The presence of these systems in urban areas (parks and gardens) is well-known because they are accessible and feasible for citizens, but the huge potential of green infrastructures is when they're implemented on the large available surface of building skins, roofs, and facades. Since the building sector contributes highly to the global environment challenges not only for its conditioning during their lifespan, but also from the indirect energy use for the production of its materials, construction, and dismantling phases, the concept of life cycle analysis (LCA) in this sector is growing (Buyle et al., 2013). In this context, LCA becomes a crucial procedure in evaluating the particular impacts of a product regarding the environmental perspectives throughout its life.

Nature Based Strategies for Urban and Building Sustainability.
DOI: https://doi.org/10.1016/B978-0-12-812150-4.00031-8

After a literature review, a large list of studies that evaluate the benefits of implementing green infrastructures, basically green roofs and vertical greenery systems on building envelopes were found. Regarding green roofs technologies, Coma et al. (2016) and Bevilacqua et al. (2016) evaluated the cooling and heating performances of different extensive green roof solutions in Mediterranean areas; at Puigverd de Lleida (Spain) and Arcavacata (Italy), respectively. On one hand, both studies experimentally demonstrated that these systems contributed in reducing the temperatures up to 12°C t at the interface with the structural roof in comparison to a typical black bituminous roof in the summer. On the other hand, the Italian study demonstrated that green roofs maintain, on average, a value that is 4°C higher in the winter. These results provided reductions of about 2.3°C of the indoor ambient temperatures for cooling and were capable to keep warm or at least maintain the same internal comfort conditions for heating compared to the reference system.

Besides contributing in reducing the heating and cooling demands of buildings in specific climatic conditions, green roof systems also provide other benefits, known as ecosystem services, at building and city scale. Many of these contributions are; the water management quality and quantity (Berndtsson, 2010), the mitigation of urban heat island effect (Coutts et al., 2013), the reduction of CO_2 concentration in urban areas (Li et al., 2010), the increment of the biodiversity and reduction of habitat losses (Brenneisen, 2006), the revalorization of the building considering the better aesthetics, the sound absorption capacity, etc.

All of these presented studies are a step forward in evaluating the aforementioned specific technical benefits of green roof systems. However, a lack of crucial information regarding the overall impact of each individual technology on the environment, economy, and society is detected and requires a precise analysis of each effect. Within this context, the most extended tool to quantify the overall impact of a product, service, manufacturing process, material or in this case, a green roof construction system is the LCA approach (ISO 14040:2006; ISO 14044:2006).

The aim of this chapter is to provide a general overview about LCA concept and its application on green roof systems. In addition, the main research findings in this topic were summarized and discussed to obtain final conclusions.

LCA GENERAL CONCEPT

Environmental life cycle assessment can be defined as a procedure or technique to obtain an environmental impact associated with products or systems throughout its life cycle. Thus, LCA provides an overall assessment that deal with the associated environmental aspects of a product from the extraction of raw materials, to the manufacture, operational, and the final management of this product at the end of its life.

Nowadays, LCA could be considered the best effective approach to perform an environmental impact analysis of a product or in that case, a green roof system, since it is based on international guidance, ISO 14040 and ISO 14044, which provides a standardized and methodological technique. In addition to the environmental LCA, economics and social aspects could be implemented on the assessment by using other combined tools, such as life cycle costing (LCC) and social life cycle assessment (S-LCA). The main targets for performing an LCA are as follows: to identify opportunities for the environmental improvement of products throughout their life cycle, to advise decision-makers of the reasonable impacts of a project proposal before a decision is made, to inform decision-makers in the industry, and to provide further information of a product or process.

Any process can be assessed by implementing the proceeding defined in the ISO 14040:2006, which establishes four principal phases for conducting an LCA (Fig. 1); goal and scope definition, inventory analysis (LCI), life cycle impact assessment (LCIA), and interpretation:

- Goal and scope definition: determines the boundary conditions of the system, the functional unit the assumptions, and the limitations of the assessment.
- Life cycle inventory analysis (LCI): comprises data collection of all processes, information on physical material and energy flows in various stages of the products lifecycle, i.e., inputs and outputs of a product system (water, energy, row material, waste, emissions, etc.).
- Life cycle impact assessment (LCIA): evaluates the significance of potential environmental impacts using the results obtained in the LCI and associating them with specific environmental impact categories.
- Interpretation: provides consistent results with a clear understanding of the uncertainty and the assumptions used in agreement with the defined goal and scope, and also should define recommendations to decision-makers.

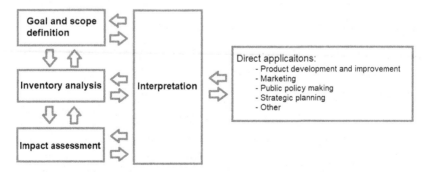

FIGURE 1   Methodology of an LCA (ISO 14040:2006).

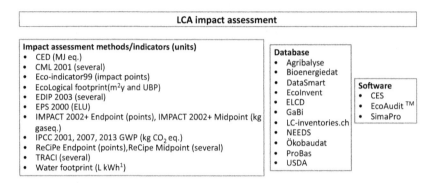

FIGURE 2   Main impact assessment methods, databases, and software available to perform an environmental life cycle assessment (Solé et al., 2017).

Fig. 2 shows many of the impact assessment methods and databases available to evaluate the environmental impacts of materials, processes, construction systems, etc. hence to identify new paths in optimizing and reducing the environmental damage during the whole process. Due to the huge collected information in all databases, several software enterprises developed friendly interfaces for the end-user to facilitate the development of an LCA such as GaBi or SimaPro, which currently are well-known software.

Regarding the building sector, Martínez-Rocamora et al. (2016) conducted a review on LCA databases containing data for building materials in order to evaluate and determine which databases accomplish more features and thus might be considered more complete and suitable than others. The following are the set of crucial features that should have appear in a LCA database according to the authors: scope, completeness, transparency, comprehensiveness, update, and licence. Fig. 3 shows comparative spider charts in which the bigger the area of the chart the more complete the database is. On one hand, GaBi and Ecoinvent databases were classified as the most complete ones, but they require a license to be used, and on other hand, ELCD was considered the best free database. It is remarkable that most of the LCA studies in the literature regarding green roofs used the EcoInvent database.

LCA FOR GREEN ROOFS

The methodology of LCA can be very useful to develop a comparative analysis between different materials or processes to obtain products such as building construction systems and even studies that consider an entire building. However, due to the precision by which the calculation of the impact indicators and the life cycles methods are organized, few opportunities in

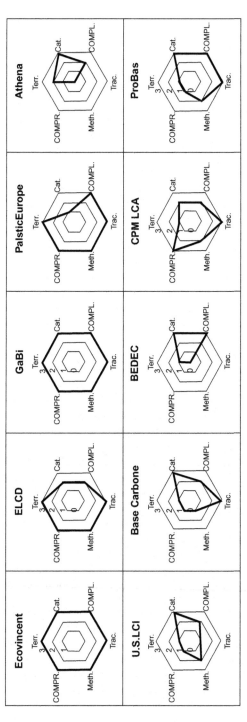

FIGURE 3 Comparative chart of LCA databases based on aforementioned features. *Terr,* Territory; *Cat,* Categories; *COMPL,* Completeness; *Trac,* Traceability; *Meth,* Methodology outlining; *COMPR,* Comprehensiveness (Martínez-Rocamora et al., 2016).

comparing LCA studies concerning green roofs when implemented on buildings are difficult to find, thus some of the most interesting findings are summarized and discussed below.

Belussi and Barozzi (2015) conducted a review about the mitigation measures (cool/green roofs, and green walls) to contain the environmental impact in urban areas from the life cycle approach. The aim of the study was to establish a general scheme to combine the main outcomes from the literature review and to provide a comparative outline of different mitigation systems supporting the decision-making phase. In order to highlight the scarce literature regarding LCA for urban mitigation measures, nineteen studies at building scale were found up to date. However, only nine of them were focused on LCA for roof systems, while the other ten refer to life cycle cost analysis (LCC) or global environment (GE). In addition to the roof construction systems, the review also included LCA, LCC, and GE for urban pavements and urban parks. The authors concluded that the analyzed studies used different databases that provided different assessment methods and impact categories, making the technical comparisons difficult or impossible. Nonetheless, the first qualitative analysis of these building construction systems (green roofs) showed an improvement of the environmental and economic results compared with the reference scenarios when evaluated using one or more specific environmental performance indicators.

Bianchini and Hewage (2012) evaluated the sustainability of green roofs by estimating the number of years that a common extensive green roof takes to balance the pollution released in its material production, with the pollution removed by the planted vegetation throughout the useful life. The considered layers for a typical green roof were the following and their thicknesses vary according to the typology of green roof (intensive/extensive): root barrier, drainage, filter, water retention, growing medium, and vegetation. Moreover, the authors established a comparative analysis of four different substances: nitrogen dioxide (NO_2), sulfur dioxide (SO_2), pzone (O_3), and particles of $10\,\mu m$ or less (PM_{10}) to analyze the air pollution created due to the production process of the existing polymers inside the aforementioned layers of a green roof. The LCA was carried out by using SimaPro 7.1 software, and the Ecoindicator (H) V2.06. method for damage categories. This software allowed performing two case studies for comparative purposes, one with recycled materials and one with nonrecycled materials. The main outcomes of this study showed that the lower environmental impacts highlight the importance of green roofs as a sustainable option for the building industry and society. However, it is essential to explore materials that can replace the current materials used as separation layers (polymers) such as reusing waste or implementing recycled materials. According to the authors, the years required in the operation phase of the green roofs to balance the air pollution released during the production of materials can be reduced more than half in the case of implementing recycled materials instead of nonrecycled materials.

Within the same topic, Rincón et al. (2014) conducted a comprehensive LCA in which the materials of two sustainable extensive green roofs without an insulation layer, neither separation layers (polymers) and with different drainage layer materials (pozzolana and recycled rubber crumbs) have environmentally compared with two conventional gravel ballasted flat roofs, with and without polyurethane as a thermal insulation layer. This LCA approach considered the production, construction, operational, and disposal phases of these roofs in agreement with UNE-EN 15643-2. The scope of the analysis was defined with a roof of 9 m^2 as a functional unit, the life cycle was estimated at 50 years, and the LCA Manager (v1.3) software and Ecoinvent database (v2.2) were used to model the systems. In addition, this study went further on calculating the operational phase by using experimental data considering heating and cooling demands for a whole year instead of assuming or using estimations of the final thermal performance.

The results presented in Fig. 4 showed the large contribution of the operational phase in comparison to the whole LCA for the existing roofing systems, the energy consumption being the determinant parameter of the total life cycle impact of the assessed systems. Approximately 85.7%−87.2% of the total impact points came from the operational phase, while only 10.5%−11.5%, 1.5%−2%, and 0.8%−1% were from construction, production, and disposal phases, respectively. The authors of the paper concluded that recycled materials, in this case rubber crumbs from out-of-use tires, can be implemented in extensive green roofs to improve both, the insulation capacity and the environmental properties in Mediterranean continental climate conditions. In addition, the extensive green roof system with recycled rubber crumbs presented similar impact points (2405) when compared with a traditional insulated flat roof (2357). Future researches will consider the benefits of carbon fixation provided by the vegetation in the operational phase as well as a LCC that can provide a better and realistic approximation of the environmental cost of a sustainable extensive green roof.

Bachawati et al. (2016) performed a cradle-to-gate LCA to obtain environmental impacts (objective parameters) about three different existing technologies; green roofs, traditional gravel ballasted roofs, and white reflective roof that are designed to enhance the sustainability of an urban area while at the same time can provide other multiple ecosystem services such as to reduce storm water run-off, and to reduce the energy consumption in buildings through the improvement of the insulation values. This study excludes the use (operational) and end-of-life (disposal) phases from the scope of the analysis since the thermal performance was not experimentally evaluated for all the studied roofs. The functional unit was a surface roof area of 834 m^2, and the lifetime was established at 45 years. EcoInvent database and SimaPro software were used to perform the LCA of the considered systems. The main outcomes of the study highlight the high environmental impact of common materials used in typical roofs such as concrete, steel bars for

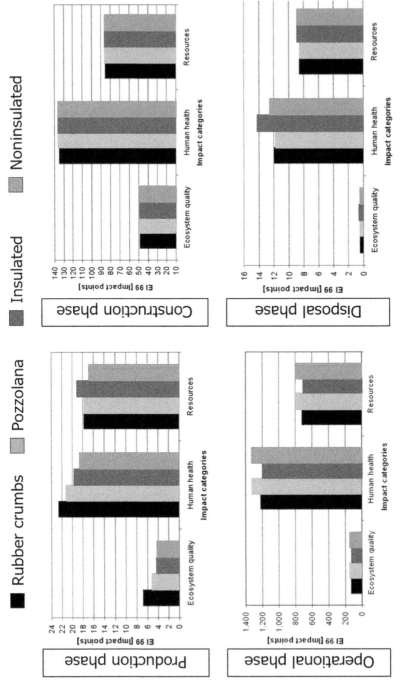

FIGURE 4 Life cycle inventory results in impact points (EI99) of each phase in the analyzed roofs (Rincón et al., 2014).

reinforcing concrete, waterproof membranes, and thermal insulation. Also, perlite showed a high impact contribution in green roofs. The authors concluded that green roofs are better, from the environmental point of view, than traditional gravel ballasted roofs, and white reflective roofs. Moreover, within the vegetated roofs the extensive ones showed the lowest environmental impacts for all impact categories.

CONCLUSION

In this chapter, a general overview LCA methodology as well as an overview of the main findings regarding LCA in green roofs are presented. From an overall point of view LCA can be used as a methodical technique to compare and evaluate the environmental properties of different construction systems focused on the same target and to provide objective parameters for decision-makers.

After conducting a literature review, GaBi and EcoInvent were classified as the most complete LCA databases for materials in the building sector despite that they require licence. Nonetheless, ELCD was considered the best free database.

Since the global building stock in Europe is currently far from the proposed new energy trends such as nearly zeroenergy building (nZEB) (European Commission, 2014), the operational phase of a LCA still plays a very important role in the overall environmental impact assessment. This fact is due to materials with high levels of embodied energy in production, dismantling, and disposal phases, but that provide higher energy savings in the operational phase, generally from 40 to 50 years, are more sustainable from the environmental point of view when compared to those materials that have less environmental impact in the production, construction, and disposal phases.

From the presented studies regarding LCA of green roofs, it can be generally concluded that these systems show less environmental impacts in comparison to white reflective roofs or typical ballasted roofs. Nowadays green roofs and other urban green infrastructures such vertical greenery systems, are standing out as promising environmental friendly systems that can be implemented on building skins to reduce the environmental impact of the whole system while at the same time they can provide multiple ecosystems services at both, building and urban scale.

ACKNOWLEDGMENTS

This study has received funding from European Union's Horizon 2020 research and innovation program under grant agreement No 657466 (INPATH-TES), from the European Commission Seventh Framework Program (FP/2007-2013) under grant agreement No PIRSES-GA-2013-610692

(INNOSTORAGE). The work is partially funded by the Spanish Government (ENE2015-64117-C5-1-R (MINECO/FEDER)). GREA is certified agent TECNIO in the category of technology developers from the government of Catalonia. The authors would like to thank the Catalan Government for the quality accreditation given to their research group (2014 SGR 123).

REFERENCES

Bachawati, E.M., Manneh, R., Belarbi, R., Dandres, T., Nassab, C., Zakhem, H., 2016. Cradle-to-gate Life Cycle Assessment of traditional gravel ballasted, white reflective, and vegetative roofs: a Lebanese case study. J. Cleaner Production 137, 833–842.

Belussi, L., Barozzi, B., 2015. Mitigation measures to contain the environmental impact of urban areas: a bibliographic review moving from the life cycle approach. Environ. Monitoring Assessment 187, 745.

Berndtsson, J.C., 2010. Green roof performance towards management of runoff water quantity and quality: a review. Ecol. Eng. 36 (4), 351–360.

Bevilacqua, P., Mazzeo, D., Bruno, R., Arcuri, N., 2016. Experimental investigation of the thermal performances of an extensive green roof in the Mediterranean area. Energy Build. 122, 63–69.

Bianchini, F., Hewage, K., 2012. How "green" are the green roofs? Lifecycle analysis of green roof materials. Build. Environ. 48, 57–65.

Brenneisen, S., 2006. Space for urban wildlife: designing green roofs as habitats in Switzerland. Urban Habitats 4, 27–36.

Buyle, M., Braet, J., Audenaert, A., 2013. Life cycle assessment in the construction sector: a review. Renew. Sustain. Energy Rev. 26, 379–388.

Coma, J., Pérez, G., Solé, C., Castell, A., Cabeza, L.F., 2016. Thermal assessment of extensive green roofs as passive tool for energy savings in buildings. Renew. Energy 85, 1106–1115.

Coutts, A.M., Daly, E., Beringer, J., Tapper, N.J., 2013. Assessing practical measures to reduce urban heat: green and cool roofs. Build. Environ. 70, 266–276.

European Commission, 2014. Technical guidance-financing the energy renovation of building with cohesion policy funding.

Green Infrastructure and the Biodiversity Strategy; Environment. European Commission, 2013. http://ec.europa.eu/environment/nature/ecosystems/index_en.htm. (Last accessed 06.07.2017).

International Green Roof Association. http://www.igra-world.com/. (Last accessed 06.07.2017).

ISO International Standard 14040:2006. Environmental Management - life cycle assessment. Principles and framework. International Organisation for Standardisation (ISO).

ISO International Standard 14044:2006. Environmental Management - life cycle assessment. Requirements and guidelines. International Organisation for Standardisation (ISO).

Li, J.-F., Wai, O.W.H., Li, Y.S., Zhan, J.-M., Ho, Y.A., Li, J., et al., 2010. Effect of green roof on ambient CO2 concentration. Build. Environ. 45, 2644–2651.

Martínez-Rocamora, A., Solís-Guzmán, J., Marrero, M., 2016. LCA databases focused on construction materials: a review. Renew. Sustain. Energy Rev. 58, 565–573.

Rincón, L., Coma, J., Pérez, G., Castell, A., Boer, D., Cabeza, L.F., 2014. Environmental performance of recycled rubber as drainage layer in extensive green roofs. A comparative Life Cycle Assessment. Build. Environ. 74, 22–30.

Solé, A., Miró, L., Cabeza, L.F., 2017. High temperature thermal storage Systems using phase change materials. Part 4: Environmental and economic approach. Chapter 10: Environmental approach. Elsevier S&T Books.

FURTHER READING

Cabeza, L.F., Rincón, L., Vilariño, V., Pérez, G., Castell, A., 2014. Life cycle assessment (LCA) and life cycle energy analysis (LCEA) of buildings and the building sector: a review. Renew. Sustain. Energy Rev. 29, 394−416.

Summary

Gabriel Pérez and Katia Perini

Nature Based Strategies for Urban and Building Sustainability provides an extensive review and discusses the most relevant research results regarding vertical greening systems, green roofs, and green streets.

International experts (engineers, architects, researchers, designers) in the field of green technologies author the chapters, which are divided in four sections to meet the general and most relevant aspects related to nature-based strategies in urban areas, regarding the main classifications and technical characteristics, qualitative and quantitative evaluation of benefits, also considering the main issues and challenges, and finally to discuss social, economic, and environmental sustainability of greening systems.

Section 1 highlights that the ecosystem services delivered by vegetation (green infrastructure) have the potential to help humanity address some of the major environmental and social challenges of our times, the urbanization process, by greening the epidermis of built structures. These services can play an important role in enhancing the resilience of cities to climate change, but only if urban planners and policy makers take into account the connectivity required by key ecological processes in order for their potential to be optimized. Engineered green infrastructure, such as green roofs, vertical greening systems and green streets, presents opportunities for providing the above mentioned connectivity even in the densest cities. Convinced by the compelling evidence of the multiplebenefits provided by nature-based solutions, city authorities are turning their attention to ways of increasing the uptake of these features, through incentives and other innovating financial products including loans at favorable rates.

From the second part of the book, regarding the available technologies, it can be concluded that these are fully consolidated, with clear classifications and descriptions of the different systems and their variants, being much clearer in the case of the green roofs and green streets features than for vertical greening systems. The different designs have evolved to meet the demanding requirements of the built environment, being the subject of water management perhaps the most determinant in order to guaranteeing the full consolidation of these systems in the market. Proper provision of related eco-system services and the possibility to design more sustainable systems, regarding to materials and water efficient use are issues to consider in future designs.

The most important researches around the world present, in the third section of the book, are the latest advances relating to the benefits assessment and the challenges overcome.

In regards to the potential of these systems to enhance the thermal performance of buildings and outdoor comfort, four main effects stand out when greenery is applied to the building skin, that are the shade effect, the cooling effect, due to evapotranspiration, the insulation effect, and the wind barrier effect. Green streets allow improving thermal outdoor comfort in cities, with effects evident at a range of scales (city, district, canyon scale). Future research must be directed towards suitable management of this potential to improve the thermal performance of buildings and cities.

Since plants naturally clean the air, acting on gaseous pollutants and particulate matter concentration, implementing nature-based solutions on a wide scale provides opportunities to utilize these typically unused surfaces to address environmental issues such as air pollution. Selecting the most suitable plant species is very important, since some species can act as a great sink of air pollutants. Attention will be paid to high performances, low allergenicity, low cost of maintenance, and climatic suitability. Since the effect of greening solutions on enhancing the air quality in urban canyons is a very complex problem, modeling allows simulating the effects of plants on pollutants' concentrations under different conditions.

The latest research highlighted that urban green infrastructure significantly contributes to buildings' acoustic insulation and noise reduction at city scale. However, research on sound transmission through urban greening systems is rather scarce and is predominantly based on measurements. Good performances have been observed not only from the vegetation layers but also from the substrate layers due to the porous structure. Future research must deal with studies regarding the types of plants, the thickness of the vegetation layer, the thickness and composition of the substrate layer, the type of support structure and materials to be used, as well as to consider insulation measures to prevent sound transmission (structural impenetrability and insulation).

Urban nature-based strategies can have a great impact on urban storm water management so that they can offer adaptable solutions to manage rainwater through minimizing the impact of development on water cycle. These systems can use natural processes to manage runoff at the source at a lower cost than traditional gray infrastructure (pipes and other water control devices). In this regard green roofs and different green streets strategies are the most suitable, whereas vertical greening remains as a complement in the whole strategy. Appropriate strategies should consider the opportunity to downsizing the sewer infrastructure, resulting in significant cost savings for municipalities and minimizing the use of potable water while enhancing the quality of urban and street landscapes. The cooling effect, due to the water evapotranspiration from plants and substrates, and the use of these systems as biofilters are other relating ecosystem services to be considered in future research.

In a context of urban greening, vertical plantings offer a great potential as habitat for biodiversity. However it is still unclear whether they act as exclusive habitats or components of urban corridors. Thus, understanding the ecological value and functioning of each system becomes critical to integrate nature into urban design and planning at the city scale. Diversity linked to these systems is mainly driven by microclimatic conditions as well as landscape configurations. When incorporated into urban green corridors, green infrastructure can provide habitat to reduce the effects of fragmentation for some plant and animal species. The general consensus is that the more habitat types and the greater proximity or connectivity included, the more potential is offered to increase biodiversity. Keeping habitats as natural as possible, incorporating local materials and creating habitats in proximity to, or linked to, other similar habitats leads to better colonization and higher biodiversity. Increasing the habitat heterogeneity by well-planned and varied management, sympathetic cutting regimes and the inclusion of pools or wetland areas can also increase biodiversity values.

Finally, the latest advances on research regarding social, economic, and environmental sustainability of urban nature-based strategies are discussed in Section 4.

Considering social and aesthetic aspects of urban green infrastructure, social perception surveys have shown that, although their environmental quality improvement potential is now recognized, resistance to their installation persists. The main fears concern construction and maintenance costs, and the possible increase of insects and pollen, as well as uncertainty that the waterproofing system will be able to adequately protect the supporting brickwork. Considering perception as an evaluation of what human senses detect, it was shown that a positive perception often goes with the experience and the knowledge of what we see. It is therefore possible to state that it is necessary to support the application of these systems with information campaigns to make people understand their qualities. On the other hand, it is demonstrated that the living environment has widely recognized the consequences reflected on the well-being of city-dwellers, so that these systems integrated in a network of green spaces can increase liveability of everyday landscapes and the health of residents, supporting and stimulating physical and psychological benefits. Moreover, they can increase social cohesion and social justice in urban contexts, counteracting health inequalities related to income deprivation and supporting the recreational value of urban contexts.

Some of the latest research relating to the economic benefits and costs of these systems showed that their lifecycle cost can be paid back by individual benefits in mature markets. If public benefits are added into the assessment, the lifecycle cost can be retrieved in most of the markets. The governments should play a key role in promoting their construction by providing incentives to transfer the social benefits into private investors, such as tax abatement, direct cash rebate, low interest loans, etc. These incentives will also expand the public benefits, and lower the lifecycle cost of urban nature-based strategies, providing a better city environment and reducing some of the negative effects generated by climate change.

From the scarce research conducted previously relating to the environmental impact of green roofs and vertical greening systems, measured by means lifecycle assessment, it can be generally concluded that these systems show less environmental impacts in comparison to traditional gray solutions. However, the impact of the materials involved and the benefits obtainable must be considered a crucial issue during the design phase. The studies conducted in this field show that the environmental benefits, mainly related to the energy savings at the building scale, can balance the environmental burden depending on the system characteristics.

Index

Printed in the United States
By Bookmasters